Notes on Numerical Fluid Mechanics and Multidisciplinary Design

Founding Editor

Ernst Heinrich Hirschel

Volume 201

Series Editor

Wolfgang Schröder, Aerodynamisches Institut, RWTH Aachen, Aachen, Germany

Editorial Board

Bendiks Jan Boersma, Delft University of Technology, Delft, The Netherlands

Kozo Fujii, Institute of Space and Astronautical Science (ISAS), Sagamihara, Kanagawa, Japan

Michael A. Leschziner, Department of Aeronautics, Imperial College, London, UK

Jacques Periaux, Paris, France

Sergio Pirozzoli, Department of Mechanical and Aerospace Engineering, University of Rome 'La Sapienza', Roma, Italy

Arthur Rizzi, Department of Aeronautics, KTH Royal Institute of Technology, Stockholm, Sweden

Bernard Roux, Ecole Supérieure d'Ingénieurs de Marseille, Marseille CX 20, France

Yurii I. Shokin, Siberian Branch of the Russian Academy of Sciences, Novosibirsk, Russia

Managing Editor

Esther Lagemann, RWTH Aachen University, Aachen, Germany

Notes on Numerical Fluid Mechanics and Multidisciplinary Design publishes state-of-art methods (including high-performance methods) for numerical fluid mechanics, numerical simulation and multidisciplinary design optimization. The series includes proceedings of specialized conferences and workshops, as well as relevant project reports and monographs.

Indexed by SCOPUS, zbMATH, SCImago.

All books published in the series are submitted for consideration in Web of Science.

Pawel Flaszynski · Holger Babinsky · Piotr Doerffer
Editors

Towards Effective Flow Control and Mitigation of Shock Effects in Aeronautical Applications

Editors
Pawel Flaszynski ⓘ
Institute of Fluid-Flow Machinery
Polish Academy of Sciences
Gdańsk, Poland

Holger Babinsky ⓘ
Department of Engineering
University of Cambridge
Cambridge, UK

Piotr Doerffer ⓘ
Institute of Fluid-Flow Machinery
Polish Academy of Sciences
Gdańsk, Poland

ISSN 1612-2909   ISSN 1860-0824   (electronic)
Notes on Numerical Fluid Mechanics and Multidisciplinary Design
ISBN 978-3-031-86604-3   ISBN 978-3-031-86605-0   (eBook)
https://doi.org/10.1007/978-3-031-86605-0

This work was supported by H2020 Research and Innovation Programme (860909).

© The Editor(s) (if applicable) and The Author(s) 2025. This book is an open access publication.

Open Access This book is licensed under the terms of the Creative Commons Attribution 4.0 International License (http://creativecommons.org/licenses/by/4.0/), which permits use, sharing, adaptation, distribution and reproduction in any medium or format, as long as you give appropriate credit to the original author(s) and the source, provide a link to the Creative Commons license and indicate if changes were made.
The images or other third party material in this book are included in the book's Creative Commons license, unless indicated otherwise in a credit line to the material. If material is not included in the book's Creative Commons license and your intended use is not permitted by statutory regulation or exceeds the permitted use, you will need to obtain permission directly from the copyright holder.
The use of general descriptive names, registered names, trademarks, service marks, etc. in this publication does not imply, even in the absence of a specific statement, that such names are exempt from the relevant protective laws and regulations and therefore free for general use.
The publisher, the authors and the editors are safe to assume that the advice and information in this book are believed to be true and accurate at the date of publication. Neither the publisher nor the authors or the editors give a warranty, expressed or implied, with respect to the material contained herein or for any errors or omissions that may have been made. The publisher remains neutral with regard to jurisdictional claims in published maps and institutional affiliations.

This Springer imprint is published by the registered company Springer Nature Switzerland AG
The registered company address is: Gewerbestrasse 11, 6330 Cham, Switzerland

If disposing of this product, please recycle the paper.

Preface

Future, sustainable, growth of the aviation industry relies on continued improvements in efficiency giving reductions in fuel consumption and increases in payload. Research efforts therefore focus on aerodynamic performance and structural weight savings. This inherently requires more highly performing wings, control surfaces and turbomachinery blades where transonic flow is common place and the formation of shock waves the key aerodynamic challenge. In particular, the interaction of shock waves with boundary-layers is one of, if not the main performance-limiting or safety critical flow phenomena across all of these flow fields. Thus, a good understanding of the interaction of shock waves with boundary layers is essential for the development of future, more efficient, air vehicles and engines. Increased aerodynamic forces can lead to flow separation and reductions in engine and airframe efficiency. In such cases, flow control is needed to maintain system performance. However, novel designs are also likely to increase the extent of laminar flow and this implies that flow control devices need to operate in a laminar or transitional regime. This requires a better understanding of their function and their interaction with flow transition.

The H2020-MSCA-ITN TEAMAero (Towards Effective Flow Control and Mitigation of Shock Effects in Aeronautical Applications) project was about training 15 Early Stage Researchers through a research-based programme to become key scientists and engineers in the aeronautics sector.

This book presents main achievements and new findings obtained within a framework of the TEAMAero project.

Gdańsk, Poland
September 2024

Pawel Flaszynski
TEAMAero Coordinator

Acknowledgements

The publication of the book is a result of a collective effort by contributors and authors of the chapters. The coordinator would like to thank all TEAMAero partners, Ph.D. students and their supervisors for the very fruitful cooperation and friendly atmosphere during the training program.

The TEAMAero project has received funding from the European Union's Horizon 2020 research and innovation programme under grant agreement No EC grant 860909 TEAMAero.

The coordinator would like to express special gratitude to the Project Officer, Szymon Sroda, who has provided the guidance along the route to make the TEAMAero project successful.

The editors would like to express their gratitude to Filip Wasilczuk for his contribution to the editing of this book.

Contents

Introduction 1
Pawel Flaszynski and Filip Wasilczuk

Numerical and Experimental Methods Development

From High-Fidelity High-Order to Reduced-Order Modeling for Unsteady Shock Wave/Boundary Layer Interactions 9
Nicolas Goffart, Benoît Tartinville, and Sergio Pirozzoli

Numerical Tools for High-Fidelity Simulation of SBLIs 33
Alessandro Ceci and Sergio Pirozzoli

Development of a PVDF Piezo-Film Sensor for Unsteady Wall-Pressure Measurements in SBLIs 57
Cosimo Corsi, Bei Wang, Julien Weiss, and Ha Duong Ngo

Transitional/Turbulent SBLI and Flow Control

Non-linearities in the Low-Frequency Dynamics of Transitional SBLI 75
Mariadebora Mauriello, Lionel Larchevêque, and Pierre Dupont

The Length and Time Scales of Transitional SBLIs 105
Nikhil Mahalingesh, Sébastien Piponniau, and Pierre Dupont

Parameter Influence on Porous Bleed Performance for Shock-Wave/Boundary-Layer Interaction Control 127
Julian Giehler, Pierre Grenson, and Reynald Bur

Unsteady Three-Dimensional Oblique Shock Wave Boundary-Layer Interactions 149
Timothy Missing and Holger Babinsky

Oblique-Shock Wave Boundary Layer Interactions Control:
Shock Control Bumps 171
Jane Bulut, Ferry Schrijer, and Bas van Oudheusden

Airfoil/Wing Configuration

Numerical Study of Unsteady Shock/Boundary Layer Interaction 193
Andrea Petrocchi, Rene Steijl, and George N. Barakos

Numerical Study and Physical Analysis of the Transonic
Interaction and Its Modification Through Morphing Around
Supercritical Wings at High Reynolds Number 221
Cesar Jimenez Navarro, Jacques Abou Khalil, Rajaa El Akoury,
Abderahmane Marouf, Jean-Baptiste Tô, Yannick Hoarau,
Jean-François Rouchon, and Marianna Braza

Transonic Compressor

Numerical Investigations of Transitional SBLI on a Highly
Loaded-Transonic Compressor 251
Selin Kahraman and Ilias Vasilopoulos

Reynolds Number Effects on Shock Wave Boundary Layer
Interaction in Highly Loaded Compressor Stator 275
Arun Joseph, Pawel Flaszynski, Michal Piotrowicz, Piotr Doerffer,
and Marcin Kurowski

Experimental and Numerical Investigations of SBLI and Flow
Control on a Transonic Compressor Cascade 297
Edwin J. Munoz Lopez and Alexander Hergt

Surface Roughness Effect on Shock Boundary Layer Interaction
on Compressor Rotor Profile 325
Ahmed H. Hanfy, Pawel Flaszyński, Piotr Kaczyński, and Piotr Doerffer

Shock Oscillation Mechanisms of Highly Separated Transitional
Shock-Wave/Boundary-Layer Interactions 345
Philipp Nel, Anne-Marie Schreyer, and Marius Swoboda

Introduction

Pawel Flaszynski and Filip Wasilczuk

Abstract This book presents the outcomes of the H2020-MSCA-ITN TEAMAero project, which addresses key aerodynamic challenges in achieving Flightpath 2050 goals for European aviation. Focusing on shock wave boundary layer interactions (SBLI), TEAMAero explores advanced flow control strategies to enhance performance in high-speed, transonic flow environments. Through a multidisciplinary collaboration of universities, research centers, and industry, the project trained 15 researchers and delivered breakthroughs in numerical methods, experimental techniques, and flow-control devices. The findings support the design of more efficient aircraft and engines by improving understanding and control of complex three-dimensional, unsteady transonic flows in both internal and external configurations.

Achieving goals defined in Flightpath 2050—Europe's Vision for Aviation require major improvements in efficiency, which for aircraft means reduced weight and better aerodynamic performance, based on high-performance wings, control surfaces and turbomachinery blades, where transonic flow is common place and the formation of shock waves is the key aerodynamic challenge. In particular, the interaction of shock waves with boundary layers is the primary performance-limiting factor across all of these flow fields. Thus, knowledge of the interaction of shock waves with boundary layers is essential for the development of more efficient aircraft and engines. The problem is that increased aerodynamic forces can lead to flow separation and reductions in engine and airframe efficiency. In such cases, flow control is needed to maintain the system's performance. Novel designs can also increase the extent of laminar flow and this means that flow-control devices need to operate in a laminar or transitional regime, which requires a better understanding of their function and their interaction with flow transition. Of course, improvements to their effectiveness will have a direct and positive impact on airframe and engine performance. It is because these modern geometries are increasingly complex that we have to really understand

---

P. Flaszynski (✉) · F. Wasilczuk
Institute of Fluid-Flow Machinery, Polish Academy of Sciences (IMP PAN), Gdańsk, Poland
e-mail: pflaszyn@imp.gda.pl

F. Wasilczuk
e-mail: fwasilczuk@imp.gda.pl

and so be able to control three-dimensional flows, especially the three-dimensional shock wave boundary layer interactions (SBLI).

The H2020-MSCA-ITN TEAMAero (Towards Effective Flow Control and Mitigation of Shock Effects in Aeronautical Applications) derives from two very successful European projects: FP6-AEROSPACE UFAST (Unsteady Effects of Shock Wave Induced Separation) and FP7-TRANSPORT TFAST (Transition Location Effect on Shock Wave Boundary Layer Interaction) and cooperation of some partners within H2020, Clean-Sky or national projects. All these projects looked at aeronautical flow fields for highly loaded compressor and turbine blades and wings, where the SBLIs are crucial.

The main objective of TEAMAero has been achieved by realising the project's four science and technology subobjectives, which are:

1. to improve our fundamental understanding of the physics of SBLI, including three-dimensionality and unsteadiness;
2. to identify the flow domains best suited to flow-control device installation;
3. to develop flow-control schemes using wall transpiration (suction/blowing), vortex generators and surface treatments to delay the onset of separation, and
4. to develop novel numerical methods for predicting the effects of SBLI.

The multi-disciplinary training platform for 15 Early Stage Researchers (ESRs) has brought together 12 Beneficiaries (universities, research centres and industrial stakeholders) and 6 Partner Organisations (Table 1), actively involved in top-level research in the fields of transonic flows with a focus on shock wave boundary layer interaction, advanced optical measurement methods, high-speed imaging, numerical methods, multi-objective optimisation and mechanical engineering (Fig. 1).

The implementation of the objectives was performed through the following Work Packages.

Work Package 1—Basic Flow Cases (Leader: AMU)

The study of fundamental, basic flow problems was be carried out in WP1, which was concerned with normal and oblique SBLIs occurring in rectilinear geometries. The WP1 was focused on investigations involving different Mach and Reynolds numbers and two upstream conditions: laminar and turbulent boundary layers. The objective of WP1 was also (1) to qualify response sensitivity of SBLIs to local perturbations and to identify best suited regions to install new control devices (passive or active) in order to influence separation state and unsteadiness of the interaction, (2) to provide a clear description of the flow topology for the different tested configurations, (3) to provide a clear description of the flow topology for the different tested configurations.

Work Package 2—Application to Internal Flows (Leader: RRD)

The second work package WP2 was focused on experimental and numerical investigations of compressor cascades, both rotor and stator. The objective of WP2 was (1) to provide detailed steady and unsteady experimental data in transitional SBLIs representative of those occurring in turbomachinery, (2) to enrich knowledge on the flow physics pertaining to laminar or transitional SBLIs in compressor representative

Table 1 Participants of TEAMAero project

No	Beneficiaries	Short name	Scientist-in-charge	Early stage researcher
1	Instytut Maszyn Przeplywowych im. Roberta Szewalskiego Polskiej Akademii Nauk	IMP PAN	Pawel Flaszynski *Project Coordinator* Filip Wasilczuk *Project Manager* Piotr Doerffer Michal Piotrowicz Piotr Kaczynski Marcin Kurowski	Arun Joseph Ahmed Hanfy
2	Rolls-Royce Deutschland Ltd & Co Kg	RRD	Marius Swoboda Christian Janke Patrick Grothe	Selin Kahraman Philipp Nel
3	Deutsches Zentrum Fuer Luft - und Raumfahrt Ev / Institute of Propulsion Technology	DLR	Alexander Hergt	Edwin Munoz Lopez
4	Universite d'aix Marseille	AMU	Pierre Dupont Lionel Larchevêque Sébastien Piponniau	Mariadebora Mauriello Nikhil Mahalingesh
5	Office National D'etudes et de Recherches Aerospatiales	ONERA	Reynald Bur Pierre Grenson Eric Garnier	Julian Giehler
6	The Chancellor Masters and Scholars of the University of Cambridge	UCAM	Holger Babinsky	Timothy Missing
7	Technische Universiteit Delft	TU Delft	Bas Van Oudheusden Ferdinand Schrijer	Jane Bulut
8	The University of Glasgow	UGLA	George N. Barakos Rene Steijl	Andrea Petrocchi
9	Cadence Design Systems Belgium	CADENCE	Benoit Tartinville Charles Hirsch	Nicolas Goffart
10	Universita Degli Studi di Roma La Sapienza	UNIROMA1	Sergio Pirozzoli	Alessandro Ceci
11	Institut National Polytechnique de Toulouse	INPT	Marianna Braza	Cesar Jimenez Navarro
12	Technische Universitat Berlin	TUB	Julien Weiss Lennart Rohlfs	Cosimo Corsi
No	Partner organisations	Short name	Person-in-charge	
1	Centro Italiano Ricerche Aerospaziali	CIRA	Antonio Schettino	
2	Politechnika Gdanska	GUT	Rafał Tylman	
3	Dassault Aviation	DAAV	Flavien Billard	

(continued)

Table 1 (continued)

No	Partner organisations	Short name	Person-in-charge	
4	Airbus Operations GmbH	AIRBUS	Bruno Stefes	
5	RWTH Aachen University	RWTH	Anne-Marie Schreyer	
6	Arts et Métiers Institute of Technology / DynFluid Laboratory	DynFluid	Jean-Christophe Robinet	

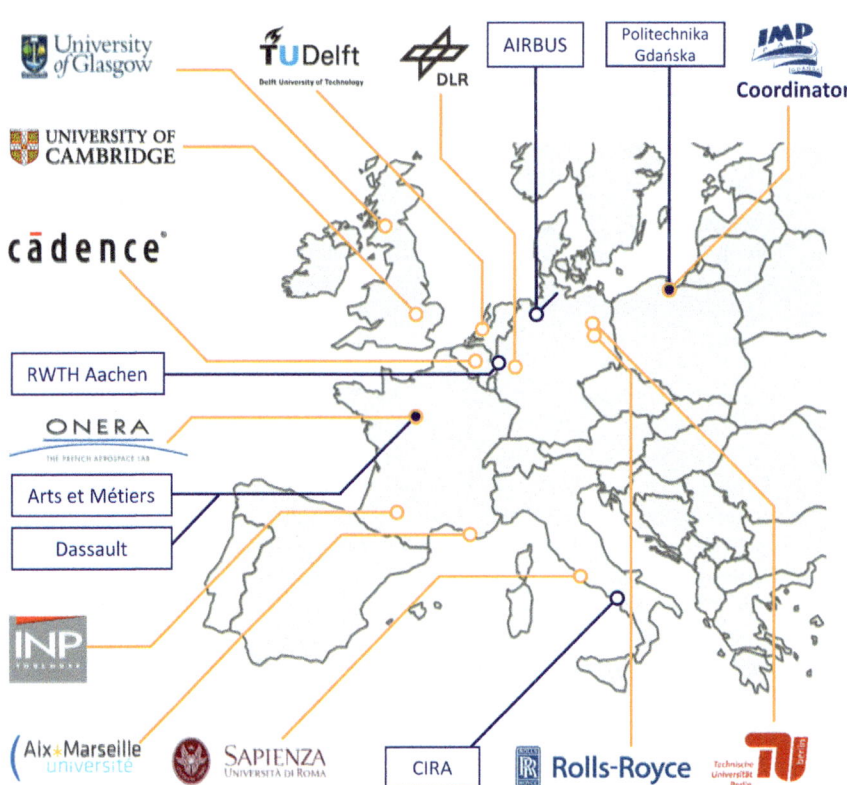

Fig. 1 Beneficiaries and Partner Organisations in TEAMAero project

configurations, (3) to develop passive and active methods to control the shock wave boundary layer interaction to improve aerodynamic performance.

Work Package 3—Application to External Flows (Leader: UGLA)

The work within WP3 concerns of external flow configurations and the installation of flow control to alleviate the adverse effects of SBLIs. WP3 consists of numerical tasks (Computational Fluid Dynamics) that are directly linked to experimental databases delivered by other EU projects as FP6 UFAST, FP7 TFAST and H2020 SMS, geometries defined by Dassault Aviation and Airbus. The key issue to address was that flow control methodology must be applicable to a very wide range of flight conditions and loadings.

Work Package 4—Research Methods—Numerical and Experimental (Leader: ONERA)

In the context of the basic and applied studies, it is important to learn and gain experience in the use of state-of-the-art measurements as well as innovative experimental and CFD methods. This was addressed in WP4. The objective of WP4 was to drive the development of novel techniques and also to help all the ESRs in the selection of the most appropriate methodology for their specific flow problem.

Work Package 5—Training, dissemination, communication and exploitation (Leader: UCAM)

WP5 ensures good coordination of training activities, dissemination and communication. The objective was to train young researchers in a wide range of specialist and transferable skills. These include selected scientific knowledge applicable to the research performed in this project and also researcher skills to enable and enhance future career progression within the wider field of science and technology research and management.

Work Package 6—Project Management (Leader: IMP PAN)

WP6 ensures the good coordination of the management.

Work Package 7—Ethics requirements (Leader: IMP PAN)

WP7 sets out the ethics requirements that the project must comply with.

The 15 Early Stage Researchers involved in the project have carried out numerical simulations and experimental investigations for shock wave boundary layer interactions in various configurations including wings, inlet ducts and transonic fan or compressor stator.

Several results have been achieved during TEAMAero project.

1. Development of numerical methods used to simulate transonic interactions using Direct Numerical Simulations, both in Cartesian and curvilinear coordinates, to improve turbulence modelling in harmonic methods for shock-induced separated flows in turbomachinery applications, to examine unsteadiness and the underlying mechanism in a transitional shock reflection with separation, to investigate the ability of the Partially-Averaged Navier- Stokes method to reproduce transonic buffet.

2. Experimental campaigns to understand unsteady flow effects and development of measurement techniques to study the time scale in a transition process of a natural laminar boundary layer and to investigate unsteady flow conditions in a turbulent shockwave-boundary layer interaction.
3. Flow control methods: a porous bleed control in mitigating the negative effects of shock-wave/boundary-layer interactions, separation-bubble-shaped shock control bumps, conical shaped artificial corner separation bodies, surface roughness, electroactive morphing concepts based on trailing-edge vibration and on travelling waves along the wall.
4. Investigations of unsteady effects due to the shock-boundary layer interactions in transonic compressor cascades. Numerical and experimental study of Reynolds number and surface roughness effect on shock wave unsteadiness in a rotor and a stator cascade. Transonic Cascade TEAMAero was optimized and its performance was validated experimentally.

The main achievements and selected results of the Individual Research Projects are presented in the following chapters.

Open Access This chapter is licensed under the terms of the Creative Commons Attribution 4.0 International License (http://creativecommons.org/licenses/by/4.0/), which permits use, sharing, adaptation, distribution and reproduction in any medium or format, as long as you give appropriate credit to the original author(s) and the source, provide a link to the Creative Commons license and indicate if changes were made.

The images or other third party material in this chapter are included in the chapter's Creative Commons license, unless indicated otherwise in a credit line to the material. If material is not included in the chapter's Creative Commons license and your intended use is not permitted by statutory regulation or exceeds the permitted use, you will need to obtain permission directly from the copyright holder.

Numerical and Experimental Methods Development

From High-Fidelity High-Order to Reduced-Order Modeling for Unsteady Shock Wave/Boundary Layer Interactions

Nicolas Goffart, Benoît Tartinville, and Sergio Pirozzoli

Abstract To design the next-generation aircraft and engines, efficient numerical tools need to be developed. Shock wave/boundary layer interactions are indeed putting the current industrial methods to the test: high-order methods lack robustness while modeling assumptions in low-fidelity methods make them not accurate enough. This work presents the first steps toward improving turbulence modeling in harmonic methods for shock-induced separated flows in turbomachinery applications, using high-fidelity data. A high-order solver based on the flux reconstruction framework is employed for performing high-fidelity simulations. Robustness is ensured by an enhanced artificial viscosity, allowing to capture shocks while not damping turbulence. A canonical oblique shock wave/boundary layer is first investigated to validate the solver and the results are in excellent agreement with the abundant existing literature. Then, the periodically forced transonic flow over a bump is considered. A study of the sensitivity to the perturbation frequency is carried out and highlights different flow regimes. The performance of the harmonic method for the bump case is finally shown to be inferior compared to the unsteady results. As a future step, the high-fidelity data generated will help to reduce the gap between the two.

Keywords Transonic flow · Shock wave—boundary layer interaction · High-fidelity simulations · Implicit Large-Eddy simulations

---

N. Goffart (✉) · B. Tartinville
Cadence Design Systems Belgium, Brussels, Belgium
e-mail: nicolas.goffart@hotmail.fr

B. Tartinville
e-mail: tartin@cadence.com

S. Pirozzoli
Department of Mechanical and Aerospace Engineering, Sapienza University of Rome, Rome, Italy
e-mail: sergio.pirozzoli@uniroma1.it

1 Introduction

The aviation industry has always been searching for improved aircraft and engine designs and, in the current context of climate change particularly, this effort is to be pursued. On the aerodynamic point of view, the challenge resides in the transonic and supersonic regimes, in which shock waves develop and interact with boundary layers. The resulting shock wave/boundary layer interaction (SWBLI) affects a wide variety of flows in aeronautical applications like aircraft wings, turbomachinery blades or supersonic air intakes to name a few. Not only can it provoke flow separation and flow distortion, but also such a phenomenon is associated with a large scale low-frequency unsteadiness. The latter is particularly critical as it is responsible for structural fatigue that can potentially lead to failure. To push the efficiency limits further and with confidence, reliable numerical design tools that can accurately predict SWBLI flow features are needed.

In Computational Fluid Dynamics (CFD), the cheapest methods are obtained by solving the Reynolds-Averaged Navier-Stokes (RANS) equations, in which the additional viscous stress term arising from turbulence is modeled. An immensity of turbulence models have been developed in CFD. With respect to shock wave/boundary layer interactions, they were found to play a significant role in the prediction of the mean flow features [1], and unsteady RANS (or URANS) simulations are inefficient at reproducing the low-frequency unsteadiness. Their defects and palliatives were also outlined in a turbomachinery context [2]. Methods relying on turbulence models are often referred to as low-fidelity methods. Nevertheless, they have been the backbone of industrial design so far, thanks to their efficiency and robustness, and will continue to be employed as such for the foreseeable future [3].

On the opposite, high-fidelity methods such as Direct Numerical Simulations (DNS) and Large-Eddy Simulations (LES) were proven to accurately predict SWBLIs [4, 5]. However, performing high-fidelity simulations is costly, even considering the rising computing power. A remedy for this is to employ high-order numerical schemes as they allow to reach the same level of accuracy as lower-order schemes but at a reduced computing time. In such schemes, a powerful shock-capturing technique is essential and the prescription of realistic turbulent inflow conditions is a challenge as well [6, 7].

High-fidelity methods are profitable for predicting features that fall in the shortcomings of RANS-based methods, such as separation, transition and heat transfer. They are therefore suitable tools to investigate the underlying physics and to generate data that would eventually help to improve the models employed in low-fidelity methods.

The idea behind the present work is to apply that framework to the Non-Linear Harmonic (NLH) method [8, 9]. Harmonic methods are popular to simulate turbomachinery flows. They indeed intend to benefit from the intrinsic flow periodicity to restrict the simulation domain to a single blade passage, while being able to predict unsteady features as in URANS. The closure of the harmonic equations derived in the NLH framework requires, however, a model for the harmonic turbulent stresses, the assumptions of which can degrade the accuracy of the solution in a configuration with shock-induced separation [10]. The development of better models would improve the predictions given by harmonic methods and widen their range of application.

This chapter is a condensed version of the work performed as part of a Ph.D. thesis [11]. The first step was to develop a high-order solver robust enough to perform Implicit Large-Eddy Simulations (ILES) of SWBLIs, and then to use it to study the transonic flow over a bump with harmonic forcing and generate high-fidelity data for harmonic turbulence modeling in the context of the NLH method. The content of this chapter is therefore the following. In the first section, all the computational methodologies employed are briefly described. Then, the results from a canonical oblique SWBLI are introduced as a validation for the high-order solver. Finally, the study of periodically forced transonic flow by the means of high-fidelity and low-fidelity simulations is presented.

2 Computational Methods

The results presented in this chapter were obtained from different flow solvers, a high-order solver dedicated to high-fidelity ILES, and an industrial second-order accurate solver for the RANS, URANS and NLH simulations. The purpose of this section is therefore to briefly describe the numerical schemes employed in each of them. The interested reader is invited to refer to [11] for a more complete description.

To carry out high-fidelity Implicit Large-Eddy Simulations, an in-house high-order solver was developed. The spatial discretization is based on the Flux Reconstruction method introduced in [12]. Similarly to Discontinuous Galerkin, the spatial order of accuracy can be directly chosen and a fourth-order of accuracy has been selected hereafter. An explicit 5-stages fourth-order accurate Runge–Kutta scheme is used for the temporal derivative following [13]. To accurately predict the shock wave/boundary layer interaction, a special attention is paid to the shock-capturing technique. The combination of a Laplacian artificial viscosity method [14] with the Ducros sensor [15] makes the approach quite efficient. Indeed, the addition of the Ducros sensor allows to discriminate the shock region from the boundary layer. It therefore avoids any over-dissipation within the boundary layer and ensures its proper development. Besides shock-capturing, the global robustness is further enhanced by a positivity-preserving limiter [16]. Turbulent inflow conditions are obtained with the digital filtering technique. Instead of the original 3D filter implementation [17], the approach of a 2D filter [18] correlated in time [19] is adopted. Velocity perturbations are finally scaled with the Lund's transformation [20]. Compressible fluctuations (for temperature and density) are also introduced based on the Strong Reynolds Analogy and the hypothesis of negligible pressure fluctuations [21].

For RANS, URANS and NLH methodologies, the simulations are performed using Cadence FINE™/Turbo flow solver, initially developed by [22]. The three-dimensional equations are discretized following a finite volume approach in which the spatial derivatives are evaluated with a centered second-order scheme in combination with a Jameson-type artificial dissipation [23], while a 4-stages explicit Runge–Kutta scheme is used for the temporal discretization. Convergence is improved by means of multigrid, local time-stepping and implicit residual smoothing. For URANS computations, a dual-time stepping method is employed, the temporal derivative being evaluated with a second-order backward Euler difference. A description of the implementation of the NLH method is given in [9].

3 Oblique Shock Wave/Boundary Layer Interaction

The first test case to be presented is a canonical oblique shock wave/boundary layer interaction. The aim is to demonstrate the ability of the high-order solver based on the flux reconstruction to handle a configuration with an unsteady shock system and to provide a realistic turbulent boundary layer. First, the particular flow conditions adopted for the simulation are introduced, together with the computational setup. Then, various results are analyzed and compared with the available literature.

3.1 Flow Conditions and Computational Setup

The flow conditions are chosen to replicate an experiment led at the IUSTI in which an oblique shock wave is generated with a flat plate inclined by 8° in a Mach 2.3 flow and impinges on the turbulent boundary layer developed on the floor of the wind tunnel [24]. The experimental characteristics of the incoming boundary layer, that are to be matched in the simulation, are given in Table 1 at a reference station upstream of the interaction. In particular, the reference boundary layer thickness $\delta_0 = 11$ mm and the Reynolds number based on the compressible momentum thickness $Re_\theta \approx$

Table 1 Boundary layer characteristics upstream of the interaction. The reference station is located at $(x - \bar{x}_{imp})/L \approx -1.66$. δ^*, θ and θ_i are the compressible displacement thickness and the compressible and incompressible momentum thicknesses, respectively

	Experiment [24, 25]	Present case
δ_0 [mm]	11	11.4
δ^* [mm]	3.4	3.42
θ [mm]	0.96	0.93
θ_i [mm]	1.28	1.28
Re_θ	5.1×10^3	5.2×10^3
Re_{θ_i}	6.9×10^3	6.9×10^3
C_f	2×10^{-3}	2.04×10^{-3}

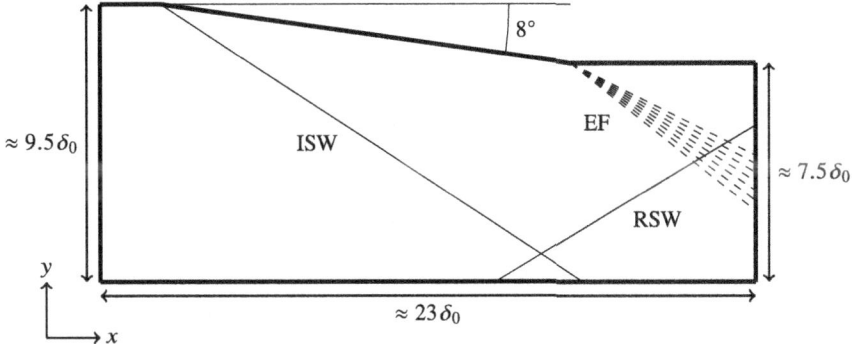

Fig. 1 Side view of the computational domain with some of the main flow features: incident shock wave (ISW), reflected show wave (RSW) and expansion fan (EF)

5.1×10^3. The upstream total temperature and total pressure are, respectively, 300 K and 50 kPa. The fluid is assumed to be a perfect gas.

A side view of the computational domain is shown in Fig. 1 together with some of the main flow features. The shock generator is modeled and is part of the top boundary, which is therefore inclined by 8°. However, the upper part of the wind tunnel, above the shock generator, is not included in the computational domain. The lengths of the domain in the streamwise and spanwise directions are, respectively, $L_x \approx 23\delta_0$ and $L_z = 5\delta_0$. The inlet is located $\approx 17\delta_0$ upstream of the inviscid impingement point of the incident shock wave so that the upstream turbulence can develop properly before reaching the interaction region. Finally, the domain height is $L_y \approx 9.5\delta_0$ at the inlet and $L_y \approx 7.5\delta_0$ at the outlet.

The mesh is entirely composed of hexahedra. The number of cells is $256 \times 97 \times 76$ in the streamwise, wall-normal and spanwise directions, respectively. The simulation is performed at polynomial order 3, leading to a total number of solution points of $N_x \times N_y \times N_z = 1024 \times 388 \times 304 \approx 121 \times 10^6$. A constant grid spacing is used in the x and z directions giving in wall units based on the upstream conditions $\Delta x^+ = 16$ and $\Delta z^+ = 12$. In the wall-normal direction, the boundary layer comprises exactly 25 cells (or 100 solution points at polynomial order 3), which are stretched according to a hyperbolic tangent law. The first cell height is imposed such that the first solution point lies below $y^+ = 1$. Outside the boundary layer, the grid spacing is kept practically constant and equal to the grid spacing in the streamwise direction. Therefore, $\Delta y^+ = 16$.

Regarding the boundary conditions, the inlet is fully supersonic with prescribed velocity components, static temperature and static pressure profiles taken from a precursor ILES of a turbulent boundary layer in the same flow conditions. To configure the digital filter, turbulence length scales in the streamwise I_x and spanwise I_z directions have been set constant and respectively equal to $0.5\delta_0$ and $0.2\delta_0$. I_y varies in the wall-normal direction such that the number of flux points constituting the filter is practically constant and around 350. Moreover, it matches I_z at the edge of the boundary layer. Reynolds stresses profiles (obtained from the same precursor simulation) are imposed to scale the filtered perturbations. The outlet boundary is supersonic but with static pressure imposed in the subsonic part of the boundary layer. The value is taken from a separate RANS simulation of the interaction, at the first supersonic point in the outlet boundary layer. The top boundary is divided into a slip wall for the shock generator and two external boundaries with Riemann invariants. The bottom wall is no-slip adiabatic and periodic boundary conditions are prescribed in the spanwise direction.

The parameters of the shock-capturing technique (see [11] for more details) are $s_0 = -4.5, \kappa = 0.5, C_T = 0.03$ and $s_{D,0} = 0.2$. Density is used as the sensor variable.

The explicit time step is $2.5 \cdot 10^{-8}$ s, giving a CFL number of around 2.5. The simulation is first restarted from an initial RANS solution for about $225\delta_0/U_\infty$ to get rid of the transient. Samples of the flow are then collected for a duration of $1000\delta_0/U_\infty$. Instantaneous span-averaged, instantaneous mid-span, as well as bottom wall data are extracted at a sampling rate of 500 kHz. The flow is also probed at various locations. The probes record the primitive variables at each time step, hence a sampling rate of 40 MHz. The frequency associated to the energy-carrying eddies in the upstream boundary layer is $\mathcal{O}(U_\infty/\delta_0) \approx 50$ kHz. Both sampling rates are consequently high enough to capture all the frequencies involved in the flow. The shock unsteadiness is expected to correspond to a frequency two orders of magnitude lower than the characteristic frequency of the upstream boundary layer. The simulation time therefore covers around 10 cycles of the reflected shock motion.

3.2 Results

A first insight into the flow field is given in Fig. 2 displaying an instantaneous view of the density gradient magnitude at mid-span and of the streamwise velocity near the bottom wall (at a distance $y^+ \approx 10$). The different flow features are easily recognized, beginning with the oblique incident shock. Its reflection as an expansion fan on the sonic line is also discerned. As expected, the reflected shock wave stands slightly upstream of its inviscid location. The weak reattachment shock wave, turning back the flow parallel to the wall after the interaction, is also captured. Turbulent streaks are clearly highlighted in the upstream boundary layer and are largely influenced by the interaction. A sudden and sharp drop in streamwise velocity occurs as soon as the boundary layer meets with the reflected shock, which provokes the separation of the flow. Further downstream, turbulent streaks develop again, progressively, as the boundary layer recovers from the interaction.

To begin with, Table 1 summarizes the characteristics of the boundary layer, at a reference station upstream of the interaction, and compares them to the experimental ones. Note that the location of the reference station is scaled to account for the different interaction length (as it will be shown later). The boundary layer thickness, the Reynolds number and the friction coefficient are all nicely matched.

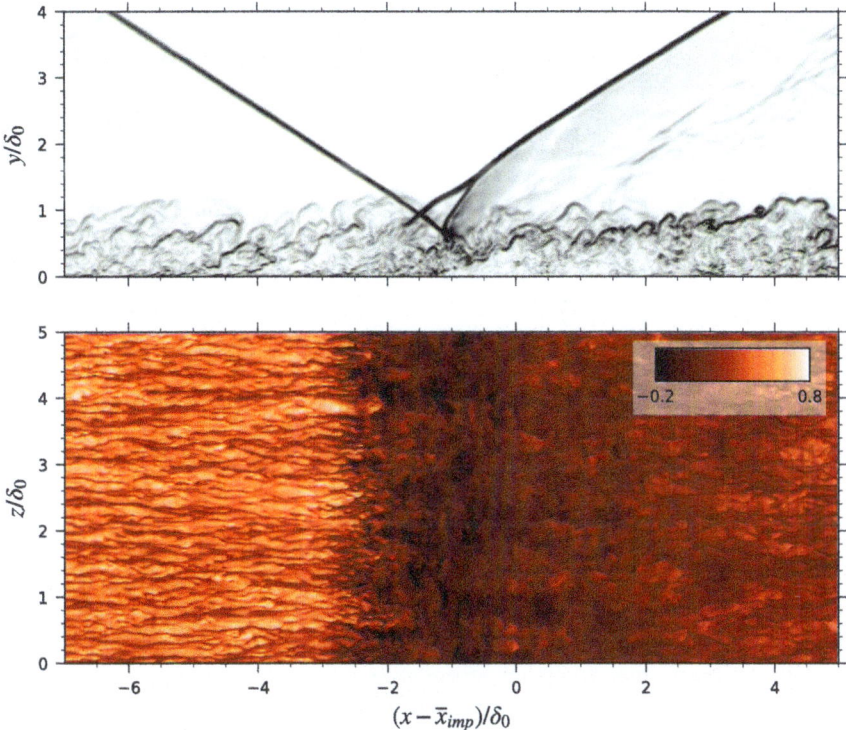

Fig. 2 Instantaneous density gradient magnitude at mid-span (*top*) and instantaneous streamwise velocity u/U_∞ near the bump wall, $y^+ \approx 10$ (*bottom*)

The boundary layer profile upstream of the interaction is shown in Fig. 3 together with PIV measurements from [24] and DNS data of [26]. Regarding the van Driest-transformed mean velocity profile (*left* figure), a very good agreement is found between the present results and the DNS. The curves are on top of each other in both the viscous sublayer and in the logarithmic region. The typical law of the wall is matched, with $\kappa = 0.41$ and $C = 5.3$. A slight departure from DNS is observed in the defect layer. With respect to experimental results, an offset is reported in the logarithmic layer. It comes from the use of a different friction velocity in the van Driest transform. The experimental value is around 25.4 m/s [21] whereas the simulation gives here 24.5 m/s. Scaling the experimental profile with the latter reduced the gap.

In Fig. 3, *right*, Reynolds stress profiles are presented and compared again with the available PIV data from [24] and the DNS results reported in [26]. A digital filtering approach was also employed to provide a turbulent inflow. A good agreement is again found between the different numerical results. In particular, the peaks of the normal stresses are very well predicted. The agreement with experimental data is satisfactory, the under-prediction of the wall normal velocity fluctuations being typical in measurements.

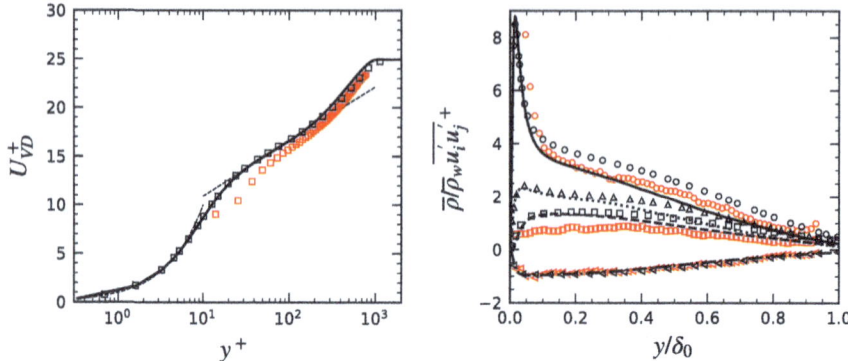

Fig. 3 Boundary layer profiles at $(x - \bar{x}_{imp})/L = -1.66$—van Driest-transformed mean velocity profile (*left*) and density-scaled Reynolds stress profiles (*right*). *red* symbols refer to experimental measurements of [24] and *black* symbols to the DNS data of [26]

A key point in the computational setup is the width of the domain, which has to be sufficiently large to guarantee the de-correlation of the turbulent structures before reaching the lateral boundary. Failing this requirement leads to a virtually confined flow and therefore mimics the presence of sidewalls. Ultimately, this can result in a stronger interaction and, as a consequence, in a larger separation bubble. The domain width is typically checked by considering the two-point streamwise velocity correlation coefficient in the spanwise direction, computed following

$$C_{uu}(x, y, \Delta z) = \overline{u'(x, y, z)u'(x, y, z + \Delta z)}/\overline{u'(x, y, z)u'(x, y, z)}. \quad (1)$$

Figure 4 presents the results at a distance $y/\delta_0 = 0.5$ from the bottom wall. The spatial distribution (*left*) indicates that C_{uu} quickly drops to zero, at $\Delta z/L_z \approx 0.1$, regardless of the streamwise location. A more detailed view at four selected stations (*right*) shows moreover that, within the separated region, the integral length scale first increases and then immediately decreases. A similar observation was reported in [27]. Upstream and downstream of the interaction region, the integral length scale is identical. In any case, the flow is de-correlated much before a distance of half the span. It is therefore concluded that the domain is wide enough. Touber [21] reached the same conclusion using the same domain width.

Figure 5 (*left*) illustrates the streamwise evolution of the friction coefficient, compared with the simulation results of [26, 27]. Upstream of the interaction, the friction coefficient is steadily decreasing, as it is expected for a developed turbulent boundary layer. Together with the boundary layer profiles showed in Fig. 3, this is comforting the idea that the inlet is located far upstream enough from the interaction region to let the boundary layer retrieve its main features but also that the digital filtering is correctly configured. Within the interaction region, two negative lobes are observed, which is in agreement with other high-fidelity predictions. Whereas the absolute

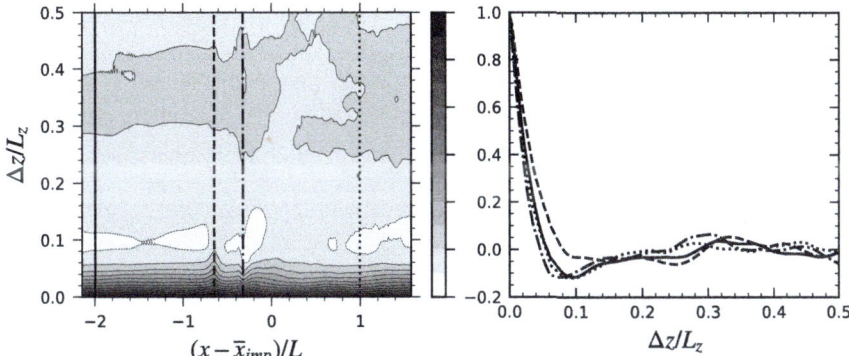

Fig. 4 Two-point streamwise velocity correlation coefficient in the spanwise direction, C_{uu}. Spatial distribution at $y/\delta_0 = 0.5$ (*left*) and spanwise evolution at four streamwise locations (*right*)—upstream of the interaction (*solid*), in the separated region (*dashed* and *dashdot*) and downstream of the interaction (*dotted*)

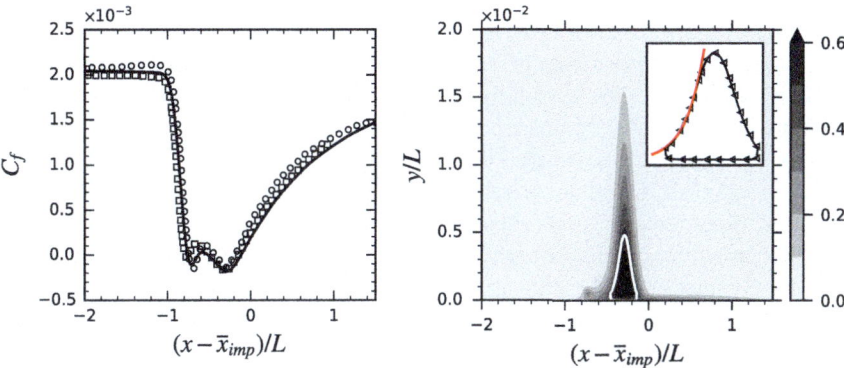

Fig. 5 Streamwise evolution of mean friction coefficient (*left*) compared with simulation results from [26] (□) and from [27] (○)—Spatial distribution of flow reversal probability (*right*), with contour of null mean streamwise velocity (*solid white*). The inset shows the contour of null streamwise velocity (*black* with ◁ symbols) with an exponential best-fit line (*red*)

magnitude of the second minimum is identical among the different results, the depth of the first lobe decreases with increasing Reynolds number. This effect was already reported in [27].

To further comment on the complex pattern of the friction coefficient in the interaction region, Fig. 5 (*right*) shows the flow reversal probability (that is to say the probability for the flow to exhibit a negative streamwise velocity component). Interestingly, the second lobe displays a high probability of reverse flow (above 60%), while the first lobe barely exceeds 30%. Besides, a clear recirculation bubble is highlighted for the second lobe only (see *white* contour) and matches the contour at 50% probability of flow reversal. The fact that the first lobe is not associated to a high probability of flow reversal means that it corresponds to rare events during which the friction coefficient is strongly negative. The maximum height of the bubble in wall unit is $h^+ \approx 11$, which is almost four times lower than the value reported in [21] ($h^+ \approx 41$) but, however, compares well with [27] ($h^+ \approx 7$). The inset reproduces the contour of the recirculation bubble and shows that its front part can be approximated by an exponential curve, a fit suggested in [21].

The streamwise evolution of mean wall pressure is depicted in Fig. 6 (*left*) together with simulation results of [26, 27] and experimental measurements given in [28]. Scaled by the length of interaction, the results are practically all on top of each other. The sharp pressure gradient imposed by the shock system in the potential flow is smeared in the boundary layer, giving a smooth increase of pressure as soon as the flow enters the interaction region. Downstream of the interaction region, pressure slowly reaches the imposed value.

On the *right* of Fig. 6 is illustrated the streamwise evolution of wall pressure fluctuations. In the present case, it was obtained from a series of probes recording static pressure at the sampling rate of 40 MHz. Upstream of the interaction, the level is approximately equal to 4% of the upstream static pressure, an offset of 1% with respect to the DNS of [26], who reported slightly less than 3%. In the interaction region and downstream of it, the same trend as the DNS is observed but the offset is still present. The reason for the additional noise might come from the relatively short domain in comparison to [26], preventing the level of fluctuation to sufficiently decrease to reach the DNS result. Moreover, the digital filter introduces additional acoustic disturbances, resulting in an over-prediction of the wall pressure variance. Measurements from [28] are known to under-estimate the pressure fluctuations because of the cutoff frequency of the pressure transducers employed to acquire the data. By integrating the PSD up to that cutoff frequency (20 kHz), a much better agreement is found between simulation and experiment, as witnessed by the *dashed*

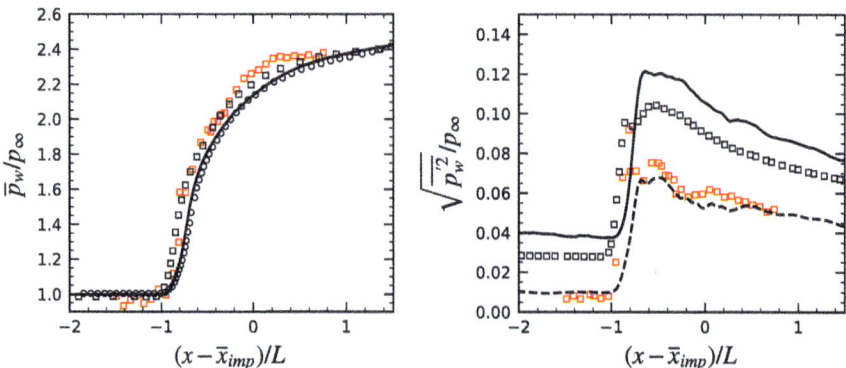

Fig. 6 Streamwise evolution of mean wall pressure (*left*) and wall pressure fluctuations (*right*). Present case (*solid*), experimental data from [28] (*red* □), simulation results from [26] (*black* □) and from [27] (*black* ○). The *dashed* line represents the variance evaluated up to a cutoff frequency of 20 kHz

Table 2 Comparison of the length of interaction and length of separation

	Re_θ	L/δ_0	L_{sep}/δ_0
Vyas et al. [31]	4.6×10^3	2.94	1.51[b]
Morgan et al. [27]	4.8×10^3	3.02	1.61
Present case	5.2×10^3	3.12	2.09
Bernardini et al. [26]	6.9×10^3	3.30	2.16
Agostini et al. [29]	5.0×10^3	3.45	2.76[a]
Experiment [32]	4.5×10^3	4.18	3.34[a]
Touber [21]	5.0×10^3	4.80	3.90
Vyas et al. [30]	4.6×10^3	5.20	4.70[b]

[a] Value estimated assuming $L_{sep} = 0.8L$ [33]
[b] Value estimated from the friction coefficient curve

line. It indicates therefore that the additional noise does not affect the low-frequency content (that is to say below the cutoff frequency) in the interaction region.

Two length scales of interest in shock wave/boundary layer interactions are the length of interaction L and the length of separation L_{sep}. The former is defined by the distance between the reflected shock foot and the incident shock foot, the positions of which result from the extrapolation of the shocks down to the wall. Table 2 compares the length scales found in the experiment and also in various high-fidelity simulations. The length of interaction in the present case is much shorter than in the experiment. It is actually consistently under-predicted by ILES and DNS when using periodic boundary conditions. The experiment suffers indeed from three-dimensional effects, strengthening the interaction at the centerline and therefore resulting in a longer length of interaction. The present result falls in line with similar simulations and L follows an increasing trend with increasing Reynolds number. Touber [21] and Agostini et al. [29] managed to obtain a value closer to the experiment but the results are affected by the choice of a particular subgrid-scale model. Including the sidewalls, [30] reported a length of interaction longer than the experiment.

For separated interactions, the length of separation L_{sep} is simply defined by the extent of the separation bubble. Figure 5 revealed, however, the existence of two smaller separated regions in the mean field. From an instantaneous perspective, these can actually merge into a single bubble, such that the length of separation is taken here as the distance between the first and the last location of null C_f. The present result is indicated in Table 2 and again conforms with ILES and DNS data. Also, following the interaction length, the separation bubble is much shorter than in the experiment and increases with increasing Reynolds number.

One of the shortcomings of low-fidelity methods when simulating oblique SWBLI is their inability to reproduce the low-frequency motion of the reflected shock, the flow feature at the origin of the research effort on SWBLI. High-fidelity methods were proven to succeed in doing so and it is therefore of prime interest to check if the high-order solver employed in this work does not escape the rule. Figure 7 shows the weighted premultiplied Power Spectral Densities (PSD) obtained from a series of

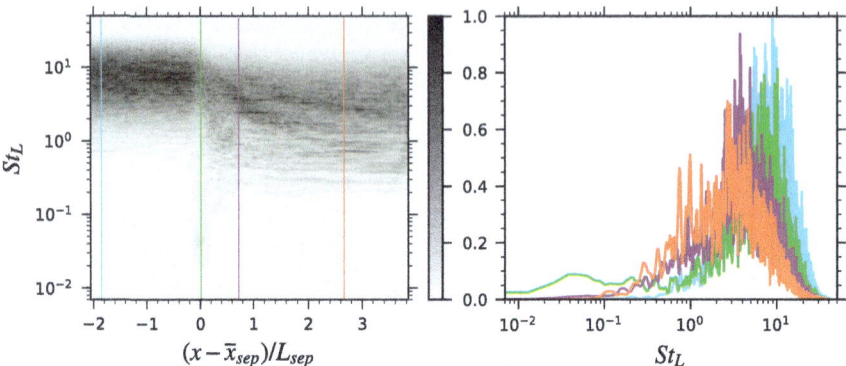

Fig. 7 Weighted premultipied Power Spectral Density of wall pressure—streamwise evolution (*left*) and four selected stations (*right*) corresponding to the upstream boundary layer (*blue*), the separation point (*green*), the minimum of friction coefficient (*purple*) and the relaxation zone (*orange*)

probes recording the wall pressure at mid-span. The PSDs are computed by using the Welch's method [34] with 7 blocks, Hamming windows and a typical overlap of 50%. The *left* figure reports the streamwise evolution of the PSDs, as a map. Several zones are discriminated beginning with the upstream boundary layer, characterized by the ridge centered at around $St_L \approx 10$. Near the mean separation point, an energetic broadband low-frequency region is observed, witnessing the low-frequency motion of the reflected shock. As expected, the associated Strouhal number is two orders of magnitude lower than in the incoming boundary layer. Then, in the interaction region, intermediate frequencies develop in the detached shear layer. Finally, this intermediate range remains in the relaxation zone, downstream of the interaction. The *right* figure illustrate the PSDs at four selected stations, corresponding to the upstream boundary layer, the mean separation point, the minimum of friction coefficient and the relaxation region. The low-frequency content of the PSD near the mean separation point clearly stands out. It is widespread, with a peak located at $St_L \approx 0.04$. The intermediate frequency range emerges for the two downstream locations at $St_L \approx 1$.

4 Periodically Forced Transonic Flow Over a Bump

The solver being validated on a canonical configuration, it can now be employed to simulate other configurations of interest. In this study, the focus is on harmonic turbulence in shock-induced separated flows. A simple yet relevant configuration is therefore the transonic flow over a two-dimensional bump. The oscillatory component of the flow is promoted by imposing a periodic forcing of the backpressure and the sensitivity of the flow to the forcing frequency is investigated. The flow conditions and the computational setup are first described. The results section then features the analysis of several high-fidelity ILES (of which the complete analysis can be found in [35]) as well as a comparison with the performance of lower-fidelity methods.

4.1 Flow Conditions and Computational Setup

The case under investigation is the transonic flow over a bump, the geometry of which is taken from the experiment of [36]. The bump length B_l is 0.184 m and its thickness B_h is 10.48 mm, whereas the wind tunnel height L_y is 0.12 m. The upstream conditions are a total pressure of 160 kPa, a total temperature of 300 K and a Mach number of 0.7. Whereas various levels of backpressure were imposed in the experiment, the focus is here for the case with a mean static pressure $\overline{p}_o$ of 106 kPa. At these conditions, a shock wave develops in the rear part of the bump, promoting flow separation. Various cases will be considered in this work depending on the way the backpressure evolves in time. The case with steady backpressure ($A_{\widetilde{p}_o} = 0$) will be referred to as the baseline case. Then, three cases with sinusoidally-varying backpressure will be presented. In the context of turbomachinery, this perturbation mimics the potential effects of a rotor/stator interaction [37]. The amplitude is fixed at 2% of the mean ($A_{\widetilde{p}_o}/\overline{p}_o = 0.02$) and the frequency is either 250, 500 or 1000 Hz. The corresponding reduced frequencies $f_r = f B_l / U_\infty$ are, respectively, ≈0.2, ≈0.4 and ≈0.8. These are values commonly met in low and high pressure turbines [2] and are consequently relevant for turbomachinery applications. The Reynolds number based on the bump length Re_{B_l} amounts to ≈1.9×10^5, which is 20 times lower in comparison to the experiment. The fluid is therefore air assumed as a perfect gas but with a reference dynamic viscosity multiplied by the same factor.

The computational domain is a rectangular box with the bump geometry as bottom boundary. With respect to the bump, the beginning of which is located at $x = 0$ m, the domain extends from $30\delta_0$ upstream to $20\delta_0$ downstream. In the spanwise direction, the domain is $4\delta_0$ wide. Following the experimental measurements of [38], the reference boundary layer thickness δ_0 is here 8.95 mm, measured at $x = -0.1$ m. This value was also considered in other numerical studies [39].

The mesh consists of hexahedra only. For ILES, using the high-order flux reconstruction approach, the target grid resolution is evaluated by considering a uniform distribution of the solution points within the cell, here with polynomial order three. In the streamwise direction, the grid spacing is initially constant, with $\Delta x^+ = 16$, in wall units based on the upstream conditions. Over the last $10\delta_0$, the mesh is progressively coarsened to $\Delta x^+ = 160$ to dampen high-frequency reflected waves. In the spanwise direction, $\Delta z^+ = 12$. The mesh is stretched in the wall-normal direction. Bottom and top boundary layers comprise 100 solution points each, the first one targeting $y^+ = 1$. The exact distribution of the solution points leads in fact to $y_w^+ \approx 0.28$. From the edge of the boundary layers and in the free stream, $\Delta y^+ = 16$. The total number of degrees of freedom rises to approximately 80 million.

For RANS, URANS and NLH, a coarser mesh ($N_x \times N_y$ = 273 × 97) is employed but with keeping $y_w^+ = 1.0$.

The inlet boundary is fully subsonic, with total pressure, total temperature and velocity direction imposed. These profiles, as well as Reynolds stress profiles (needed for the digital filtering approach for the ILES), are taken from the averaged solution of a precursor simulation of a turbulent boundary layer in the same flow conditions. The top and bottom boundaries are no-slip adiabatic walls and periodic boundary conditions are imposed in the spanwise direction. A spatially constant static pressure is imposed along the fully subsonic outlet boundary. For ILES and URANS, it consists in a sinusoidally varying temporal signal with the appropriate amplitude and frequency, while for the NLH, the frequency and the imaginary part of the perturbation are prescribed.

The explicit time step for the ILES is 4×10^{-8} s and corresponds to a CFL number of around 2.5. The simulation is initially restarted from an initial RANS solution for a duration of around 45 convective time units (CTU) to set up the flow. One CTU is evaluated here with respect to the bump length and is therefore equal to B_l/U_∞. After the transient phase, the ILES is run over approximately 25 CTUs. In RANS and NLH, the simulations are run until reaching convergence. Finally, in URANS, data is accumulated over two periods after the transient.

4.2 Results

4.2.1 High-Fidelity ILES

To start with an overall description of the flow field, Fig. 8 shows instantaneous contours of density gradient magnitude at mid-span and streamwise velocity near the bump wall, at $y^+ \approx 10$. A fully turbulent boundary layer is observed upstream of the bump, with its characteristic streaks. Approaching the bump, the flow slightly decelerates on the concave part and then quickly accelerates as it evolves on its convex part. The boundary layer undergoes partial re-laminarization due to the favorable pressure gradient there, which is also witnessed as the structures are widening in the spanwise direction. An oblique compression wave is generated when the flow separates and forms a large lambda pattern as it joins the normal shock standing downstream, responsible for the remaining compression. The separated shear layer is unstable, breaks down to turbulence and as a consequence, additional weak oblique compression waves are observed at the root of the normal shock. Finally, the boundary layer slowly recovers its initial, unperturbed state while reaching the end of the domain as thin and elongated structures appear again.

Figure 9 (*left*) illustrates the mean friction coefficient with its Probability Density Function (PDF). As the incoming flow is subsonic, the bump has an upstream influence that is observed up to $x/B_l \approx -0.45$. Nevertheless, the friction coefficient steadily decreases already before, further confirming that the inflow is far enough

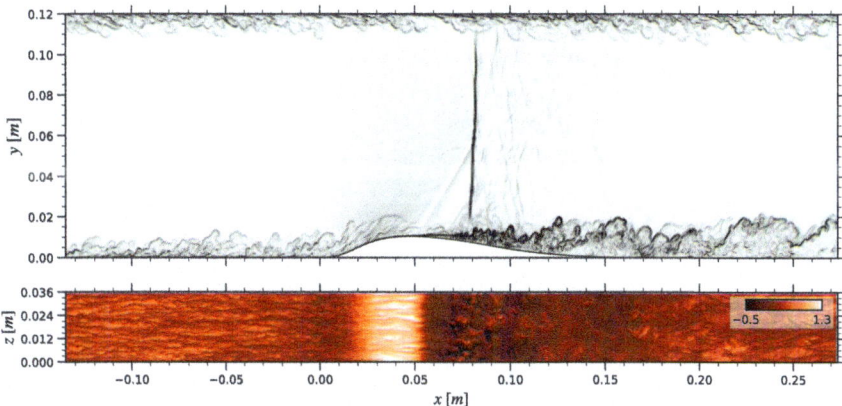

Fig. 8 Instantaneous density gradient magnitude at mid-span (*top*) and instantaneous streamwise velocity u/U_∞ near the bump wall, $y^+ \approx 10$ (*bottom*)

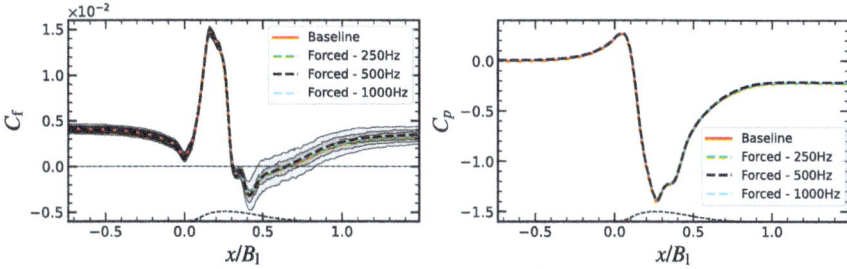

Fig. 9 Comparison of mean friction coefficient (*left*) with superimposed PDF obtained in the unperturbed case and mean wall pressure coefficient (*right*) on the bump wall between the baseline and the forced cases. The *dashed* line represents the geometry

from the bump for the turbulence to develop properly. Over the bump, the friction coefficient reaches its maximum 1.47×10^{-2} at $x/B_l \approx 0.15$, in the favorable pressure gradient region. The flow then separates at $x/B_l \approx 0.3$ and reattaches at $x/B_l \approx 0.65$. It is noted that the first location is slightly downstream the section throat ($x/B_l \approx 0.26$). In between, the distribution is typical of thin separated zones [47] with first a short region over which the skin friction is barely negative, and a second, longer region with larger negative values. The minimum is -3.2×10^{-3} and is found at $x/B_l \approx 0.4$. The superimposed PDF shows moreover that the first part is associated with a low variance. It is actually referred to as the region of stable recirculation [48]. On very rare occasions, the flow almost reattaches (at $x/B_l \approx 0.35$). The second part exhibits a much higher variance that is linked to the vortex shedding occurring at the breakdown of the shear layer. Similar descriptions are provided in previous studies on the same bump geometry, even though the flow conditions were different [39, 48] but also for other configurations with [47] or without shock wave interaction [49–51]. A comparison is also provided between the baseline and the

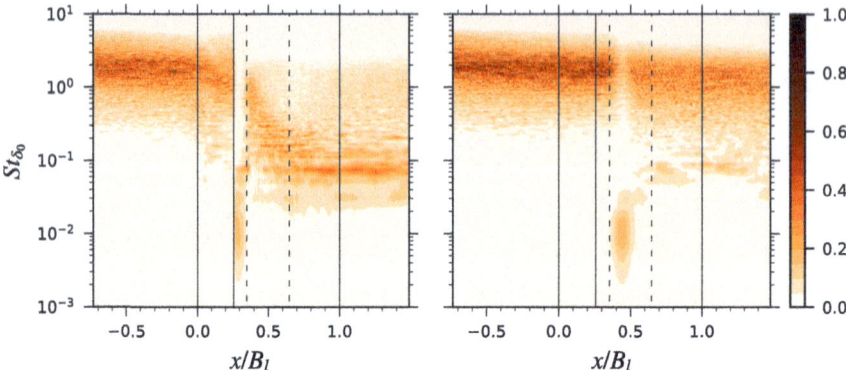

Fig. 10 Weighted premultiplied PSD maps of bottom (*left*) and top (*right*) wall pressure for the baseline case

three forced cases. No distinction can be made. It indicates first that the harmonic disturbance has no effect on these mean quantities, which is explained by the low perturbation amplitude that has been prescribed. Moreover, the frequency has no impact neither.

To summarize how the flow responds to the forcing, Power Spectral Density maps of wall pressure are now considered. Figure 10 shows the weighted premultiplied Power Spectral Density map of wall pressure, normalized by its global maximum. The PSDs have been evaluated using the Welch periodogram method [34]. Various locations are highlighted by vertical lines to ease the analysis. Solid lines refer to geometrical stations whereas dashed lines are related to physical phenomena. These locations are, from left to right, the beginning of the bump, the bump throat, the end of the region of stable recirculation, the reattachment point and the end of the bump.

The upstream boundary layer is characterized by the ridge centered at $St_{\delta_0} \approx 1$. A broadband low-frequency energetic contribution is observed at both walls, around $St_{\delta_0} = 0.01$. At the bottom wall, this contribution starts from the separation point and is consequently associated to the front leg of the lambda shock. It is moreover contained within the region of stable recirculation. At the top wall, this contribution is located at $x/B_l \approx 0.45$, and is therefore related to the normal shock (see Fig. 8). These results indicate that the entire shock system is naturally oscillating. Actually, observing a low-frequency unsteadiness at a Strouhal number that is two orders of magnitude lower than the incoming boundary layer is typical for shock wave/boundary layer interactions with separation [40]. The spurious contribution of the shock motion, at $St_{\delta_0} \approx 0.03$ and originating from the reflective outlet, is not influencing the wall pressure significantly. At the bottom wall, the upstream ridge is progressively shifted toward $St_{\delta_0} \approx 0.1$ in the interaction region. These intermediate frequencies develop as a consequence of the vortex shedding occurring at the breakdown of the shear layer and persist in the downstream boundary layer, from the reattachment point onward. Some contributions at intermediate frequencies are also

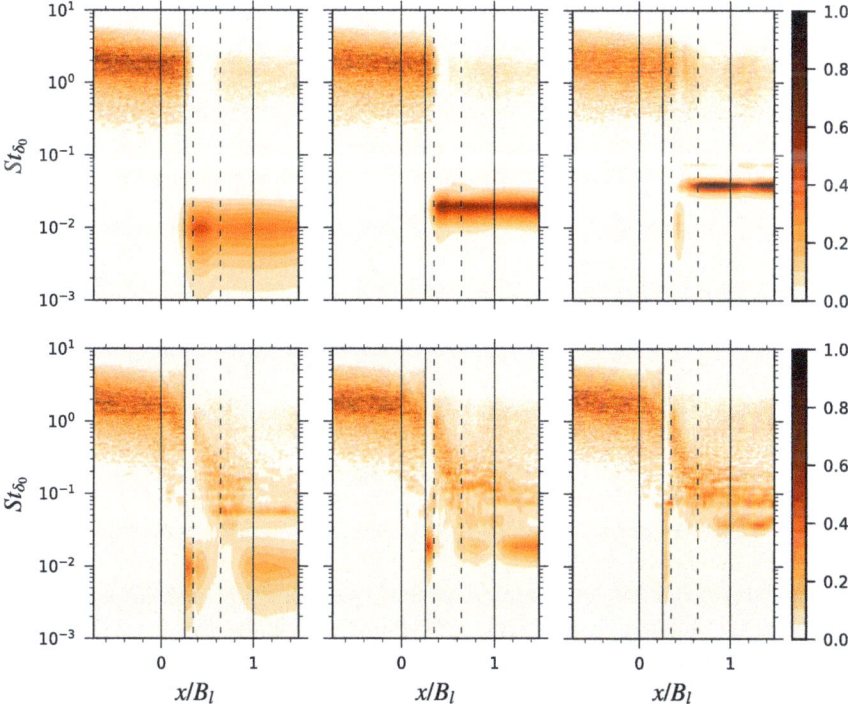

Fig. 11 Weighted pre-multiplied PSD maps of top (*top*) and bottom (*bottom*) wall pressure for the forced cases—250 Hz (*left*), 500 Hz (*center*) and 1000 Hz (*right*)

captured at the top wall, downstream of the interaction, but most of the variance of the signal is due to the barely perturbed boundary layer.

Figure 11 shows the weighted premultiplied Power Spectral Density maps of wall pressure for each forced case. As for the baseline case, the PSDs have been obtained by using the Welch method. When the flow is perturbed at 250 or 500 Hz (*left* and *center* figures, respectively), a strong influence of the forcing is perceived at the as soon as the flow separates at the bottom wall. In particular, the region of stable recirculation only receives a contribution corresponding to the forcing frequency. This influence persists further downstream and at some locations it conceals the contribution from the intermediate frequencies related to vortex shedding. At the top wall, the perturbation frequency is virtually the sole contributor in the region downstream of the interaction. The extent of the gap between the ridges corresponding to the upstream and downstream boundary layers (the latter being barely detectable) is larger at lower forcing frequency, and is directly reflecting the amplitude of the shock motion. When the flow is forced at 1000 Hz (see *right* figures), the similarity with the baseline case is striking (see Fig. 10). At the bottom wall, the contribution from the forcing frequency is almost indistinguishable from the vortex shedding contribution. Approaching the separation bubble, its influence vanishes and the broadband

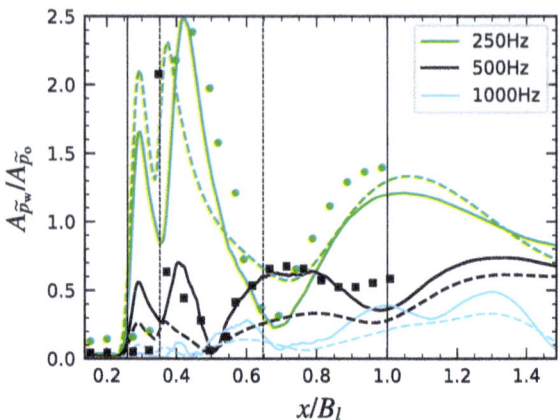

Fig. 12 Bottom wall pressure amplification factor—ILES (*solid*), URANS (*dashed*) and experiments from [36] (*symbols*)

low-frequency energetic region is retrieved at $St_{\delta_0} \approx 0.01$ in the region of stable recirculation. At the top wall, the perturbation frequency stands out but the contribution from the natural shock oscillation is also detected.

The streamwise evolution of the bottom wall pressure amplification factor ($A_{\tilde{p}_w}/A_{\tilde{p}_o}$) at the forcing frequencies is illustrated in Fig. 12, *left*. For all the cases, the three first local extrema are co-located. The first and second amplification peaks are positioned at $x/B_1 \approx 0.3$ and $x/B_1 \approx 0.4$ and are caused by the oscillation of the weak oblique compression wave emanating from the separation point and of the normal shock, respectively. In between, the first attenuation peak is related to the end of the region of stable recirculation ($x/B_1 \approx 0.35$). Further downstream, in the subsonic boundary layer, a succession of lobes is observed. With increasing frequency, these lobes are shrunk and shifted toward more upstream locations, indicating upstream traveling waves. The ratio between the size of the first lobe (equal to half of the wavelength) and the period is constant for all frequencies and gives a propagation velocity of $\approx$87.5 m/s. The frequency insensitivity of the pattern under the shock region compared to the downstream boundary layer is in line with the conclusions of [41]. Acoustic waves are damped as they propagate upstream, because of viscous effects, and therefore the strong pressure amplification is due to the oscillation of the shock system in the region beneath. The shock system position being, on average, independent of the frequency, so are the locations of the three first extrema.

The available experimental results from [36] at the reference Reynolds number are reported as well. Because of the different shock structure and separation bubble topology, no agreement is expected for the first extrema which are related to these features. Nonetheless, the first downstream lobe is reasonably well captured, revealing that the upstream propagation of pressure waves inside the boundary layer is not subject to Reynolds number effects. These comparisons give further confidence in the computational setup for the investigation performed in this work.

While the pattern beneath the shock region is independent of the frequency, the magnitude of the amplification factor is strongly affected and clearly decreases with increasing frequency. It actually reflects the extent of the shock system displacement. For a larger displacement (and therefore for a lower forcing frequency), a bigger portion of the pressure gradient will be felt by a fixed point on the wall, resulting in a higher pressure amplitude.

4.2.2 Comparison with RANS, URANS and NLH

In this section, results from RANS, URANS and NLH simulations are presented. In all these methodologies, the eddy viscosity is evaluated with the Spalart–Allmaras model [42]. To begin with, Fig. 13 shows the mean density contours for the case forced at 500 Hz, obtained from URANS and NLH. The first observation is that the NLH method is able to reproduce the same mean flow as URANS, the latter being identical to RANS (not shown here). Secondly, similar features are reported compared to ILES, that is to say an early separation of the boundary layer, leading to the development of a weak oblique compression wave and the appearance of a large lambda shock system.

The agreement between RANS, URANS and NLH for the mean flow is further demonstrated in Fig. 14, comparing the mean friction and pressure coefficients on the bottom wall. A perfect match between all the methodologies is reported. With respect to ILES (see Fig. 9), the separation point is rather well located, whereas the reattachment point is predicted much more downstream, leading to a larger separation bubble in low-fidelity methods.

The larger separation bubble obtained in RANS, URANS and NLH compared to ILES is actually an effect of the turbulence model employed. In Table 3, the locations of the separation and reattachment points are compared all together for various turbulence models such as the Spalart-Allmaras model [42], the $k - \omega$ model [43], the $k - \omega$ SST model [44] and the EARSM implemented in [45]. The results from the baseline, unperturbed, ILES are also reported. None of the models agree on

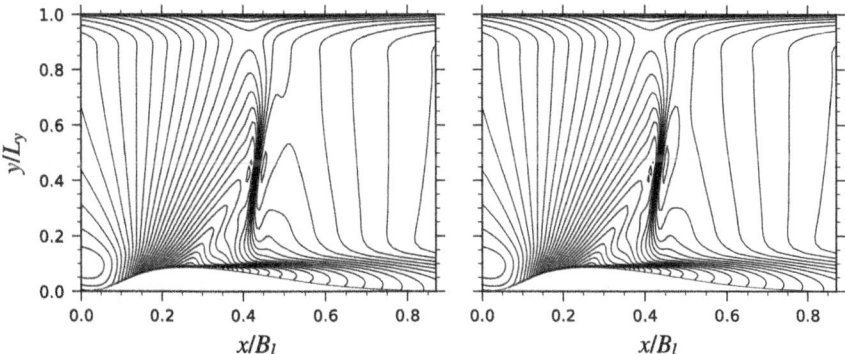

Fig. 13 Mean ρ/ρ_∞ from URANS (*left*) and NLH (*right*) simulations at $Re_{B_l} = 1.91 \times 10^5$ and $f = 500$ Hz, 40 equally-spaced contours between 0.4 and 1.1

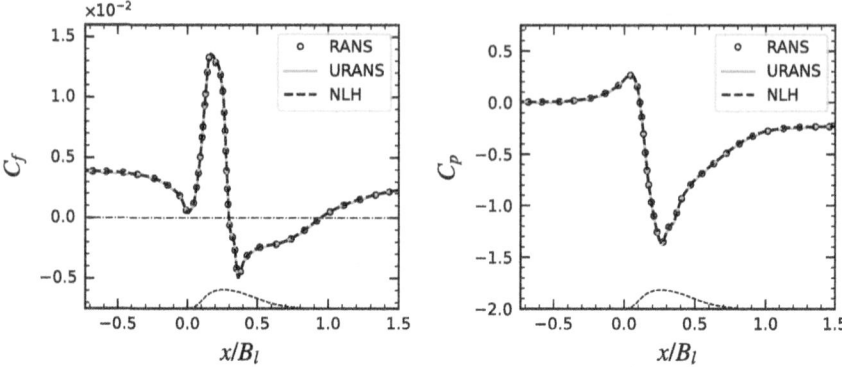

Fig. 14 Comparison of mean friction coefficient (*left*) and mean wall pressure coefficient (*right*) on the bump wall between RANS, URANS and NLH simulations at $Re_{B_l} = 1.91 \times 10^5$. For the forced cases, $f = 500\,\text{Hz}$. The *thin dashed* line represents the bump geometry

Table 3 Comparison of separation and reattachment points location as well as shock position at mid-height for various turbulence models

	x_{sep}/B_l	x_{rea}/B_l
Spalart–Allmaras	0.298	0.947
$k - \omega$	0.337	0.731
$k - \omega$ SST	0.296	0.872
EARSM	0.310	0.663
Baseline ILES	0.306	0.640

the same location of separation and reattachment. However, the EARSM shows the smallest separation bubble with, in particular, the most upstream reattachment, which makes it the closest to the ILES. On the opposite lies the Spalart-Allmaras model, which exhibits the longest separation bubble and the most downstream reattachment. Finally, the $k - \omega$ SST model displays the earliest separation.

The attention is now focused on the prediction of the unsteady flow and therefore, obviously, only URANS and NLH results are reported. The evolution of the first harmonic amplitude of pressure at the bottom wall is depicted in Fig. 15, for both the 500 and 1000 Hz cases. The mismatch between NLH and URANS results is clear, as the lobes downstream of the shock are shifted in the NLH and therefore not properly reproduced. Debrabandere [46] ruled out several possible explanations for this. The inclusion of higher-order cross coupling terms, the non-reflecting treatment at the outlet boundary and the neglect of the deterministic stresses were indeed not found to yield better results. Such an inconsistency was also pointed out by [10] on the same configuration, using a fully-linearized harmonic solver. The $k - \omega$ turbulence model of [43] was employed, while the Spalart-Allmaras model is used here, showing that changing the type of turbulence closure does not solve the issue neither. However,

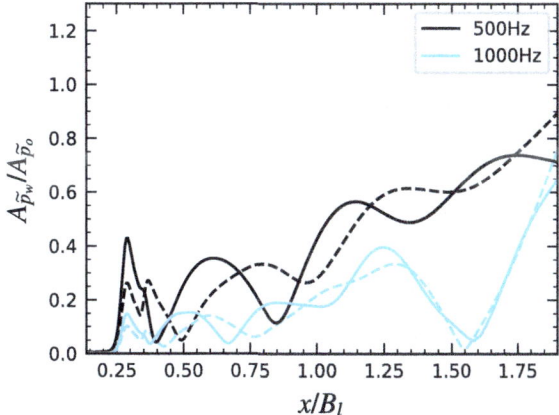

Fig. 15 Normalized wall pressure amplitude at the forcing frequencies—comparison of streamwise evolution between URANS (*dashed*) and NLH (*solid*) simulations at $Re_{B_l} = 1.91 \times 10^5$

their investigation demonstrated that taking into account harmonic turbulence allows to recover URANS results. In other words, the reason for the discrepancy lies in the freezing of the turbulence.

To compare with ILES, the results of URANS simulations are also shown in Fig. 12. The trends predicted from the URANS corroborate qualitatively those provided by ILES. However, large discrepancies are noticed in terms of amplitude, and the first extrema are not correctly located, which can be explained as a shortcoming of URANS in resolving time-dependent phenomena.

5 Conclusion

The numerical prediction of shock wave/boundary layer interactions is a challenge regardless of the method employed. On one hand, high-order methods are suitable for high-fidelity simulations but lacks robustness with respect to shock waves. On the other hand, lower-fidelity methods are robust but not accurate enough because of their modeling assumptions. A natural complementary use of these methods is to generate high-fidelity data from the former to improve models in the latter.

This chapter intended to summarize the work undertaken in [11] toward improving turbulence models for harmonic methods employing data obtained from high-fidelity simulations, and in particular for shock-induced separated flows.

The high-order solver developed for the high-fidelity simulations was briefly introduced. It is based on the flux reconstruction framework, with a Laplacian artificial viscosity to capture shock waves. Turbulence is injected by a digital filtering technique.

Two cases featuring shock wave/boundary layer interactions were then discussed. The first case is a well-known canonical oblique shock wave/boundary layer interaction, for validation purposes. The basic aspects of the flow have been detailed and, by a systematic comparison, were shown to be consistent with the rich existing literature. In particular, the typical low-frequency unsteadiness of the reflected shock could be captured.

Then, the transonic flow over a two-dimensional bump was addressed. It is indeed a relevant case for shock-induced separation, and a periodic forcing of the backpressure was further prescribed to generate the harmonic component of the flow. Results from ILES highlighted a large lambda shock system with a massive flow separation on the downstream part of the bump. The shock system was found to oscillate naturally at low-frequency. When subjected to the harmonic forcing, the flow behavior ranges from a fully locked configuration at the forcing frequency, for low forcing frequencies, to a decoupling between the natural and the forced flow, when increasing the forcing frequency.

Finally, a comparison between low-fidelity methods and ILES was performed for the bump case. It showed a good agreement for the mean flow but not for its unsteady features. URANS gives the closest results to the high-fidelity ILES, while the NLH method is not as accurate. The latter is attributed to the treatment of turbulence in the NLH method, which is the target for improvement using the high-fidelity data, in a future work.

Acknowledgements This work is part of a project that has received funding from the European Union's Horizon 2020 research and innovation program under grant agreement no. 860909 (TEAMAero-Towards Effective Flow Control and Mitigation of Shock Effects In Aeronautical Applications).

References

1. DeBonis, J.R., Oberkampf, W.L., Wolf, R.T., Orkwis, P.D., Turner, M.G., Babinsky, H., Benek, J.A.: AIAA J. **50**, 891 (2012)
2. Tucker, P.G.: Prog. Aerosp. Sci. **63**, 1 (2013)
3. Rumsey, C.L., Coleman, G.N.: NASA Symposium on Turbulence Modeling: Roadblocks, and the Potential for Machine Learning (2022)
4. Garnier, E., Sagaut, P., Delville, M.: AIAA J. **40**, 1935 (2002)
5. Pirozzoli, S., Grasso, F.: Phys. Fluids **18**, 065113 (2006)
6. Wang, Z.J.: Phil. Trans. R. Soc. A **372**, 20130318 (2014)
7. Tyacke, J., Vadlamani, N.R., Trojak, W., Watson, R., Ma, Y., Tucker, P.G.: Prog. Aerosp. Sci. **110**, 100554 (2019)
8. He, L., Ning, W.: AIAA J. **36**, 2005 (1998)
9. Vilmin, S., Lorrain, E., Hirsch, C., Swoboda, M.: In: ASME Turbo Expo: Power for Land, Sea, and Air, Volume 6: Turbomachinery, Parts A and B, 1227 (2006)
10. Philit, M., Ferrand, P., Labit, S., Chassaing, J.C., Aubert, S., Fransson, T.: In: 28th International Congress of Aeronautical Sciences (2012)
11. Goffart, N.: From high-fidelity high-order to reduced-order modeling for unsteady shock wave/boundary layer interactions. Ph.D. Thesis, Sapienza University of Rome (2024)
12. Huynh, H.T.: In: 18th AIAA Computational Fluid Dynamics Conference, 4079 (2007)
13. Carpenter, M.H., Kennedy, C.A.: Fourth-Order 2N-Storage Runge-Kutta Schemes (1994)
14. Persson, P.-O., Peraire, J.: In: 44th AIAA Aerospace Sciences Meeting and Exhibit, 112 (2006)

15. Ducros, F., Ferrand, V., Nicoud, F., Weber, C., Darracq, D., Gacherieu, C., Poinsot, T.: J. Comput. Phys. **152**, 517 (1999)
16. Wang, C., Zhang, X., Shu, C., Ning, J.: J. Comput. Phys. **231**, 653 (2012)
17. Klein, M., Sadiki, A., Janicka, J.: J. Comput. Phys. **186**, 653 (2003)
18. Adler, M., Gonzalez, D., Stack, C., Gaitonde, D.: Comput. & Fluids **165**, 127 (2018)
19. Xie, Z., Castro, I.: Flow Turbul. Combust. **81**, 449 (2008)
20. Lund, T., Wu, X., Squires, K.: J. Comput. Phys. **140**, 233 (1998)
21. Touber, E.: Unsteadiness in shock-wave/boundary layer interactions. Ph.D. Thesis, University of Southampton (2010)
22. Rizzi, A., Eliasson, P., Lindblad, I., Hirsch, C., Lacor, C., Haeuser, J.: Comput. & Fluids **22**, 341 (1993)
23. Jameson, A., Schmidt, W., Turkel, E.: In: 14th Fluid and Plasma Dynamics Conference, 1259 (1981)
24. Dupont, P., Piponniau, S., Sidorenko, A., Debiève, J.-F.: AIAA J. **46**, 1365 (2008)
25. Piponniau, S., Dussauge, J.-P., Debiève, J.-F., Dupont, P.: J. Fluid Mech. **629**, 87 (2009)
26. Bernardini, M., Della Posta, G., Salvadore, F., Martelli, E.: J. Fluid Mech. **954**, A43 (2023)
27. Morgan, B., Duraisamy, K., Nguyen, N., Kawai, S., Lele, S.K.: J. Fluid Mech. **729**, 231 (2013)
28. Dupont, P., Haddad, C., Debiève, J.-F.: J. Fluid Mech. **559**, 255 (2006)
29. Agostini, L., Larchevêque, L., Dupont, P., Debiève, J.-F., Dussauge, J.-P.: AIAA J. **50**, 1377 (2012)
30. Vyas, M.A., Yoder, D.A., Gaitonde, D.V.: In: AIAA Scitech 2019 Forum, 1890 (2019)
31. Vyas, M.A., Yoder, D.A., Gaitonde, D.V.: AIAA J. **57**, 4698 (2019)
32. Dupont, P., Haddad, C., Ardissone, J.-P., Debiève, J.-F.: Aerosp. Sci. Technol. **9**, 561 (2005)
33. Clemens, N.T., Narayanaswamy, V.: In: 39th AIAA Fluid Dynamics Conference, 3710 (2009)
34. Welch, P.D.: IEEE Trans. Audio Electroacoust. **15**, 70 (1967)
35. Goffart, N., Tartinville, B., Pirozzoli, S.: AIAA J. **62**, 940 (2024)
36. Bron, O.: Etude Expérimentale et Numérique de l'interaction Onde de Choc-Couche Limite en Ecoulement Transsonique Instationnaire. PhD Thesis, Ecole Centrale de Lyon (2004)
37. Korakianitis, T.: J. Turbomach. **115**, 118 (1993)
38. Sigfrids, T.: Hot wire and PIV studies of transonic turbulent wall-bounded flows. Licenciate Thesis, Royal Institute of Technology, Sweden (2003)
39. Wollblad, C., Davidson, L., Eriksson, L.-E.: AIAA J. **44**, 2340 (2006)
40. Clemens, N.T., Narayanaswamy, V.: Annu. Rev. Fluid Mech. **46**, 469 (2014)
41. Bur, R., Benay, R., Galli, A., Berthouze, P.: Aerosp. Sci. Technol. **10**, 265 (2006)
42. Spalart, P.R., Allmaras, S.R.: Flows. In: 30th Aerospace Sciences Meeting and Exhibit, 439 (1992)
43. Wilcox, D.C.: Turbulence Modeling for CFD. DCW Industries (2006)
44. Menter, F.R.: Improved two-equation $k - \omega$ turbulence models for aerodynamic flows (1992)
45. Mehdizadeh, O.Z., Temmerman, L., Tartinville, B., Hirsch, C.: In: ASME Turbo Expo: Power for Land, Sea, and Air, Volume 8: Turbomachinery, Parts A, B, and C, 2079 (2012)
46. Debrabandere, F.: Comparison of NLH and full unsteady methods on the 2D bump, 2013 (unpublished)
47. Sandham, N.D., Yao, Y.F., Lawal, A.A.: Int. J. Heat Fluid Flow **24**, 584 (2003)
48. Brouwer, J.: A study of transonic shock-wave/boundary-layer interactions using conservative, skew-symmetric finite-differences. Ph.D. Thesis, TU Berlin (2016)
49. Laval, J.P., Marquillie, M.: In: Progress in Wall Turbulence: Understanding and Modeling: Proceedings of the WALLTURB International Workshop, 203 (2011)
50. Schiavo, L., Jesus, A.B., Azevedo, J., Wolf, W.R.: Int. J. Heat Fluid Flow **56**, 137 (2015)
51. Schiavo, L., Wolf, W.R., Azevedo, J.: Phys. Fluids **29**, 115108 (2017)

Open Access This chapter is licensed under the terms of the Creative Commons Attribution 4.0 International License (http://creativecommons.org/licenses/by/4.0/), which permits use, sharing, adaptation, distribution and reproduction in any medium or format, as long as you give appropriate credit to the original author(s) and the source, provide a link to the Creative Commons license and indicate if changes were made.

The images or other third party material in this chapter are included in the chapter's Creative Commons license, unless indicated otherwise in a credit line to the material. If material is not included in the chapter's Creative Commons license and your intended use is not permitted by statutory regulation or exceeds the permitted use, you will need to obtain permission directly from the copyright holder.

Numerical Tools for High-Fidelity Simulation of SBLIs

Alessandro Ceci and Sergio Pirozzoli

Abstract Shock-wave/turbulent boundary layer interactions (SBLIs) are a typical hallmark of high-speed aerodynamics. Common examples of SBLIs can be found both in external flows, such as transonic/supersonic airfoils, wing-body junctions, aircraft control surfaces, and in internal flows such as engine supersonic inlets, compressors and turbines [1]. More broadly, SBLIs occur whenever a shock wave encounters a turbulent boundary layer developing on a solid surface. The impact of a shock on a boundary layer typically results in substantial flow separation, which can lead to significant decreases in performance. This chapter concentrates on the numerical aspects of SBLIs. In the first part, we elucidate the numerical framework utilized to simulate these interactions using Direct Numerical Simulations (DNS), both in Cartesian and curvilinear coordinates. Special emphasis has been placed on the handling of convective terms, which have been reformulated into a convenient split form ensuring the discrete preservation of total kinetic energy. Following this, the chapter presents qualitative and quantitative outcomes from a classical time-reversibility test. Subsequently, it delves into a more practical scenario, detailing the results of a fully turbulent supersonic compression corner.

Keywords Transonic flow · Shock wave—boundary layer interaction · High-fidelity simulations · Direct numerical simulations

1 Governing Equations

The physical model of the present work is based on the Navier-Stokes equation for a Newtonian, non-reacting, calorically perfect ideal gas. In the present section we recall the system of equations upon which the numerical framework is developed.

A. Ceci (✉) · S. Pirozzoli
DIMA, Sapienza University of Rome, Rome, Italy
e-mail: alessandro.ceci@uniroma1.it

S. Pirozzoli
e-mail: sergio.pirozzoli@uniroma1.it

An accurate description of the discretization strategy, both in space and time, is given to the reader, with particular emphasis on the treatment of convective fluxes, which is of crucial importance for a genuine representation of wall-bounded turbulence.

The Navier-Stokes equations for the aforementioned non-reacting, calorically perfect, ideal gas (say air), when projected onto a Cartesian reference frame, are

$$\frac{\partial \rho}{\partial t} + \frac{\partial \rho u_i}{\partial x_i} = 0 \qquad (1)$$

$$\frac{\partial \rho u_i}{\partial t} + \frac{\partial \rho u_i u_j}{\partial x_j} = -\frac{\partial p}{\partial x_i} + \frac{\partial \sigma_{ij}}{\partial x_j} \qquad (2)$$

$$\frac{\partial \rho E}{\partial t} + \frac{\partial \rho u_j H}{\partial x_j} = -\frac{\partial q}{\partial x_i} + \frac{\partial \sigma_{ij} u_j}{\partial x_j}, \qquad (3)$$

where u_i is the ith velocity component. These velocity components are denoted as u, v, w, in their respective x, y, z direction, ρ is the density field, p the pressure, $E = c_v T + u_i u_i / 2$ the total energy per unit mass (with c_v being the isochoric specific heat and T the temperature field) and $H = E + p/\rho$ the total enthalpy per unit mass. The components of the heat flux vector and viscous stress tensor are q_i, σ_{ij} respectively, which are defined as

$$q_i = -k \frac{\partial T}{\partial x_i} \qquad (4)$$

$$\sigma_{ij} = \mu \left(\frac{\partial u_i}{\partial x_j} + \frac{\partial u_j}{\partial x_i} - \frac{2}{3} \frac{\partial u_k}{\partial x_k} \delta_{ij} \right). \qquad (5)$$

The dependence of the molecular viscosity (μ) on temperature is accounted either by a power law of type $\mu/\mu_{\text{ref}} = T/T_{\text{ref}}^{0.76}$ or by Sutherland's law. The thermal conductivity, on the other hand, is expressed as $k = c_p \mu / Pr$, where c_p is the isobaric specific heat and Pr is the Prantl number, which we set to be constant and equal to 0.72. Finally, the equation of state linking pressure, density and temperature is the ideal gas law

$$p = \rho RT, \qquad (6)$$

where R is the ideal gas constant.

2 A High-Fidelity Framework for High-Speed Flows

Throughout the chapter, we follow a very robust framework for high-fidelity computations of high-speed flows as accurately described in the review work by Pirozzoli [2]. The convective, non-linear terms in the Navier-Stokes equations in Cartesian coordinates are discretized in space using accurate and stable approximations delivered by locally conservative, energy-preserving formulas of arbitrary order of accuracy. These schemes guarantee discrete conservation of the total kinetic energy in the limit case of inviscid flow [2, 3].

2.1 Spatial Discretization

We now introduce the foundational components of the analysis let us consider a candidate Eulerian flux of the type

$$\frac{\partial \rho u_k \varphi}{\partial x_k}, \tag{7}$$

where φ stands as the transported quantity in the convective terms of the Navier-Stokes equations (1), (2) and (3), i.e. $\varphi = 1$ for the mass equation, $\varphi = u_i (i = 1, 2, 3)$ in the momentum equations and $\varphi = H$ in the total energy equation. For the purpose of a straightforward analysis, we assume only one spatial dimension and an equally spaced grid stencil with nodes $x_j = j \cdot h$. A sketch of the numerical stencil is depicted in Fig. 1. We now look at finite-difference approximations, cast in conservative form, of the type

$$\left. \frac{\partial \rho u_k \varphi}{\partial x_k} \right|_{x_j} \approx \frac{1}{h} \left(\hat{f}_{j+1/2} - \hat{f}_{j-1/2} \right), \tag{8}$$

in which $\hat{f}_{j+1/2}$ is the numerical flux. Kennedy and Gruber [4] suggested that additional robustness for flows with strong density variations can be gained when the flux in Eq. (7) has the triple products fully expanded in the following generalized form

$$\frac{\partial \rho u \varphi}{\partial x} = \alpha \frac{\partial \rho u \varphi}{\partial x} + \beta \left(u \frac{\partial \rho \varphi}{\partial x} + \rho \frac{\partial u \varphi}{\partial x} + \varphi \frac{\partial \rho u}{\partial x} \right) + (1 - \alpha - 2\beta) \left(\rho u \frac{\partial \varphi}{\partial x} + \rho \varphi \frac{\partial u}{\partial x} + u \varphi \frac{\partial \rho}{\partial x} \right). \tag{9}$$

It can be shown [3] that conservative approximations of the split form can be recovered only if $\alpha = \beta = 1/4$. An energy-preserving numerical flux at the interface $j + 1/2$ can be obtained recasting in conservative form the split formulation of the Eulerian fluxes

$$\hat{f}_{j+1/2} = 2 \sum_{l=1}^{L} a_l \sum_{m=0}^{l-1} \widetilde{(\rho, u, \varphi)}_{j-m,l}, \tag{10}$$

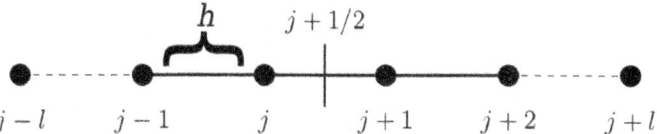

Fig. 1 Computational stencil in one direction. Nodes are marked by bullet points

where the a_l are the standard coefficients for central finite-difference approximations of the first derivative, yielding order of accuracy $2L$, and by defining the three-point averaging operator [3]

$$\widetilde{(f,g,h)}_{j,l} = \frac{1}{8}(f_j + f_{j+l})(g_j + g_{j+l})(h_j + h_{j+l}). \tag{11}$$

This discretization is applied to smooth (shock-free) regions, yielding excellent stability and accuracy performance for the description of wall-bounded turbulence in compressible flows, with no numerical dissipation. In fact, Fig. 2 shows the instantaneous density field from a DNS calculation by Ceci et al. [5] depicting an adiabatic, zero-pressure-gradient boundary layer at Mach 2. Such simulation was run with the fully central, energy-preserving, finite difference framework described above.

However, in complex compressible flows of practical interest, the flow can very often experience the presence of discontinuities. Among them, shock waves are a typical hallmark of high-speed flows. Across a shock, the flow undergoes sharp gradients in a tiny region of space proportional to the molecular mean free path. The discretization of such small regions of space, would result in a prohibitively large number of points across the discontinuity, leading to prohibitive computational cost. Despite that, the framework of locally conservative spatial schemes allows straightforward hybridization of the convective flux with classical shock-capturing reconstructions. We present now the fundamental elements that make up the analysis in "shocked" regions. For shock capturing we rely on Lax–Friedrichs flux vector splitting, where positive and negative characteristic fluxes components are reconstructed at the grid interfaces using a weighted essentially non-oscillatory (WENO) reconstruction [6]. In order to judge the local smoothness of the solution, we use a tailored version of the Ducros [7] shock sensor for effective hybridization between the energy preserving and the shock-capturing discretization

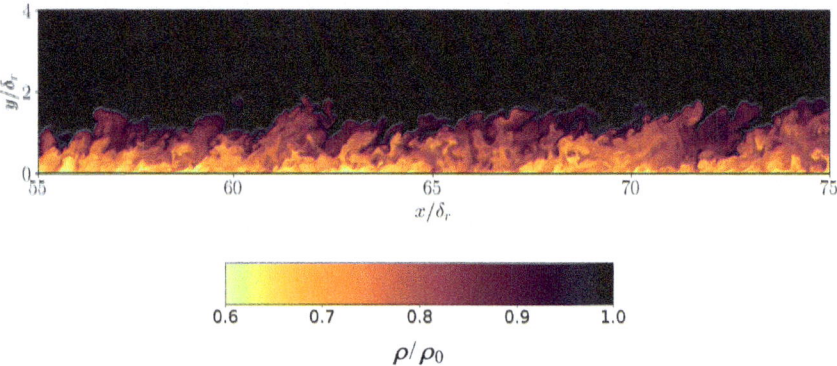

Fig. 2 Instantaneous density field of an adiabatic, zero pressure gradient turbulent boundary layer at Mach 2 [5]: Here δ_r is a reference boundary layer thickness taken at $x/\delta_0 = 200$, where δ_0 is the boundary layer thickness at the inflow. ρ_0 is the free-stream density

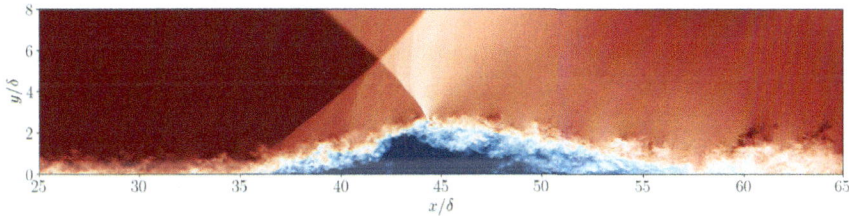

Fig. 3 Instantaneous snapshot of the streamwise velocity field u for an oblique SBLI [8]. Contour levels are $0 < u < 1$, blue to white to red $\leftrightarrow$ low to high. δ is the boundary layer thickness at a reference location upstream of the interaction

$$\Theta = \max\left(\frac{-\partial u_i/\partial x_i}{\sqrt{(\partial u_i/\partial x_i)^2 + \omega_i\omega_i + \varepsilon^2}}, 0\right) \in [0, 1], \qquad (12)$$

where ω_i is the vorticity vector. The sensor is designed to be $\Theta \approx 0$ in smooth flow regions and $\Theta \approx 1$ in the presence of shock-waves. The variable ε has the dimension of a velocity gradient, say $\varepsilon = u_0/L_0$, where suitable velocity (u_0) and length scales (L_0) must be carefully chosen so that the sensor is inactive in flow region where use of low-dissipative discretizations is of crucial importance. Effective choices of u_0, L_0 for spatially developing zero pressure gradient turbulent boundary layers are the free-stream velocity and the boundary layer thickness at the inflow. The results for such effective hybridization can be appreciated in the instantaneous field of Fig. 3 from a DNS calculation of an oblique SBLI at Mach 2.28 by Ceci et al. [8], where the shock sensor is only active across shock wave discontinuities.

Finally, the viscous fluxes of Eqs. (2) and (3), are expanded in Laplacian form in order to benefit of a higher dissipation in the wavenumber space and avoid odd-even decoupling phenomena, and are approximated with central finite-difference formulas, e.g.

$$\left.\frac{\partial}{\partial x}\left(\mu\frac{\partial u}{\partial x}\right)\right|_j = \left.\frac{\partial \mu}{\partial x}\right|_j \left.\frac{\partial u}{\partial x}\right|_j + \mu\left.\frac{\partial^2 u}{\partial x^2}\right|_j = \frac{1}{h^2}\sum_{l=-L}^{L} a_l^2 \mu_{j+l} u_{j+l} + \frac{\mu_j}{h^2}\sum_{l=-L}^{L} b_l u_{j+l}, \qquad (13)$$

in which b_l are the finite difference coefficients for the second derivative of order $2L$.

2.2 Time Integration

The numeric discretization of the spatial derivatives leads to a system of semi-discrete ordinary differential equations of the type

$$\frac{d\mathbf{w}}{dt} = \mathbf{R}(\mathbf{w}), \qquad (14)$$

where, **w** is the vector of conservative variables ($\rho, u_j, \rho E$), and **R** is the residual vector. A three-stage, third-order Runge-Kutta scheme is used to advance the solution in time [9]

$$\mathbf{w}^{(l+1)} = \mathbf{w}^{(l)} + \alpha_l \Delta t \mathbf{R}^{(l-1)} + \beta_l \Delta t \mathbf{R}^{(l)}, \quad l = 0, 1, 2 \tag{15}$$

$$\mathbf{w}^{(0)} = \mathbf{w}^n \tag{16}$$

$$\mathbf{w}^{n+1} = \mathbf{w}^{(3)}, \tag{17}$$

with integration coefficients being $\alpha_l = (0, 17/60, -5/12)$ and $\beta_l = (8/15, 5/12, 3/4)$.

One can note that the solution of the compressible Navier-Stokes equations can be, in general, more computationally demanding than the incompressible counterpart. This is due to the presence of additional terms and equations, non-uniform thermodynamic properties but also due to the acoustic time step limitation, absent in the incompressible case. Considering the non linear system of advection equations arising from the Euler equations in characteristic form, the greatest eigenvalues in the ith direction is $\lambda_i = u_i + c$ (with c being the local speed of sound). Thus, it embeds an acoustic (c) and convective (u_i) part. The acoustic contribution is suppressed in the incompressible Navier-Stokes equation, where the time-step limitation in wall-bounded turbulence is generally governed by the streamwise convective one. Hence, the time-step for explicit time advancement of compressible flows is always more limiting than the incompressible case, especially at low Mach number.[1]

3 Generalized Curvilinear Coordinates

Geometric complexities give rise to further complications, when dealing with computational problems. These are typically addressed using local mesh refinement in Cartesian frameworks or unstructured meshes, which, in turn, cause high computational cost or decreased accuracy, respectively [10]. In recent years, the Immersed Boundary Method (IBM) has gained popularity and it has been successfully used in conjunction with turbulence models to simulate high-Reynolds-number flows [11]. Within DNS, though, the IBM struggles to provide a sufficiently accurate description of the boundary layer, especially in presence of separated flow [12]. A different approach to deal with complex geometries relies on the use of body-fitted meshes in a generalized curvilinear coordinate framework. This technique ensures most accurate simulation of the fluid dynamics near the wall and straightforward implementation of high-order Finite-Difference (FD) schemes.

---

[1] We are now disregarding viscous and conductive time step limitations, which are typically less restrictive than the convective one, provided $\Delta y^+ \approx 1$, and the wall is adiabatic.

A general curvilinear coordinate system (ξ, η, ζ) is defined with reference to a Cartesian system (x, y, z) by three scalar functions

$$\begin{cases} \xi = \xi(x, y, z) \\ \eta = \eta(x, y, z) \\ \zeta = \zeta(x, y, z). \end{cases} \tag{18}$$

Our aim is to obtain a representation of the position vectors in terms of tangential or normal unit base vectors. If we consider the inverse function of the coordinate transformation

$$\mathbf{r} = x(\xi, \eta, \zeta)\mathbf{i} + y(\xi, \eta, \zeta)\mathbf{j} + z(\xi, \eta, \zeta)\mathbf{k}, \tag{19}$$

where $\mathbf{i}, \mathbf{j}, \mathbf{k}$ are the unit vectors in the three space directions, by differentiation we get

$$d\mathbf{r} = \frac{\partial \mathbf{r}}{\partial \xi} d\xi + \frac{\partial \mathbf{r}}{\partial \eta} d\eta + \frac{\partial \mathbf{r}}{\partial \zeta} d\zeta, \tag{20}$$

and the vector $\partial \mathbf{r}/\partial \xi_i$ is tangent to ξ_i at any point, with $i = 1, 2, 3$ and $\xi_1 = \xi, \xi_2 = \eta, \xi_3 = \zeta$. The three unit vectors tangent to ξ_i, called fundamental vectors, are defined as

$$\mathbf{e}_i = \frac{\partial \mathbf{r}/\partial \xi_i}{||\partial \mathbf{r}/\partial \xi_i||} = \frac{1}{h_i}\frac{\partial \mathbf{r}}{\partial \xi_i}, \quad h_i = \left|\left|\frac{\partial \mathbf{r}}{\partial \xi_i}\right|\right|, \tag{21}$$

where h_i is called metric coefficient. In a generic system, if we increase the coordinate ξ_i by an infinitesimal quantity $d\xi_i$, without changing the other two coordinates, a point will move by an elementary arc of length ds_i, which is proportional to $d\xi_i$, but not equal to it (as it would be in Cartesian coordinates). The proportionality coefficients are precisely the metric coefficients, hence $ds_i = h_i d\xi_i$. The complete elemental arc results

$$ds = ||d\mathbf{r}|| = \sqrt{h_1^2 d\xi^2 + h_2^2 d\eta^2 + h_3^2 d\zeta^2}, \tag{22}$$

and the volume of the elementary parallelepiped is given by

$$dV = ds_1 ds_2 ds_3 = h_1 h_2 h_3 d\xi d\eta d\zeta. \tag{23}$$

The product of the metric coefficients is equal to the inverse determinant of the Jacobian matrix, J^{-1}, associated to the inverse function of the coordinate transformation:

$$h_1 h_2 h_3 = J^{-1} = \left|\frac{\partial(x, y, z)}{\partial(\xi, \eta, \zeta)}\right|. \tag{24}$$

We now restrict our attention on configurations in which geometric complexities are confined in two coordinate directions, say in the $x - y$ plane. The third dimension is orthogonal to the other two and the ζ coordinate axis is always aligned with the z

direction. This implies that $\partial \xi/\partial z = \partial \eta/\partial z = 0$ and $\partial \zeta/\partial x = \partial \zeta/\partial y = 0$.[2] In this simplified case the determinant of the Jacobian matrix, J, is

$$J = \left|\frac{\partial(\xi,\eta)}{\partial(x,y)}\right| = \frac{\partial \xi}{\partial x}\frac{\partial \eta}{\partial y} - \frac{\partial \xi}{\partial y}\frac{\partial \eta}{\partial x} = \frac{1}{J^{-1}} = \left(\frac{\partial x}{\partial \xi}\frac{\partial y}{\partial \eta} - \frac{\partial x}{\partial \eta}\frac{\partial y}{\partial \xi}\right)^{-1} = \frac{1}{h_1 h_2}. \tag{25}$$

3.1 Spatial Discretization in Two-Dimensional Generalized Curvilinear Coordinates

Regarding the use of generalized curvilinear coordinates, the aforementioned framework can be applied in a straightforward manner to the calculations of the spatial derivatives in the computational space (ξ, η, ζ). Particular attention has to be given to the correct accounting of the metric terms, which plays an important role, especially in the convective terms.

This specific approach to the Navier-Stokes equations enables us to calculate the temporal evolution of the conservative variables, in Cartesian coordinates, by computing the spatial derivatives of the fluxes in a curvilinear reference system. We consider the particular case

$$\begin{cases} \xi = \xi(x,y) \\ \eta = \eta(x,y) \\ \zeta = \zeta(z). \end{cases} \tag{26}$$

We now introduce the contravariant velocity vector $\tilde{u}_j$ and the Jacobian determinant-normalized contravariant velocity vector $\hat{u}_j$, defined as

$$\hat{u} = \frac{\tilde{u}}{J} = \frac{1}{J}\left(\frac{\partial \xi}{\partial x}u + \frac{\partial \xi}{\partial y}v\right), \quad \hat{v} = \frac{\tilde{v}}{J} = \frac{1}{J}\left(\frac{\partial \eta}{\partial x}u + \frac{\partial \eta}{\partial y}v\right), \quad \hat{w} = \frac{\tilde{w}}{J} = \frac{1}{J}\left(\frac{\partial \zeta}{\partial z}w\right), \tag{27}$$

together with the normalized metric terms

$$\begin{aligned} \hat{\xi}_x &= \frac{\xi_x}{J} = \frac{1}{J}\frac{\partial \xi}{\partial x} & \hat{\xi}_y &= \frac{\xi_y}{J} = \frac{1}{J}\frac{\partial \xi}{\partial y} \\ \hat{\eta}_x &= \frac{\eta_x}{J} = \frac{1}{J}\frac{\partial \eta}{\partial x} & \hat{\eta}_y &= \frac{\eta_y}{J} = \frac{1}{J}\frac{\partial \eta}{\partial y} \\ \hat{\zeta}_z &= \frac{\zeta_z}{J} = \frac{1}{J}\frac{\partial \zeta}{\partial z}. \end{aligned} \tag{28}$$

---

[2] One must recall that in principle $\partial^2 \zeta/\partial z^2 \neq 0$, since the computational grid can be stretched in the spanwise direction z.

If the computational grid is stationary,[3] then

$$\frac{\partial}{\partial t}\left(\frac{\mathbf{w}}{J}\right) = \frac{1}{J}\frac{\partial \mathbf{w}}{\partial t}. \tag{29}$$

In order to present the treatment of convective terms, we begin with the Euler equations in a system of curvilinear coordinates determined by the mapping $x_k = x_k(\xi_j)$, with $j, k = 1, \ldots, d$ (d is the number of spatial dimensions). These equations are presented in the strong conservation form

$$\frac{1}{J}\frac{\partial \mathbf{w}}{\partial t} + \frac{\partial}{\partial \xi_j}\left(\frac{\mathbf{f}_j}{J}\right) = 0, \quad \mathbf{f}_j = \begin{bmatrix} \rho \tilde{u}_j \\ \rho u_i \tilde{u}_j + p \tilde{J}_{ji} \\ \rho \tilde{u}_j H \end{bmatrix}, \quad i = 1, \ldots, d, \tag{30}$$

where $\mathbf{w}$ is once more the vector of conservative variables in Cartesian coordinates, $\tilde{u}_j = \tilde{J}_{jk} u_k$ are the contravariant velocity components and $\tilde{J}_{jk} = \partial \xi_j / \partial x_k$. Building upon Pirozzoli's [3] work, we employ a central skew-symmetric finite difference splitting of convective derivatives. This method is not depending on explicit upwinding or filtering of the physical variables and ensures the conservation of kinetic energy under semi-discrete, low Mach number conditions. Temporarily neglecting the impact of pressure forces, any element of the vector in Eq. (30) can be expressed as

$$\frac{1}{J}\frac{\partial \rho \varphi}{\partial t} + \frac{\partial}{\partial \xi_j}\left(\frac{\rho \tilde{u}_j \varphi}{J}\right) = 0, \tag{31}$$

where φ stands for the usual generic transported scalar property, i.e. unity for the continuity equation, u_i for the momentum equations and H for the total energy equation. For the purpose of discretization, we make the assumption, without loss of generality, that the spacing in the computational space defined by the variables ξ_j are all set to one. We are examining explicit central approximations for the derivative of the general function f in the jth direction at the grid node $N \equiv (\xi_1, \ldots, \xi_d)$ of type

$$D_j f_N = \sum_{l=1}^{L} a_l (f_{j;l} - f_{j;-l})_N, \tag{32}$$

where the shorthand notation

$$(f_{j;l})_N = f_{\xi_1, \ldots, \xi_j + l, \ldots, \xi_d} \tag{33}$$

is introduced to denote a shift by l nodes in the positive-j coordinate direction about N. The coefficients a_l that define the discrete derivative operator's values are chosen in a way that maximizes the formal order of accuracy of the approximation.

---

[3] In this particular case the time integration is identical to the Cartesian framework.

Consequently, for a stencil with a given half-width L the accuracy order of the derivative approximation is $2L$.

Our goal is to devise a partitioning of the convective derivatives in Eq. (31) in a way that ensures the preservation of kinetic energy at the semi-discrete level when pressure forces are not considered. To achieve this objective, we commence with the scenario of Cartesian coordinates, where kinetic energy-preserving stabilization techniques that rely on splitting the convective terms are at our disposal [3]. To achieve this objective, we initiate the process by starting from the strong conservation form of the convective terms as described in Eq. (31). It is important to note that analogues to the Cartesian split forms [3, 13] can potentially be derived by incorporating the Jacobian determinant into either ρ, $\tilde{u}_j$, φ.

Nevertheless, a straightforward check reveals that incorporating the Jacobian determinant J into φ (i.e. by defining $\hat{\varphi} = \varphi/J$) does not yield an equivalent result to that of the Cartesian case, since à for $\varphi = 1$, $\hat{\varphi} = 1/J \neq 1$, which implies that the discrete split form of the continuity equation would differ from

$$\frac{d\rho_N}{dt} + D_j(\rho u_j)_N = 0, \quad (34)$$

and it would have an additional term when combined with the momentum equation. On the other hand, it can be shown [13] that discrete kinetic energy preservation, as for the Cartesian case, is attained if J is incorporated with the density ($\hat{\rho} = \rho/J$, RHO/J formulation) or with the contravariant velocity ($\hat{u} = \tilde{u}/J$, U/J formulation). In our case, as already presented in Eq. (30), we adhere to the U/J formulation due to its greater robustness over the RHO/J formulation [13]. Adopting the U/J formulation, Eq. (31) simply becomes

$$\frac{1}{J}\frac{\partial \rho\varphi}{\partial t} + \frac{\partial}{\partial \xi_j}\left(\rho\hat{u}_j\varphi\right) = 0. \quad (35)$$

In the footsteps of the research by Pirozzoli [13], we consider the following generalized split form of the convective terms

$$\frac{1}{J}\frac{\partial \rho\varphi}{\partial t} + \frac{1}{4}\frac{\partial \rho\hat{u}_j\varphi}{\partial \xi_j} + \frac{1}{4}\left(\rho\frac{\partial \hat{u}_j\varphi}{\partial \xi_j} + \hat{u}_j\frac{\partial \rho\varphi}{\partial \xi_j} + \varphi\frac{\partial \rho\hat{u}_j}{\partial \xi_j}\right) + \frac{1}{4}\left(\rho\hat{u}_j\frac{\partial \varphi}{\partial \xi_j} + \rho\varphi\frac{\partial \hat{u}_j}{\partial \xi_j} + \hat{u}_j\varphi\frac{\partial \rho}{\partial \xi_j}\right) = 0. \quad (36)$$

A computationally effective implementation of convective derivatives cast in split form has been proposed by Pirozzoli [3]. In relation to the Cartesian scenario, it was discovered that finite-difference approximations with local conservation properties for the convective derivative in the jth direction

$$\left(\frac{\partial \rho\hat{u}_j\varphi}{\partial x_j}\right)_N \approx \frac{1}{h}\left(\hat{f}_{j;1/2} - \hat{f}_{j;-1/2}\right)_N, \quad (37)$$

are available for particular split convective forms, like Eq. (9) [4]. In Eq. (37), $\hat{f}_{j;1/2}$ is the numerical flux in the jth direction taken at the intermediate node $(\xi_1, \ldots, \xi_{j+1/2}, \ldots, \xi_d)$ (consistent with the notation in Eq. (33)). Specifically, the subsequent expression for the numerical flux was identified for the split form proposed by Kennedy and Gruber [4]

$$(\hat{f}_{j;1/2})_N = 2 \sum_{l=1}^{L} a_l \sum_{m=0}^{l-1} \left(\widetilde{(\rho, u_j, \varphi)}_{j;-m,l} \right)_N, \tag{38}$$

where the two-point, three-variable averaging operator in the jth direction is defined as

$$\widetilde{(f, g, h)}_{j;n,l} = \frac{1}{8}(f_{j;n} + f_{j;n+l})(g_{j;n} + g_{j;n+l})(h_{j;n} + h_{j;n+l}). \tag{39}$$

Porting the following notation to a system of curvilinear coordinates (the only change being the substitution of u_j with $\hat{u}_j$), the numerical flux becomes

$$(\hat{f}_{j;1/2})_N = 2 \sum_{l=1}^{L} a_l \sum_{m=0}^{l-1} \left(\widetilde{(\rho, \hat{u}_j, \varphi)}_{j;-m,l} \right)_N. \tag{40}$$

This identical formula can also be utilized to compute a conservative approximation for the pressure gradient term when presented in the split form, i.e.

$$(\hat{f}^p_{j;1/2})_N = 2 \sum_{l=1}^{L} a_l \sum_{m=0}^{l-1} \left(\widetilde{(p, J_{ji}, 1)}_{j;-m,l} \right)_N. \tag{41}$$

which needs to be added to the i-th component of the momentum equation.

In conclusion, we propose the following semi-discrete, locally conservative approximation of the Euler equations in curvilinear coordinates

$$\frac{d\mathbf{w}_N}{dt} = -J_N \sum_{j=1}^{d} \left(\hat{\mathbf{f}}_{j;1/2} - \hat{\mathbf{f}}_{j;-1/2} \right)_N, \tag{42}$$

with

$$\left(\hat{\mathbf{f}}_{j;1/2} \right)_N = 2 \sum_{l=1}^{L} a_l \sum_{m=0}^{l-1} \left[\left(\widetilde{\rho, \hat{u}_j, u_i} \right)_{j;-m,l} + \begin{pmatrix} \widetilde{(\rho, \hat{u}_j, 1)}_{j;-m,l} \\ \widetilde{(p, J_{ji}, 1)}_{j;-m,l} \\ \widetilde{(\rho, \hat{u}_j, H)}_{j;-m,l} \end{pmatrix} \right]_N, \quad i = 1, \ldots, d. \tag{43}$$

Finally, it is important to underline the threefold relevance of locally conservative approximations (i) they immediately guarantee global conservation of mass, momentum and total energy (from telescopic property); (ii) their implementation in existing compressible flow solvers is generally simple since they are typically based on conservative approximations; (iii) they lead to straightforward hybridization with shock-capturing schemes, that are also formulated in conservative form.

As regards the hybridization part, the adopted shock-capturing schemes rely on the use of the Lax–Friedrichs flux vector splitting, whereby the components of the positive and negative characteristic fluxes are reconstructed at the interfaces using a WENO reconstruction, as for the Cartesian case. Once more, we rely on the modified version of the Ducros [7] sensor in Eq. (12) in order to judge the local smoothness of the numerical solution. For the purpose of ensuring clarity, we are not presenting the entire exposition of the WENO schemes in curvilinear coordinates. Instead, we are outlining the fundamental concepts in the hope that this will enable the reader to understand our methodology.

The first step is to express the Euler equations in their characteristic form

$$\frac{1}{J}\frac{\partial \mathbf{w}}{\partial t} + \mathbf{A}_j \frac{\partial \mathbf{w}}{\partial \xi_j} = 0, \qquad (44)$$

where $\mathbf{A}_j = \partial \mathbf{f}_j/\partial \mathbf{w} = \mathbf{R}_j \Lambda_j \mathbf{L}_j$ is the flux Jacobian matrix, $\mathbf{R}_j$, $\mathbf{L}_j$ are the associated matrices of the right and left eigenvectors and Λ_j is the diagonal matrix of the eigenvalues. In this case, we propose a different formulation for

$$\mathbf{f}_j = \begin{bmatrix} \rho \breve{u}_j \\ \rho u_i \breve{u}_j + p \breve{J}_{ji} \\ \rho \breve{u}_j H \end{bmatrix}, \qquad (45)$$

where $\breve{J}_{jk}$ are the metric terms normalized by m_j (e.g. $\breve{J}_{1k} = J_{1k}/m_1$)

$$m_1 = \left|\frac{\partial \xi}{\partial x_i}\right|, \qquad m_2 = \left|\frac{\partial \eta}{\partial x_i}\right|, \qquad m_3 = \left|\frac{\partial \zeta}{\partial x_i}\right| \qquad (46)$$

and $\breve{u}_j = \breve{J}_{jk} u_k$ is the projection of the Cartesian velocity along the ortho-normal contravariant base vectors. The left and right eigenvector matrices are evaluated at some average state (defined by the substript $_{j;1/2}$ consistent with the notation in Eq. (33)) for both flow variables and metric terms. The average state for flow variables is Roe's state, while it is the arithmetic mean for the metric terms. Positive and negative flux components, consistently normalized, are then projected along the characteristic directions using a local Lax–Friedrichs flux splitting

$$(\hat{\mathbf{g}}^{\pm}_{j;1/2})_N = \frac{1}{2}[\mathbf{L}_{j;1/2}(\mathbf{f}_j \pm |\lambda_{\max}|\mathbf{w})]_N \frac{(m_j)_N}{J_N}, \qquad (47)$$

in which $|\lambda_{\max}|$ is the maximum eigenvalue of the Jacobian matrix $\mathbf{A}_j$ over the entire WENO sub-stencil. Then, through the classical WENO interpolation, positive and negative characteristic fluxes, labeled ($\hat{\mathbf{g}}^{\pm,\text{WENO}}_{j;1/2}$), are reconstructed at the midpoint and they are brought back into the physical reference system, obtaining the numerical flux at the cell interface

$$(\hat{\mathbf{f}}_{j;1/2})_{\mathcal{N}} = [\mathbf{R}_{j;1/2}(\hat{\mathbf{g}}_{j;1/2}^{+,\text{WENO}} - \hat{\mathbf{g}}_{j;1/2}^{-,\text{WENO}})]_{\mathcal{N}}. \qquad (48)$$

Finally, the viscous terms of the Navier-Stokes equations are expanded to Laplacian form to avoid odd-even decoupling phenomena. Specifically, viscous terms of the momentum equation in the three Cartesian directions

$$\sigma_i = \frac{\partial \sigma_{ij}}{\partial x_j} = \frac{\partial}{\partial x_j}\left[\mu\left(\frac{\partial u_i}{\partial x_j} + \frac{\partial u_j}{\partial x_i} - \frac{2}{3}\frac{\partial u_k}{\partial x_k}\delta_{ij}\right)\right], \quad i=1,2,3, \qquad (49)$$

are expanded as follows

$$\sigma_i = \mu\left[\frac{\partial^2 u_i}{\partial x_j \partial x_j} + \frac{1}{3}\frac{\partial}{\partial x_i}\left(\frac{\partial u_j}{\partial x_j}\right)\right] + \varepsilon_{ij}\frac{\partial \mu}{\partial x_j}, \qquad (50)$$

where $\varepsilon_{ij} = \sigma_{ij}/\mu$.

Finally, the viscous terms of the energy equation are expressed as

$$\sigma_q = \frac{\partial(\sigma_{ij}u_i - q_j)}{\partial x_j} = \mu\varepsilon_{ij}\frac{\partial u_i}{\partial x_j} + u_i\sigma_i + \frac{c_p}{Pr}\left(\mu\frac{\partial^2 T}{\partial x_j \partial x_j} + \frac{\partial \mu}{\partial x_j}\frac{\partial T}{\partial x_j}\right). \qquad (51)$$

When focusing solely on two-dimensional orthogonal and non-orthogonal curvilinear coordinate systems, first and second derivatives terms are reconstructed in the physical space by simply applying chain-rule differentiation. In the computational space, they are approximated by using central finite-difference formulas. Let us consider the ξ direction, then they are respectively expressed, at node $\mathcal{N}$, as

$$\left.\frac{\partial \varphi}{\partial \xi}\right|_{\mathcal{N}} = \sum_{l=-L}^{L} a_l(\varphi_l)_{\mathcal{N}}, \qquad \left.\frac{\partial^2 \varphi}{\partial \xi^2}\right|_{\mathcal{N}} = \sum_{l=-L}^{L} b_l(\varphi_l)_{\mathcal{N}}, \qquad (52)$$

where a_l and b_l are the finite difference coefficients for the first and second derivative of order $2L$, respectively.

The expansion of the Laplacian terms is computed using the elements of the contravariant metric tensor g^{ij}, being

$$g^{11} = \left(\frac{\partial \xi}{\partial x}\right)^2 + \left(\frac{\partial \xi}{\partial y}\right)^2, \qquad g^{22} = \left(\frac{\partial \eta}{\partial x}\right)^2 + \left(\frac{\partial \eta}{\partial y}\right)^2. \qquad (53)$$

In the interest of a lucid exposition, we present the viscous terms only along the streamwise x−direction

$$\sigma_x = \frac{\partial}{\partial x}\left[\mu\left(\frac{4}{3}\frac{\partial u}{\partial x} - \frac{2}{3}\frac{\partial v}{\partial y} - \frac{2}{3}\frac{\partial w}{\partial z}\right)\right] + \frac{\partial}{\partial y}\left[\mu\left(\frac{\partial u}{\partial y} + \frac{\partial v}{\partial x}\right)\right] + \frac{\partial}{\partial z}\left[\mu\left(\frac{\partial u}{\partial z} + \frac{\partial w}{\partial x}\right)\right]. \tag{54}$$

Defining

$$\varepsilon_{11} = \frac{4}{3}\frac{\partial u}{\partial x} - \frac{2}{3}\left(\frac{\partial v}{\partial y} + \frac{\partial w}{\partial z}\right) \quad \varepsilon_{12} = \frac{\partial u}{\partial y} + \frac{\partial v}{\partial x} = \varepsilon_{21}$$

$$\varepsilon_{22} = \frac{4}{3}\frac{\partial v}{\partial y} - \frac{2}{3}\left(\frac{\partial u}{\partial x} + \frac{\partial w}{\partial z}\right) \quad \varepsilon_{13} = \frac{\partial u}{\partial z} + \frac{\partial w}{\partial x} = \varepsilon_{31} \tag{55}$$

$$\varepsilon_{11} = \frac{4}{3}\frac{\partial w}{\partial z} - \frac{2}{3}\left(\frac{\partial u}{\partial x} + \frac{\partial v}{\partial y}\right) \quad \varepsilon_{23} = \frac{\partial v}{\partial z} + \frac{\partial w}{\partial y} = \varepsilon_{32},$$

the viscous fluxes in the streamwise direction can be rewritten as

$$\sigma_x = \mu\left(\frac{\partial \varepsilon_{11}}{\partial x} + \frac{\partial \varepsilon_{12}}{\partial y} + \frac{\partial \varepsilon_{13}}{\partial z}\right) + \varepsilon_{11}\frac{\partial \mu}{\partial x} + \varepsilon_{12}\frac{\partial \mu}{\partial y} + \varepsilon_{13}\frac{\partial \mu}{\partial z}. \tag{56}$$

It can be easily shown that the viscous fluxes in the three Cartesian directions can be written as

$$\sigma_x = \mu\left[\nabla^2 u + \frac{1}{3}\frac{\partial}{\partial x}(\nabla \cdot \mathbf{u})\right] + \varepsilon_{11}\frac{\partial \mu}{\partial x} + \varepsilon_{12}\frac{\partial \mu}{\partial y} + \varepsilon_{13}\frac{\partial \mu}{\partial z} \tag{57}$$

$$\sigma_y = \mu\left[\nabla^2 v + \frac{1}{3}\frac{\partial}{\partial y}(\nabla \cdot \mathbf{u})\right] + \varepsilon_{21}\frac{\partial \mu}{\partial x} + \varepsilon_{22}\frac{\partial \mu}{\partial y} + \varepsilon_{23}\frac{\partial \mu}{\partial z} \tag{58}$$

$$\sigma_z = \mu\left[\nabla^2 w + \frac{1}{3}\frac{\partial}{\partial z}(\nabla \cdot \mathbf{u})\right] + \varepsilon_{31}\frac{\partial \mu}{\partial x} + \varepsilon_{32}\frac{\partial \mu}{\partial y} + \varepsilon_{33}\frac{\partial \mu}{\partial z}, \tag{59}$$

and finally each term of Eq. (57) is expanded from Cartesian to curvilinear coordinates,[4,5] where the elements of ε_{ij} can be discretized adapting the relations of Eq. (63) to the velocity components v and w:

$$\nabla^2 u = J\left[\frac{\partial}{\partial \xi}\left(\frac{g^{11}}{J}\right)\frac{\partial u}{\partial \xi} + \frac{g^{11}}{J}\frac{\partial^2 u}{\partial \xi^2} + \frac{\partial}{\partial \eta}\left(\frac{g^{22}}{J}\right)\frac{\partial u}{\partial \eta} + \frac{g^{22}}{J}\frac{\partial^2 u}{\partial \eta^2}\right] + \frac{\partial^2 u}{\partial \zeta^2}\left(\frac{\partial \zeta}{\partial z}\right)^2 + \frac{\partial u}{\partial \zeta}\frac{\partial^2 \zeta}{\partial z^2} \tag{60}$$

$$\frac{\partial}{\partial x}(\nabla \cdot \mathbf{u}) = \frac{\partial}{\partial \xi}(\nabla \cdot \mathbf{u})\frac{\partial \xi}{\partial x} + \frac{\partial}{\partial \eta}(\nabla \cdot \mathbf{u})\frac{\partial \eta}{\partial x} \tag{61}$$

$$\nabla \cdot \mathbf{u} = \frac{\partial u}{\partial \xi}\frac{\partial \xi}{\partial x} + \frac{\partial u}{\partial \eta}\frac{\partial \eta}{\partial x} + \frac{\partial v}{\partial \xi}\frac{\partial \xi}{\partial y} + \frac{\partial v}{\partial \eta}\frac{\partial \eta}{\partial y} + \frac{\partial w}{\partial \zeta}\frac{\partial \zeta}{\partial z} \tag{62}$$

$$\frac{\partial u}{\partial x} = \frac{\partial u}{\partial \xi}\frac{\partial \xi}{\partial x} + \frac{\partial u}{\partial \eta}\frac{\partial \eta}{\partial x}, \quad \frac{\partial u}{\partial y} = \frac{\partial u}{\partial \xi}\frac{\partial \xi}{\partial y} + \frac{\partial u}{\partial \eta}\frac{\partial \eta}{\partial y}, \quad \frac{\partial u}{\partial z} = \frac{\partial u}{\partial \zeta}\frac{\partial \zeta}{\partial z}. \tag{63}$$

---

[4] Not all terms are shown, but all derivatives in the Cartesian directions are expanded.

[5] This is also valid for Eqs. (58), (59) but their expansion is not shown herein.

In case of derivatives of a composite variable, say the dynamic viscosity function of temperature ($\mu(T)$), the expressions can be simply rearranged as

$$\frac{\partial \mu}{\partial x} = \frac{\partial \mu}{\partial T}\frac{\partial T}{\partial x}, \quad \frac{\partial \mu}{\partial y} = \frac{\partial \mu}{\partial T}\frac{\partial T}{\partial y}, \quad \frac{\partial \mu}{\partial z} = \frac{\partial \mu}{\partial T}\frac{\partial T}{\partial z} \quad (64)$$

$$\frac{\partial T}{\partial x} = \frac{\partial T}{\partial \xi}\frac{\partial \xi}{\partial x} + \frac{\partial T}{\partial \eta}\frac{\partial \eta}{\partial x}, \quad \frac{\partial T}{\partial y} = \frac{\partial T}{\partial \xi}\frac{\partial \xi}{\partial y} + \frac{\partial T}{\partial \eta}\frac{\partial \eta}{\partial y}, \quad \frac{\partial T}{\partial z} = \frac{\partial T}{\partial \zeta}\frac{\partial \zeta}{\partial z}. \quad (65)$$

If the curvilinear system is non-orthogonal, the only different term is the Laplacian one, which must be written using the non zero metrics tensor cross terms g^{12} and g^{21}, making the resulting formulation more cumbersome [14]. Using the aforementioned expressions we can also transform the viscous fluxes energy equation

$$\sigma_q = \mu \left(\frac{\partial u}{\partial x}\varepsilon_{11} + \frac{\partial u}{\partial y}\varepsilon_{12} + \frac{\partial u}{\partial z}\varepsilon_{13} + \frac{\partial v}{\partial x}\varepsilon_{21} + \frac{\partial v}{\partial y}\varepsilon_{22} + \frac{\partial v}{\partial z}\varepsilon_{23} + \frac{\partial w}{\partial x}\varepsilon_{31} + \frac{\partial w}{\partial y}\varepsilon_{32} + \frac{\partial w}{\partial z}\varepsilon_{33} \right) + \\ + u\sigma_x + v\sigma_y + w\sigma_z + \lambda \nabla^2 T + \frac{\partial \lambda}{\partial x}\frac{\partial T}{\partial x} + \frac{\partial \lambda}{\partial y}\frac{\partial T}{\partial y} + \frac{\partial \lambda}{\partial z}\frac{\partial T}{\partial z} \quad (66)$$

from Cartesian to curvilinear coordinates.

4 Numerical Results

In this section, we present the outcomes derived from employing the advanced high-fidelity framework discussed earlier in the chapter, which is implemented within the GPU-accelerated community code STREAmS [15, 16], developed at Sapienza University of Rome. Our investigation begins with the assessment of the energy-preserving properties of the solver, when adopting two-dimensional curvilinear coordinates. We then transition towards a case study of greater practical significance, namely DNS of a supersonic compression corner. This segment serves to demonstrate the solver adaptability within a two-dimensional energy-preserving framework employing generalized curvilinear coordinates.

4.1 Inviscid Taylor-Green Vortex

The Euler equations in a d-dimensional unbounded space have a variety of exact integrals. First, direct integration of the conservation equations yields (as obvious) conservation of the total mass, momentum, and energy. Under the assumption of smooth flow, combining the continuity and the momentum equations and integrating yields a balance equation for the kinetic energy.

$$\frac{d}{dt}\int_V \rho \frac{u_k u_k}{2} dV = -\int_{\partial V}\left(\rho \frac{u_k u_k}{2} + p\right) u_i n_i dS + \int_V p \frac{\partial u_i}{\partial x_i} dV. \tag{67}$$

This equation shows that the total kinetic energy only varies because of momentum flux through the boundary or to volumetric work of pressure forces (which is zero for incompressible flow), whereas the convective terms do not cause any net variation.[6] Hence, assuming an incompressible flow ($\partial u_i/\partial x_i \approx 0$) and an unbounded or periodic domain Eq. (67) becomes

$$\frac{d}{dt}\int_V \rho \frac{u_k u_k}{2} dV = 0. \tag{68}$$

This property has inspired numerical schemes based on the attempt to enforce "kinetic energy preservation" in the discrete sense. The energy-preserving properties of the solver are now tested for the case proposed by Duponcheel et al. [17], namely the time reversibility of the inviscid Taylor-Green [18] flow. This flow is widely studied as a model for turbulence formation from ordered initial conditions, exhibiting rapid creation of small-scale structures with incurred growth of vorticity [19].

We now describe the formulation of the problem in curvilinear coordinates. The computational domain is a $(2\pi/k_0)^3$ triply-periodic cuboid (with $k_0 = 2\pi/\lambda_{\text{ref}} = 1$, being λ_{ref} a reference wavelength), where two dimensions, say x and y, are described by the coordinate transformations

$$\begin{cases} x(\xi,\eta) = x_{\min} + \frac{\xi-1}{N_\xi-1}L_x + A_\xi(2\pi\frac{\eta-1}{N_\eta-1}) \\ y(\xi,\eta) = y_{\min} + \frac{\eta-1}{N_\eta-1}L_x + A_\eta(2\pi\frac{\xi-1}{N_\xi-1}) \\ z(\zeta) = z_{\min} + \frac{\zeta-1}{N_\zeta-1}L_z, \end{cases} \tag{69}$$

where $x_{\min} = y_{\min} = z_{\min} = 0$, $L_x = L_y = L_z = 2\pi/k_0$. The intensities of the harmonic functions are $A_\xi = 0.4/k_0$ and $A_\eta = 0.2/k_0$. The number of points in each direction is $N_\xi \times N_\eta \times N_\zeta = (32)^3$. A two-dimensional section of the grid in the $x-y$ plane is shown in Fig. 4.

The solution is initialized as follows

$$\begin{cases} \rho(x,y,z,0) = \rho_0 \\ u(x,y,z,0) = u_0 \sin(k_0 x)\cos(k_0 y)\cos(k_0 z) \\ v(x,y,z,0) = -u_0 \cos(k_0 x)\sin(k_0 y)\cos(k_0 z) \\ w(x,y,z,0) = 0 \\ p(x,y,z,0) = p_0 + \frac{\rho_0 u_0^2}{16}[\cos(2k_0 x) + \cos(2k_0 y)](\cos(2k_0 z) + 2), \end{cases} \tag{70}$$

in which $u_0 = M_0 c_0$ is s the reference velocity (here $M_0 = 0.01$) and M_0, c_0, p_0, T_0 and ρ_0 are the reference Mach number, speed of sound, pressure, temperature and

---

[6] Additional conservation laws can be derived from the Euler equations for smooth flows, namely for arbitrary, differentiable function of the thermodynamic entropy, but they are not shown herein.

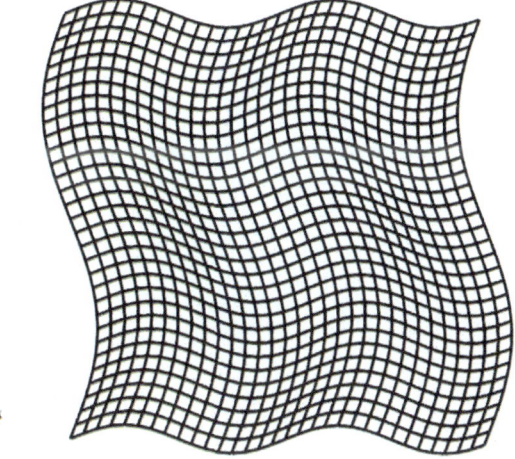

Fig. 4 Two-dimensional cross section in the $x - y$ plane of the curvilinear computational domain adopted for the inviscid Taylor-Green vortex simulation [20]

density. The solution is advanced in time up to time $tu_0k_0 = 8$, at which all velocity vectors are reversed, and then further advanced in time up to $tu_0k_0 = 16$. Based on the time-reversibility properties of the Euler equations, the initial conditions should be exactly recovered [17]. This will not be the case if numerical diffusion is present, since it spoils time reversibility.

In Fig. 5 where we report the time evolution of turbulence kinetic energy and of the total enstrophy, defined as

$$K = \frac{1}{2}\sum_i (u_k u_k)_i V_i, \qquad \Omega = \sum_i (\omega_k \omega_k)_i V_i, \qquad (71)$$

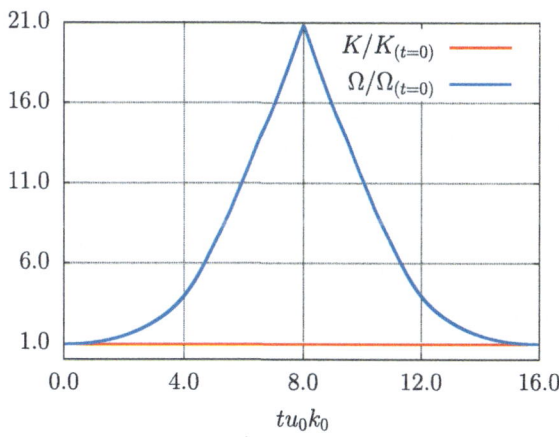

Fig. 5 Time evolution of normalized total kinetic energy (red line) and normalized enstrophy (blue line) [20]. All quantities are normalized by their value at the beginning of the simulation

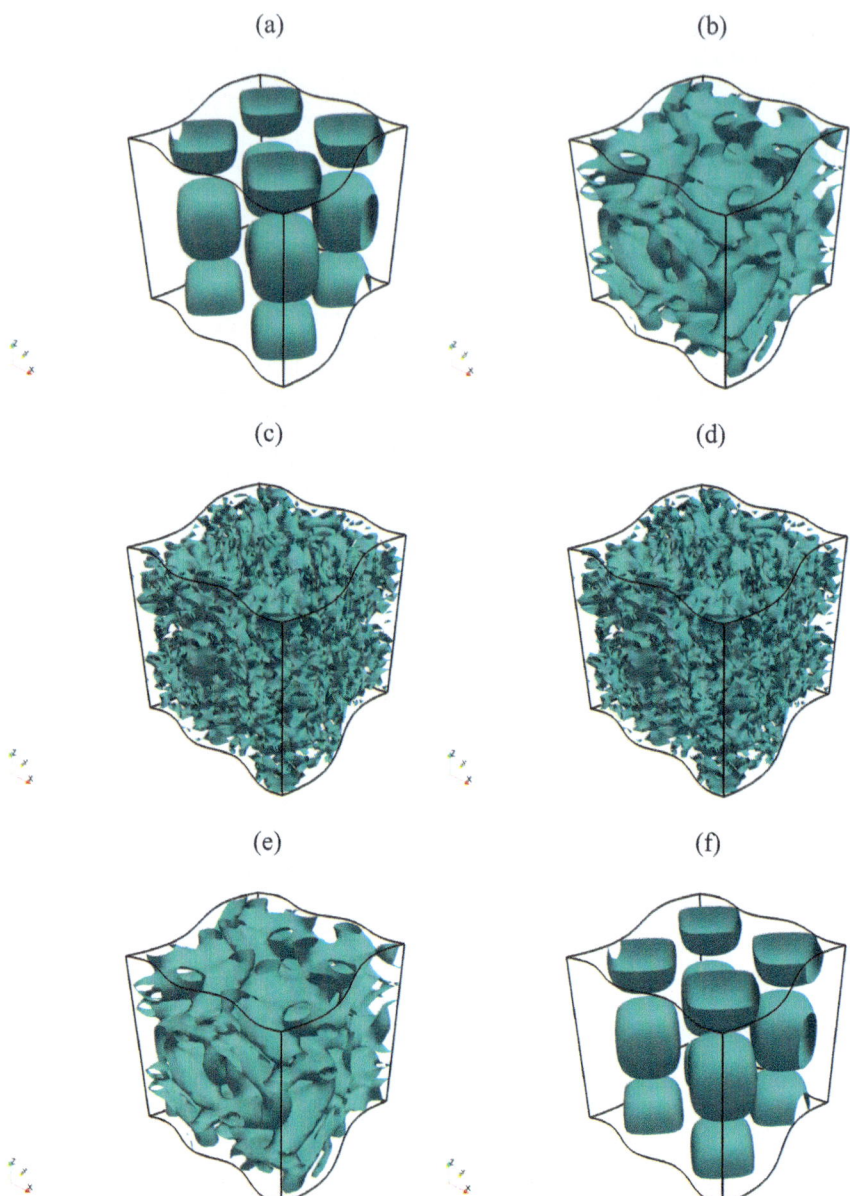

Fig. 6 Snapshots of the inviscid Taylor-Green vortex simulation at times [20]: **a** $tu_0k_0 = 0$; **b** $tu_0k_0 = 3.2$; **c** $tu_0k_0 = 6.4$; **d** $tu_0k_0 = 9.6$; **e** $tu_0k_0 = 12.8$ **f** $tu_0k_0 = 16$

where $\sum_i$ denotes the sum over all computational nodes and V_i is the volume of the ith computational element. We can see that kinetic energy is perfectly retained and that the initial conditions are perfectly recovered, as shown by the total enstrophy trend. In Fig. 6, instead, a sequence of snapshots is displayed, each corresponding to different time points in the simulations. These snapshots offer a qualitative insight into the observation that the coherent structures present in the initial conditions are faithfully reproduced.

With no doubt, we can confidently assert that the benchmark test has been successfully passed.

4.2 Supersonic Turbulent Compression Corner

In Fig. 7 we depict a typical SBLI configuration on a compression corner. The flow is forced to turn at the location $x = x_c$ by the ramp angle ϕ. Since the upstream flow is supersonic, this would lead to a formation of a shock-wave in the proximity of the corner, which would most certainly lead to flow separation with formation of a separation bubble, provided the shock is strong enough. The aim of the present setup is to validate our new code formulation, which adopts generalized curvilinear coordinates in two directions, say in the $x - y$ plane.

A synthetic turbulent boundary layer is injected through the inflow plane $x = 0$ with a free-stream Mach number M_0 and a free-stream temperature T_0. The inflow velocity is determined by combining two components. The first component is derived from an inverse Van Driest transformation applied to the incompressible mean turbulent profiles from the Musker family [21]. The second component involves fluctuations that are obtained through recycling-rescaling. The ramp angle is set to $\phi = 24°$

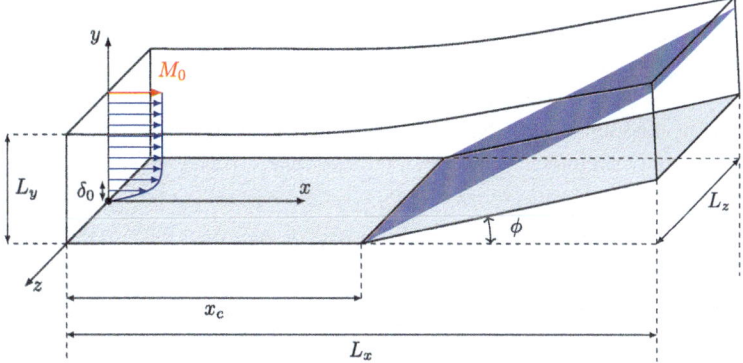

Fig. 7 Numerical setup for the analysis of a SBLI on a compression corner. δ_0 is the inflow boundary layer thickness, x_c is the corner location, M_0 is the free-stream Mach number and ϕ is the corner deflection angle with respect to the flat wall. L_x, L_y, L_z are the extent of the computational box in the streamwise, wall-normal and spanwise direction

Table 1 Incoming boundary layer properties for the supersonic compression corner validation. M_0 and T_0 are the free-stream Mach number and temperature, respectively, Re_θ and C_f are the Reynolds number based on momentum thickness and the friction coefficient upstream the interaction, and T_w is the wall temperature

	M_0	Re_θ	C_f	T_0(K)	T_w(K)
Experimental [24]	2.9	2400	0.00225	108.1	–
Reference DNS [23]	2.9	2300	0.00217	108.1	307
Present DNS	2.9	2400	0.00219	108.1	307

and the wall is isothermal, at a temperature $T_w = 1.14 T_r$, being T_r the recovery one. The dependence of the viscosity on temperature is accounted by using Sutherland's law. Finally, the flow is considered homogeneous in the spanwise direction and characteristic relaxation [22] is adopted at the top and outlet boundaries. The flow configuration reproduces the numerical simulation conducted by [23]. The dimensions of the computational domain are $L_x \times L_y \times L_z) = (29 + 18\cos\phi)\delta_0 \times 6\delta_0 \times 2.2\delta_0$ with a number of grid points $N_x \times N_y \times N_z = 2432 \times 256 \times 160$, which lead to a grid resolution upstream of the corner equal to $\Delta x^+ = 4.3$, $\Delta y_w^+ = 0.6$ and $\Delta z^+ = 3.3$.

The primary objective of this study is to validate our code formulation in curvilinear coordinates. This validation will be carried out in the context of a turbulent SBLI setup. It's noteworthy that experimental data [24] and numerical data [23] are both accessible for this particular SBLI configuration, and they pertain to a moderate Reynolds number. The properties of the incoming boundary layer are provided in Table 1. The two-dimensional curvilinear grid is generated as described in the work of Wu and Martín [25]. In our subsequent discussion of the results, we will use δ as the reference boundary layer thickness to compare our results with experimental [24] and numerical [23] data, accordingly to the work by Wu and Martín [23]. In our case, this reference location is obtained at $x/\delta_0 \approx 20$, that is where the incoming boundary layer properties match the experimental value of $Re_\theta = 2400$, corresponding to the inflow of the simulations by Wu and Martín [23]. The ratio of boundary layer thicknesses is $\delta/\delta_0 = 1.2$.

Figure 8 displays a qualitative instantaneous snapshot of the flow field. The approaching supersonic boundary layer is deflected due to the presence of the ramp, the shock system originates at the corner and the coherent flow structures, highlighted by the colored Q-criterion iso-contours, clearly show the presence of an extensive separated flow region.

Flow statistics are gathered using 400 flow fields with time intervals equal to $\Delta t = 3.43\delta_0/u_0$. We define the separation and reattachment points as the points where the mean friction coefficient C_f changes sign. Centering the streamwise location in the corner, the predicted separation and reattachment points are registered at $x - x_c = -3.2\delta$ and $x - x_c = 1.3\delta$, respectively. In the experiment of Bookey et al. [24], the separation and reattachment points are at $x - x_c = -3.2\delta$ and $x - x_c = 1.6\delta$, respectively, while they are at $x - x_c = -3\delta$ and $x - x_c = 1.3\delta$ in the calculations of Wu and Martín [23].

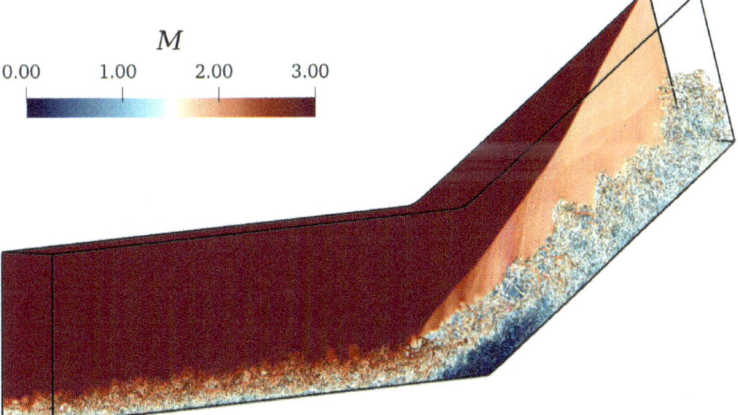

Fig. 8 Instantaneous Mach number field in a cross-stream slice of the present numerical simulation, together with Mach number-colored iso-contours highlighting vortical structures. Contour levels range is from 0 to 3 (blue to white to red ↔ low to high)

To asses the accuracy of our simulation, we compare mean wall pressure ($\overline{p}_w$) and mean velocity ($\tilde{u}$) distributions across the interaction. Figure 9, panel (a), shows the mean wall-pressure distribution. The experimental uncertainty of 5%. Both our and Wu and Martín [23] DNS data predict the wall-pressure distribution within the experimental uncertainty. In Fig. 9, panel (b), we show the velocity profiles in the incoming boundary layer and at a location 4δ downstream of the corner. These profiles are derived from our simulation, from the DNS conducted by Wu and Martín [23], and from the experiments carried out by Bookey et al. [24]. The pressure is normalized by the free-stream value p_0 while the velocity is normalized by the value at the

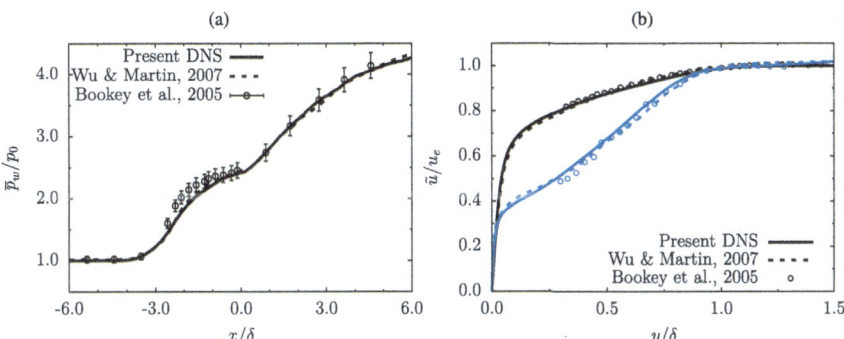

Fig. 9 Comparison of mean wall pressure ($\overline{p}_w$) and mean velocity ($\tilde{u}$) distributions across the interaction [20]. Solid lines represent the present DNS, dashed lines the DNS by Wu and Martín [23] and open circles the experimental results of Bookey et al. [24]

boundary layer edge u_e. For both the upstream and downstream velocity profiles, the agreement all simulations and experiments is within 5%.

Based on these results, it can be concluded that our calculations predict the experimental findings of Bookey et al. [24] and the numerical data from the DNS by Wu and Martín [23] with a satisfactory level of accuracy. This is particularly evident in terms of the upstream boundary layer, mean wall-pressure distribution, the size of the separation bubble, and the velocity profile downstream of the interaction.

5 On the Applicability of Energy-Consistent Schemes for SBLI

Owing to their great computational efficiency, finite difference methods are preferable for large-scale computations whenever the physical problem allows the use of Cartesian or curvilinear meshes. There are two classes of methods specifically designed for application to smooth and shocked flows. With regard to the former class, several mathematical principles can be exploited for the design of stable, low-dissipative schemes without reverting to upwinding or filtering. In particular, nonlinearly stable schemes can be designed by enforcing preservation of the total kinetic energy at the discrete level in Euler equations, in the absence of shocks.

In particular, the variant of convective splitting proposed by Kennedy and Gruber [4] seems to be promising for smooth flows with strong density variations. Regarding methods for shocked flows, WENO schemes (and their variants) seem to have superseded other shock-capturing methods in the past decade, having proven to be extremely accurate and robust in the presence of strong shock waves and complex shock interactions. WENO schemes, however, are quite computationally intensive and suffer from excessive numerical damping in smooth zones of the flow field. Therefore, their application to LES and DNS is suggested only in hybrid form, i.e., in conjunction with a non-dissipative algorithm to treat smooth flow zones. Effective strategies for coupling shock-capturing and non-dissipative methods include hybridization and nonlinear filtering. Both strategies rely on shock sensors that have to be as simple and effective as possible. One such choice is the Ducros sensor [7], which was found to perform reasonably well in many shock/turbulence interaction problems. With regard to the hybrid methods, a suggested best practice is marking critical nodes using a shock sensor and then padding a sufficient number of nodes around the critical ones to make sure that the stencil of the underlying non-dissipative scheme does not cross shocked zones.

Some open issues remain. First, one must be aware that the global order of accuracy of shock- capturing schemes in unsteady problems is always reduced to unity, and shock-capturing is the cause of spurious oscillations, especially downstream of slowly moving shocks. These limitations, related to the misrepresentation of discontinuities on a mesh with finite spacing, can only be overcome by some form of shock-fitting. A detailed study of the effect of shock-capturing oscillations on the prediction of shock/sound and shock/turbulence interactions is lacking and would be highly desirable. Second, even though hybrid schemes are frequently used, a systematic quantitative analysis of the coupling between shock-capturing and non-dissipative schemes has not been carried out. Third, a comparative efficiency analysis of numerical algorithms for problems involving shock waves is not available at present, and cost figures are seldom reported in computational studies. Fourth, it appears that efficient, low-dissipative methods suitable for compressible turbulence simulation on unstructured meshes are lacking in the literature. In this respect, alternative avenues to finite volume methods are also worth exploring.

Further efforts are needed before computational gasdynamics can reach a fully mature stage and cope with the growing demand for DNS and LES of high-speed turbulent flows for configurations of technological relevance.

Competing Interests The authors have no conflicts of interest to declare that are relevant to the content of this chapter.

Acknowledgements If you want to include acknowledgments of assistance and the like at the end of an individual chapter please use the `acknowledgement` environment – it will automatically be rendered in line with the preferred layout.

References

1. Smits, A.J., Dussauge, J.-P.: Turbulent Shear Layers in Supersonic Flow. Springer Science & Business Media (2006)
2. Pirozzoli, S.: Numerical methods for high-speed flows. Ann. Rev. Fluid Mech. **43**, 163–194 (2011)
3. Pirozzoli, S.: Generalized conservative approximations of split convective derivative operators. J. Comput. Phys. **19**, 7180–7190 (2010)
4. Kennedy, C.A., Gruber, A.: Reduced aliasing formulations of the convective terms within the Navier-Stokes equations. J. Comput. Phys. **227**, 1676–1700 (2008)
5. Ceci, A., Palumbo, A., Larsson, J., Pirozzoli, S.: Numerical tripping of high-speed turbulent boundary layers. Theor. Comput. Fluid Dyn. **36**, 865–886 (2022)
6. Jiang, G.S., Shu, C.W.: Efficient implementation of weighted ENO schemes. J. Comput. Phys. **126**, 202 (1996)
7. Ducros, F., Ferrand, V., Nicoud, F., Weber, C., Darracq, D., Gacherieu, C., Poinsot, T.: Large-eddy simulation of the shock/turbulence interaction. J. Comput. Phys. **152**, 517–549 (1999)
8. Ceci, A., Palumbo, A., Larsson, J., Pirozzoli, S.: On low-frequency unsteadiness in swept shock wave-boundary layer interactions. J. Fluid Mech. **956**, R1 (2023)
9. Wray, A.A.: Minimal storage time advancement schemes for spectral methods. NASA Ames Research Center, California, Report No. MS 202 (1990)
10. Piquet, A., Zebiri, B., Hadjadj, A., Safdari Shadloo, M.: A parallel high-order compressible flows solver with domain decomposition method in the generalized curvilinear coordinates system. Int. J. Numer. Methods Heat Fluid Flow **30**, 2–38 (2020)
11. Bernardini, M., Modesti, D., Pirozzoli, S.: On the suitability of the immersed boundary method for the simulation of high-Reynolds-number separated turbulent flows. Comput. Fluids **130**, 84–93 (2016)

12. Johnson, J.P., Iaccarino, G., Chen, K.-H., Khalighi, B.: Simulations of high reynolds number air flow over the NACA-0012 airfoil using the immersed boundary method. J. Fluids Eng. **136**, 040901 (2014)
13. Pirozzoli, S.: Stabilized non-dissipative approximations of Euler equations in generalized curvilinear coordinates. J. Comput. Phys. **230**, 2997–3014 (2011)
14. Regenstreif, E.: On the use of a non-orthogonal system of coordinates in potential theory. CERN Report PS-DL-76-15 (1976)
15. Bernardini, M., Modesti, D., Salvadore, F., Pirozzoli, S.: STREAmS: a high-fidelity accelerated solver for direct numerical simulation of compressible turbulent flows. Comput. Phys. Commun. **263** (2021)
16. Bernardini, M., Modesti, D., Salvadore, F., Sathyanarayana, S., Della Posta, G., Pirozzoli, S.: STREAmS-2.0: supersonic turbulent accelerated Navier-Stokes solver version 2.0. Comput. Phys. Commun. **285** (2023)
17. Duponcheel, M., Orlandi, P., Winckelmans, G.: Time-reversibility of the Euler equations as a benchmark for energy conserving schemes. J. Comput. Phys. **227**, 8736–8752 (2008)
18. Taylor, G.I., Green, A.W.: Mechanism of the production of small eddies from large ones. Proc. R. Soc. Lond. Ser. A **158**, 799–521 (1937)
19. Modesti, D., Pirozzoli, S.: A low-dissipative solver for turbulent compressible flows on unstructured meshes, with OpenFOAM implementation. Comput. Fluids **152**, 14–23 (2017)
20. Soldati, G., Ceci, A., Pirozzoli, S.: FLEW: A DNS Solver for Compressible Flows in Generalized Curvilinear Coordinates. Aerotec, Missili Spaz (2024)
21. Musker, A.: Explicit expression for the smooth wall velocity distribution in a turbulent boundary layer. AIAA J. **17**(6), 655–657 (1979)
22. Pirozzoli, S., Colonius, T.: Generalized characteristic relaxation boundary conditions for unsteady compressible flow simulations. J. Comput. Phys. **248**, 109–126 (2013)
23. Wu, M., Martín, M.P.: Direct numerical simulation of supersonic turbulent boundary layer over a compression ramp. AIAA J. **45**, 879–889 (2007)
24. Bookey, P., Wyckham, C., Smits, A.J., Martín, M.P.: New experimental data of STBLI at DNS/LES accessible Reynolds numbers. AIAA paper. 2005-309 (2005)
25. Wu, M., Martín, M.P.: Direct numerical simulation of shockwave/turbulent boundary layer interaction. AIAA Paper 2004-2145 (2005)

Open Access This chapter is licensed under the terms of the Creative Commons Attribution 4.0 International License (http://creativecommons.org/licenses/by/4.0/), which permits use, sharing, adaptation, distribution and reproduction in any medium or format, as long as you give appropriate credit to the original author(s) and the source, provide a link to the Creative Commons license and indicate if changes were made.

The images or other third party material in this chapter are included in the chapter's Creative Commons license, unless indicated otherwise in a credit line to the material. If material is not included in the chapter's Creative Commons license and your intended use is not permitted by statutory regulation or exceeds the permitted use, you will need to obtain permission directly from the copyright holder.

Development of a PVDF Piezo-Film Sensor for Unsteady Wall-Pressure Measurements in SBLIs

Cosimo Corsi, Bei Wang, Julien Weiss, and Ha Duong Ngo

Abstract An innovative, flexible wall-pressure sensor array for unsteady flow conditions has been developed and evaluated in a turbulent shockwave-boundary layer interaction (SBLI) setup at Mach 2. Compared to the previous version, the new sensor's flexibility makes it easier to fit on different surfaces, while offering enhanced durability and improved sensitivity. The array comprises 18 circular sensors, each with a diameter of 3 mm, fabricated using screen printing techniques from a thin piezoelectric PVDF film (thickness: 110 μm). Remarkably, this sensor array achieves excellent spatial resolution while minimizing flow interference, all at a fraction of the cost associated with traditional dynamic pressure transducers. To validate its performance, the sensor array underwent dynamic calibration using a ball-drop impact test device. Subsequently, it was rigorously tested in a supersonic wind tunnel, demonstrating strong agreement with reference measurements obtained using a state-of-the-art Kulite pressure sensor. The resulting premultiplied power spectral density $f \cdot PSD$ distributions align closely with findings reported in existing literature. Notably, the low-frequency unsteadiness region beneath the separation shock foot ($X^* = 0$) exhibits a Strouhal range of $St = 0.03 - 0.05$.

Keywords Transonic flow · Shock wave—boundary layer interaction · Unsteady wall-pressure sensor · Piezofoil sensor array

1 Introduction

SBLIs continue to be a subject of active research due to their intricate flow patterns and the growing interest in supersonic flow applications. A crucial aspect of these phenomena lies in their inherent unsteadiness, which manifests as shockwave

---

C. Corsi (✉) · J. Weiss
Technische Universität Berlin, Berlin, Germany
e-mail: cosimo.corsi@tu-berlin.de

B. Wang · H. D. Ngo
Hochschule für Technik und Wirtschaft Berlin, Berlin, Germany

oscillations at frequencies significantly lower than the turbulent fluctuations within the incoming boundary layer. This unsteadiness becomes particularly pronounced during boundary layer separation caused by the strong adverse pressure gradient induced by the incident shock [9]. The high-amplitude, low-frequency unsteadiness associated with SBLIs plays a critical role in various industrial applications. It underlies several challenges, including structural failures, panel flutter, and self-induced instabilities such as buffet and inlet buzz [5, 11, 17]. Despite extensive research on turbulent SBLIs over the past decades [6], the mechanisms driving this low-frequency unsteadiness remain incompletely understood. Currently, there exists an ongoing debate between two major potential causes:

- **Upstream Mechanism**: This hypothesis attributes the unsteadiness to low-frequency flow structures present in the incoming turbulent boundary layer [8].
- **Downstream Mechanism**: An alternative perspective links the fluctuations of the separation shock to periodic contractions and expansions within the separation bubble, similar to a breathing motion [18].

A commonly employed experimental approach to study the low-frequency unsteadiness in separated shockwave-boundary layer interactions (SBLIs) involves measuring wall-pressure fluctuations beneath the separated shock foot. Researchers have predominantly utilized piezo-resistive pressure transducers, often from the *XCQ series Kulites*® [17]. These transducers are renowned for their high accuracy and sensitivity. However, when it comes to capturing spatial wall-pressure correlations, Kulites face a significant limitation: they necessitate simultaneous acquisition of multiple sensors. Consequently, deploying a large number of sensors in the experimental facility increases the overall setup cost. Another option is to use unsteady pressure-sensitive paint, which has the advantage of providing two-dimensional unsteady pressure fields, though at the expense of complex calibration procedures and lower frequency-response range and signal-to-noise ratio than conventional pressure transducers.

To overcome the described issues, we adopted an intermediary concept, where local, pointwise measurements may be obtained at many positions on a test surface but without any significant increase of instrumentation cost. We developed a novel noninvasive and cost-effective technique for unsteady wall-pressure measurements in shockwave-boundary layer interactions (SBLIs) using a piezo-film sensor array [4]. These film sensors are based on the organic polymer *polyvinylidene fluoride* (PVDF), whose piezoelectric properties were first discovered by Kawai [10]. The piezoelectric behavior (change in surface charge due to mechanical loading of the film) arises from the specific crystalline arrangement that the polymer acquires after artificial polarization. The PVDF layer is double-sided coated with a thin layer of metal, allowing charges generated by external forces through the piezoelectric effect to be collected on both faces of the film. By partially etching the metal coating layer into custom shapes and patterns, specific arrays of active sensors can be formed from a single foil. Figure 1 illustrates the schematic setup of a PVDF sensor, along with examples of etched arrays obtained by removing the coating from one side (Fig. 1b) or both sides (Fig. 1c). In Fig. 1a, the signal chain is depicted: external forces cause

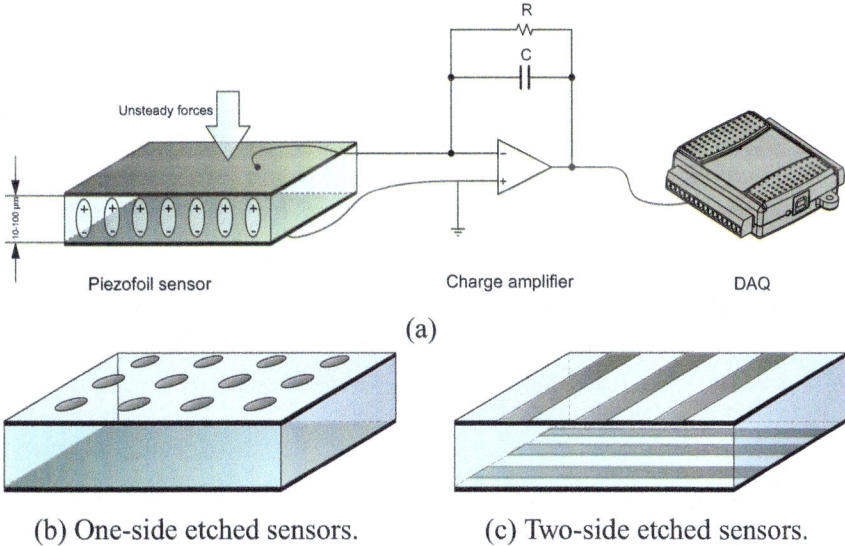

Fig. 1 Piezofoil sensor setup [4]

a charge buildup in the foil, which is acquired via a charge amplifier. The amplifier transforms the charges collected on the metal faces into a voltage signal detectable by the data acquisition system (DAQ). Notably, due to the capacitive behavior of the PVDF material, charges generated on the two electrodes gradually cancel each other. Consequently, this type of sensor is capable of detecting pressure fluctuations (AC signal) but not the mean static pressure field.

Given the appealing properties of PVDF piezo-films, which include their relatively low cost, the ability to install them on curved surfaces, wide frequency and dynamic ranges, high sensitivity to pressure variations, and robustness to mechanical stress and external environmental factors, the objective of this study is to design and test a cost-effective, non-invasive PVDF unsteady-pressure sensor array. This array is specifically intended for use in turbulent shockwave-boundary layer interactions (SBLIs) to address the low-frequency unsteadiness associated with separation shocks. The structure of this paper will be as follows: in Sect. 2, we introduce the experimental setup used in this work, including details of the supersonic facility employed for testing and a comprehensive description of the sensor designed and constructed for this study. Section 3 presents the results of fluctuating pressure measurements obtained using the new PVDF sensor array during an incident shock interaction at Mach 2. We place particular emphasis on comparing and validating these results against those obtained using a conventional piezo-resistive pressure sensor (Kulite). Finally, we discuss the advantages and limitations of the new sensor array for SBLI research.

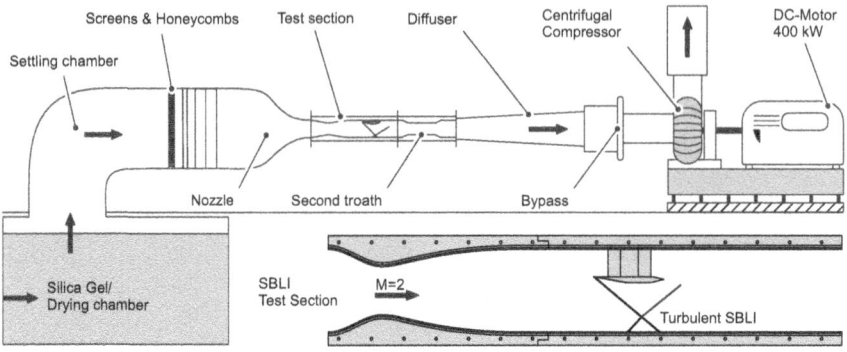

Fig. 2 Schematic representation of the supersonic wind tunnel present at *TU Berlin* with a focus on the turbulent SBLI test section [16]

Table 1 Supersonic wind tunnel parameters [16]

M	Re_U	δ_0	Shock generator angle	Test section dimensions
2	$12.5 \times 10^6 \, \text{m}^{-1}$	7.2 mm	$66° - 10°$	150 mm x 150 mm

2 Experimental Setup

2.1 Test Facility and Test Case Arrangement

The experimental investigations were carried out at the Chair of Aerodynamics at *Technische Universität Berlin* using a supersonic wind tunnel. A comprehensive overview of the facility is provided by Rohlfs et al. [16], and a schematic representation is depicted in Fig. 2. The tunnel operates continuously with an indraft mechanism: air is drawn in by a centrifugal compressor, which is driven up to 16 000 rpm by a 400 kW electric DC motor. The air then traverses a heated drying chamber located in the laboratory's basement. After passing through the settling chamber and screens designed for turbulence reduction, it enters the test section. The test section comprises a Mach 2 supersonic nozzle and a shock generator. This setup is specifically used to create an incident shockwave-boundary layer interaction (SBLI) with the (turbulent) floor boundary layer. The key characteristics of the wind tunnel and the test section are summarized in Table 1.

The study investigates the turbulent shockwave boundary layer interaction produced using a shock generator with a deflection angle of 10 degrees. This setup creates an oblique shock that interacts with a fully turbulent boundary layer of thickness $\delta_0 = 7.2$ mm on the wind-tunnel floor. The flow topology of interest is depicted in Fig. 3: an oblique shockwave impinges on a fully turbulent boundary layer, resulting in a strong adverse pressure gradient. This leads to the formation of a separation bubble with reversed flow between the *separation point S* and the *reattachment point R*, characterized by a length L_{sep}. The curvature of the bubble induces converging

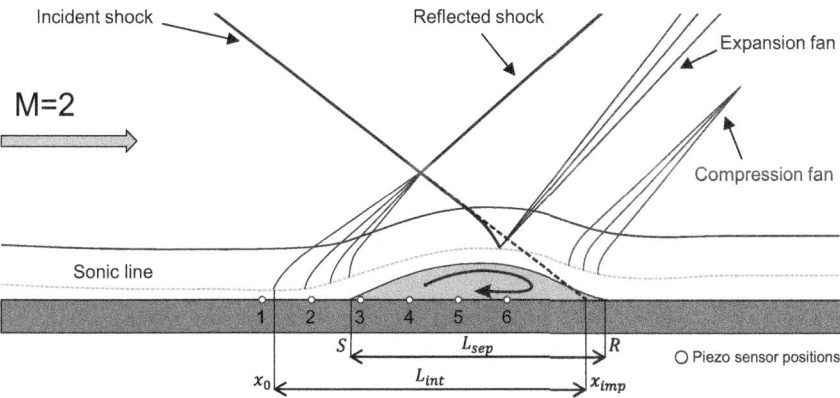

Fig. 3 Representation of a separated shockwave-boundary interaction, adapted from [9]. The white circles are representing the positions of the PVDF sensors

compression waves in the incoming flow, which merge into a *separation shock*. This shock interacts with the incident shock, forming the characteristic λ-foot of the separated shockwave boundary layer interaction (SBLI). Another important parameter for subsequent investigations is the interaction length L_{int}, defined as the distance between the extrapolated impingement point x_{imp} (predicted by inviscid shock theory) and the interaction onset x_0, where the wall pressure begins to rise.

2.2 Piezofoil Sensor Array

2.2.1 Piezofoil Design

For the purpose of this study, a clean polarized PVDF film with a thickness of 110 μm was utilized. The key properties of the film are presented in Table 2 [15]. It is essential to recognize that while the frequency and dynamic ranges listed in the table strictly pertain to the material, the actual values of the final sensors depend on the entire signal chain-from the charge amplifier to the DAQ system-as we will discuss later.

As previously introduced, the operational principle of this type of piezoelectric sensor relies on the charge generated within the semicrystalline structure of PVDF due to external mechanical stress. The resulting charge density D, in the absence of an external electric field, can be expressed as:

Table 2 Typical characteristics of PVDF piezo-films [15]

t	d_{33}	E	ϵ/ϵ_0	Freq. range	Dynamic range	T Range
110 μm	33×10^{-12} C/N	4×10^9 Pa	12	10^{-3}–10^{10} Hz	10^{-5}–10^9 Pa	−40 to 80 °C

$$D = \Sigma d_{3n}\sigma_{3n} \qquad (n = 1, 2, 3) \qquad (1)$$

The piezoelectric constant d_{3n} and the mechanical stress σ_{3n} correspond to the direction n, where 1 represents length, 2 represents width, and 3 represents thickness. In our study, we focus on the pressure fluctuation applied to the sensor, so we consider $n = 3$ (normal stress). Additionally, we assume that shear stress fluctuations ($n = 2, 3$) can be neglected [12]. By considering only the 3rd direction and multiplying Eq. 1 by the active area A (the sensitive part of the sensor where positive and negative electrodes overlap), we obtain the charge Q generated by the sensor:

$$D = d_{33}\sigma_{33} \quad \Longrightarrow \quad Q = DA = d_{33}\sigma_{33}A \qquad (2)$$

Evaluations regarding spatial resolution can be derived from Eq. 2. Ideally, the smallest possible sensor size should be chosen for optimal resolution. However, a sensing area that is too small may not generate sufficient charges, leading to a poor signal-to-noise ratio. Additionally, the sensor size impacts the frequency response. The high-frequency range of the spectrum is influenced by the smallest fluid structures. If the sensor size exceeds these flow structures, it will filter their contribution, resulting in attenuated high-frequency response.

Figure 4 illustrates the final design. The sensor array comprises 18 circular sensors, with 6 arranged in the streamwise direction and 3 positioned along the span. Each sensing element has a diameter of 3 mm, covering a total surface area of 50×18 mm^2. The pitch between consecutive sensing points along the streamwise direction is 10 mm, while the spacing across the spanwise direction is 6 mm. Figure 4a depicts the developed array, which consists of a multilayered structure including the following components: a Flex-PCB, a 3M™ 9703 anisotropic conductive adhesive, and the PVDF piezoelectric active layer with its silver electrode array. The Flex-PCB provides support for the entire sensor and is custom-patterned with Ag-ink to create contact pads and wires for connecting the sensors to the acquisition system. The 3M™ 9703 layer serves as a special anisotropic conductive adhesive containing dispersed silver nanoparticles, facilitating electrical conductivity between the Flex-PCB and the PVDF layer. The PVDF layer itself constitutes the active part of the sensor, comprising both the piezoelectrically active layer and the silver electrode array.

The sensor production process for the PVDF and FlexPCB layers is conducted in the laboratory of *Hochschule für Technik und Wirtschaft Berlin*. The manufacturing process begins with a clean substrate, which can be either polarized PVDF or a PE sheet. Both sides of the substrate are then patterned using screen printing techniques with silver ink. This involves creating a stencil of the desired pattern, which is placed on the substrate film. The silver ink is applied through the stencil onto the film using a squeegee, resulting in a precise and repeatable pattern on the film's surface. After patterning, the material is heated to allow the metallic ink to cure and enhance its electrical and mechanical properties. This adaptable process can be used to create patterns of varying sizes and shapes, making it a versatile method for producing sensors across a wide range of applications.

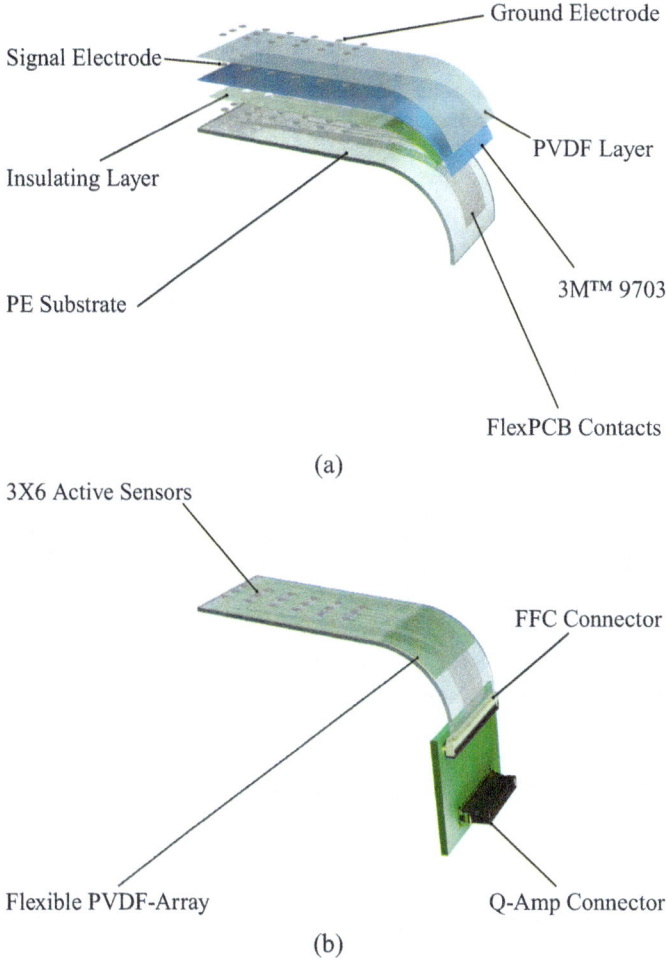

Fig. 4 Flexible piezofoil array: **a** structure of the proposed sensor; **b** final setup with connection interface for charge amplifier

A critical aspect of sensor instrumentation involves selecting an appropriate connection setup. This step is crucial because it determines the interface between the metal coatings of the piezofilm and the signal conditioning system, aiming to prevent unwanted noise and signal degradation. Various methods exist for connecting to the PVDF's electrodes. In this study, we utilize a Flexible Flat Cable (FFC) connector. FFC connectors resemble ribbon-like structures, composed of flexible plastic, polymers, films, or engineered rubber, with a metallic connector at one end. These connectors are embedded parallel to the base, allowing seamless integration with the signal conditioning system (see Fig. 4b).

The final sensor is securely adhered flush to the test section floor. Thanks to the foil's flexibility and low thickness, the connection interface is positioned outside the test section, transmitting the signal to the charge amplifier for conditioning.

The newly developed piezoelectric PVDF flexible sensor array offers several advantages over the previous rigid design presented by Corsi et al. [4]. Firstly, the new sensor exhibits significantly greater flexibility, enhancing versatility in experimental setups. Its pliability allows easy bending and shaping to conform precisely to the surface where it is placed, thereby improving pressure measurement accuracy in complex flow fields. Secondly, the improved robustness of the new sensor is a noteworthy benefit. The adoption of an FFC (Flexible Flat Cable) connector ensures a secure and reliable connection between the sensor and the data acquisition system, minimizing the risk of signal loss and interference. This enhanced connection setup also simplifies array usage and maintenance, as the FFC connector facilitates quick and effortless connection and disconnection from the amplification and data acquisition system.

2.2.2 Signal Conditioning and Acquisition System

The direct output voltage measured from piezoelectric sensors is typically only a few millivolts (mV). To obtain a readable signal with good signal-to-noise ratio (SNR), an appropriate amplification system is essential. Due to the high impedance of PVDF (polyvinylidene fluoride), a high-impedance amplifier is necessary. In this work, a charge amplifier has been selected. Figure 5 illustrates the electrical circuit of a typical charge amplifier and its frequency response.

One advantage of using a charge amplifier is that the output voltage depends solely on the feedback capacitance C_f (i.e., $V_o = -Q/C_f$) and not on the cable capacitance (C_c) or sensor capacitance (C_p). This design allows for the use of longer cables while minimizing charge leakage [15]. To optimize the performance of our PVDF sensors, we designed the charge amplifier using the components specified in Table 3.

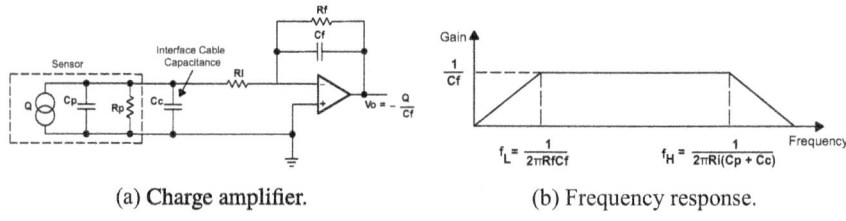

(a) Charge amplifier. (b) Frequency response.

Fig. 5 Electrical circuit of a charge amplifier and its frequency response [2]

Table 3 Charge amplifier design parameters

Op-Amp	C_f	R_f	R_i	C_p	f_L
TLC-272	100 pF	100 MΩ	56 Ω	10 pF	15.9 Hz

The central component of the system is the *Texas Instruments*™ TLC-272 operational amplifier. This CMOS single-supply amplifier features low offset voltage drift, high input impedance, and low noise behavior. The feedback capacitor C_f balances the charges injected into the negative input of the op-amp, while the resistor R_f prevents the amplifier from drifting into saturation by gradually discharging the feedback capacitor. The values of C_f and R_f determine the low cutoff frequency of the charge amplifier, which can be calculated as $f_L = \frac{1}{2\pi R_f C_f}$. In our case, these values result in a cutoff frequency of $f_L = 15.9$ Hz. The high cutoff frequency $f_H = \frac{1}{2\pi R_i (C_p + C_c)}$ is determined by the resistor R_i, which also provides electrostatic discharge (ESD) protection. Although the capacitance of the sensor (C_p) and cable (C_c) contribute to this frequency, it lies well beyond the frequency range of interest, typically in the megahertz (MHz) range [2].

The signal acquisition is carried out by means of a *National Instruments*™ NI-USB 6353 data acquisition card, in order to obtain PSDs up to 100 kHz a sample rate of $f_s = 200$ kHz for $N = 200000$ samples with a resulting period of $T = N/f_s = 1$ s was used.

2.2.3 Frequency Response Evaluation

To assess the dynamic response of the sensor in conjunction with the signal chain, a known-frequency response source signal is essential. An ideal calibration source should apply a unidirectional force to a localized region of the test specimen and generate a smooth, wide range of frequencies. For this study, an impact ball-drop test was conducted-an impulsive (step-like) source that produces a signal spanning a broad frequency range.

The calibration setup, depicted in Fig. 6c, involves a 3D-printed support holding the plug instrumented with the PVDF sensor and a 30 mm tube. The support features a traverse system, enabling testing of each sensor in the array. A small steel ball with a diameter of $R_1 = 2.4$ mm is dropped onto the specimen from a height of $h = 30$ mm. The ball's impact generates an impulse-like force, which can be precisely calculated using Hertzian theory and accurately modeled by a forcing function $f(t)$ [13, 14].

$$f(t) = \begin{cases} F_{max} sin(\pi t/t_c)^{3/2}, & t \in [0, t_c] \\ 0, & t \in (t_c, \infty) \end{cases} \quad (3)$$

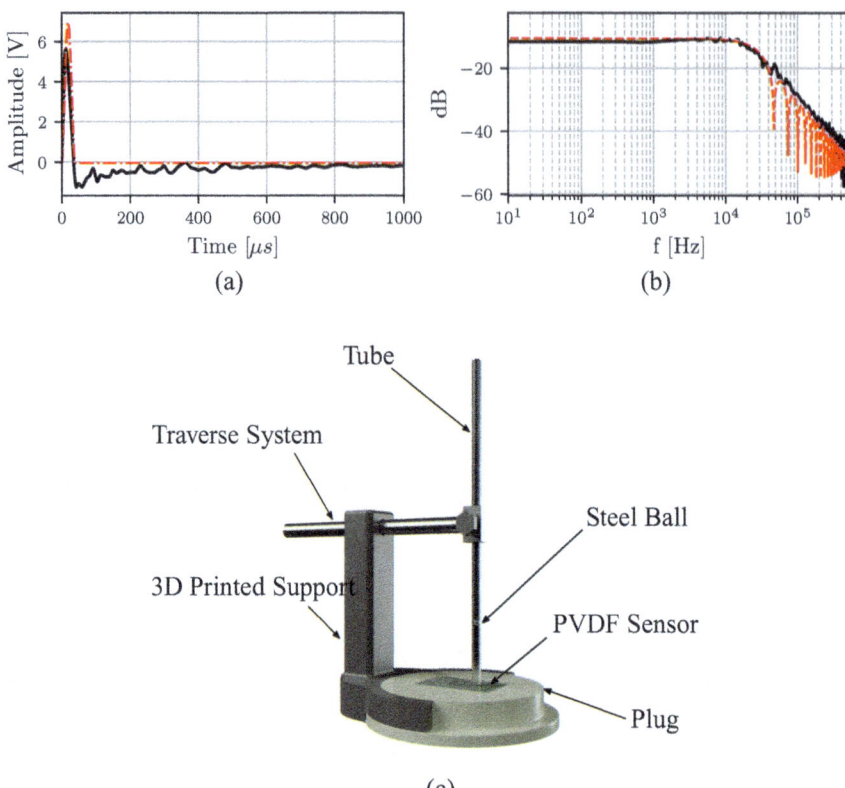

Fig. 6 Ball-drop impact test results: **a** theoretical impulse generated by the impact (red dashed line) and signal measured by the PVDF sensor (black solid line); **b** FFT spectra of the acquired signal (black solid line) compared to the theoretical one; **c** setup adopted for the ball drop impact test

The maximum force F_{max} can be expressed as:

$$F_{max} = 1.917 \rho_1^{3/5} (\delta_1 + \delta_2)^{-2/5} R_1^2 v_0^{6/5}$$

The contact time t_c during the ball impact is given by:

$$t_c = 4.53 \left(\frac{4}{3\rho_1 \pi (\delta_1 + \delta_2)} \right)^{-2/5} R_1 v_0^{-1/5}$$

Here, $\delta_i = \frac{1-\mu_i^2}{\pi E_i}$ represents the material properties, where E_i and μ_i are Young's modulus and Poisson's ratio for the respective materials. Subscripts $i = 1$ correspond to the steel ball, and $i = 2$ corresponds to the more massive test specimen (the instrumented plug). Finally, R_1 represents the radius, and $v_0 = \sqrt{2gh}$ is the impact speed of the steel ball.

In Fig. 6a, the impulsive time signal resulting from the ball impact is shown for both experimental (black) and theoretical (red) cases. Figure 6b displays the amplitude of the Fourier transform spectra. The red spectra correspond to the source function $f(t)$, characterized by a series of lobes separated by zeros at higher frequencies and remaining flat at low frequencies. Due to the low Young's modulus E_2 of the plug material, the contact time t_c is relatively high, resulting in a cutoff frequency of the steel ball spectra around $f_0 = 25\,000$ Hz. The experimental results obtained with the PVDF sensor closely match the theoretical results, validating the assumption of the piezofoil's flat frequency response, at least up to f_0. Future experiments with stiffer plug materials will further validate the sensor response at higher frequencies.

3 Results

In the following section, we present the results obtained with the novel flexible sensor. The sensor arrangement under the shockwave boundary layer interaction (SBLI) flow is depicted in Fig. 3:

- Positions 1–2: These sensors are located in proximity to the interaction onset location x_0, directly under the λ-shock foot.
- Positions 3–6: These sensors are placed beneath the separation bubble.

Figure 7a displays the time signals obtained at each position, as described earlier. Due to limitations in the data acquisition (DAQ) system, we present only one sensor out of the three per row instead of the full array of 18 piezoelectric sensors. Notably, there is an observable amplitude variation along the streamwise direction. Specifically, the root-mean-square (RMS) of the signal increases as the position moves toward the end of the interaction region, attributed to the developing shear layer behind the separation bubble. This observation aligns with the findings from a previous experiment conducted by Corsi et al. [3] in the same facility using Kulite transducers.

The Power Spectral Density (PSD) plots in Fig. 7b were obtained using Welch's method on the time traces, employing 20 Hanning windows with 75% overlap. However, interpreting the PSD spectra alone does not provide a clear understanding of flow unsteadiness. To better characterize the flow, we computed the normalized premultiplied PSD ($PSD(f)f/\sigma^2$). The normalization was performed using the variance σ^2 of the time signal. Additionally, we normalized the frequency f with the Strouhal number ($St_L = fL_{int}/u_\infty$), where L_{int} represents the length of the separation region, and u_∞ is the free-stream velocity. Finally, the streamwise length was normalized with respect to the separation length as $X^* = (x - x_0)/L_{int}$, where x denotes the streamwise position, and x_0 corresponds to the position of the interaction onset.

The piezofoil array's results are depicted in Fig. 8 as normalized premultiplied power spectral density (PSD) distributions. These illustrations are instrumental in

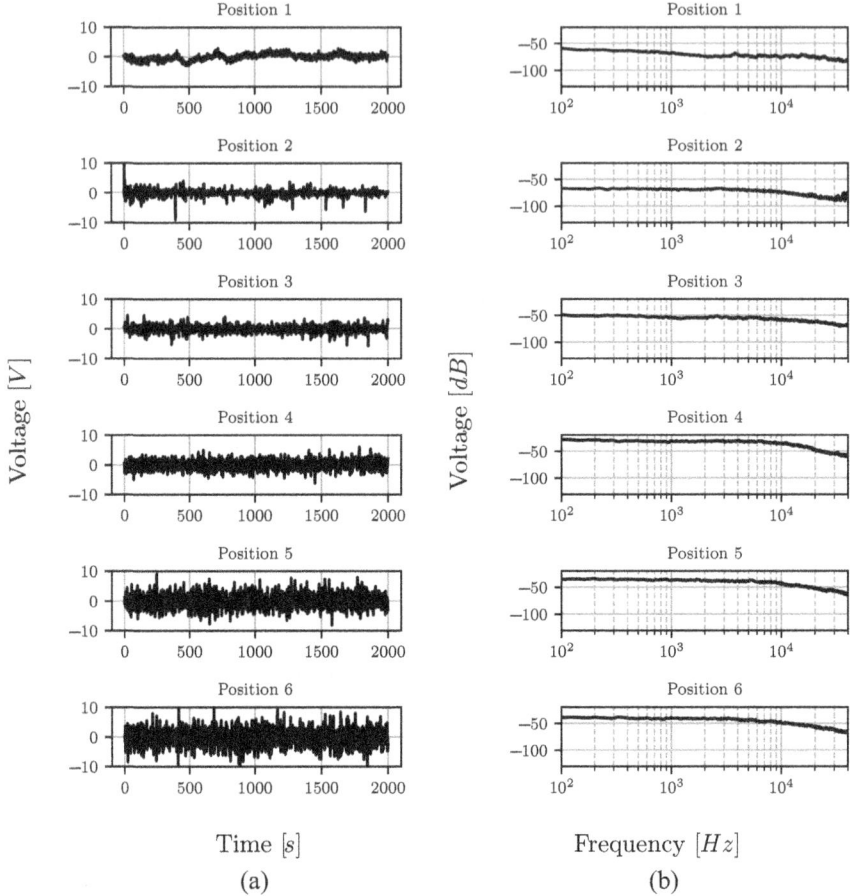

Fig. 7 a Time signals acquired by the PVDF array across the turbulent SBLI; **b** relative Power Spectral Densities for each streamwise position obtained from the PVDF measurements

elucidating the flow's energy distribution across various frequencies and SBLI locations. In line with established observations in the field [1, 3, 4, 7, 16, 17, 19], the expected patterns emerge, showcasing distinct frequency bands that are segmented into zones, each characterized by unique temporal dynamics.

- The first zone, known as the low frequency zone (LFZ), is located at the beginning of the interaction region near the foot of the reflected shock, around $X^* \approx 0$. The LFZ exhibits fluctuations of low frequency, distinguished by a pronounced peak centered at ($St_L \approx 0.04 - 0.05$). This hump is a clear indication of the large-scale, low-frequency motion of the reflected shock.
- The subsequent zone, referred to as the intermediate frequency zone (IFZ), aligns with the interaction zone $-0.2 \leq X^* \leq {'}1$, with spectra indicating a medium-

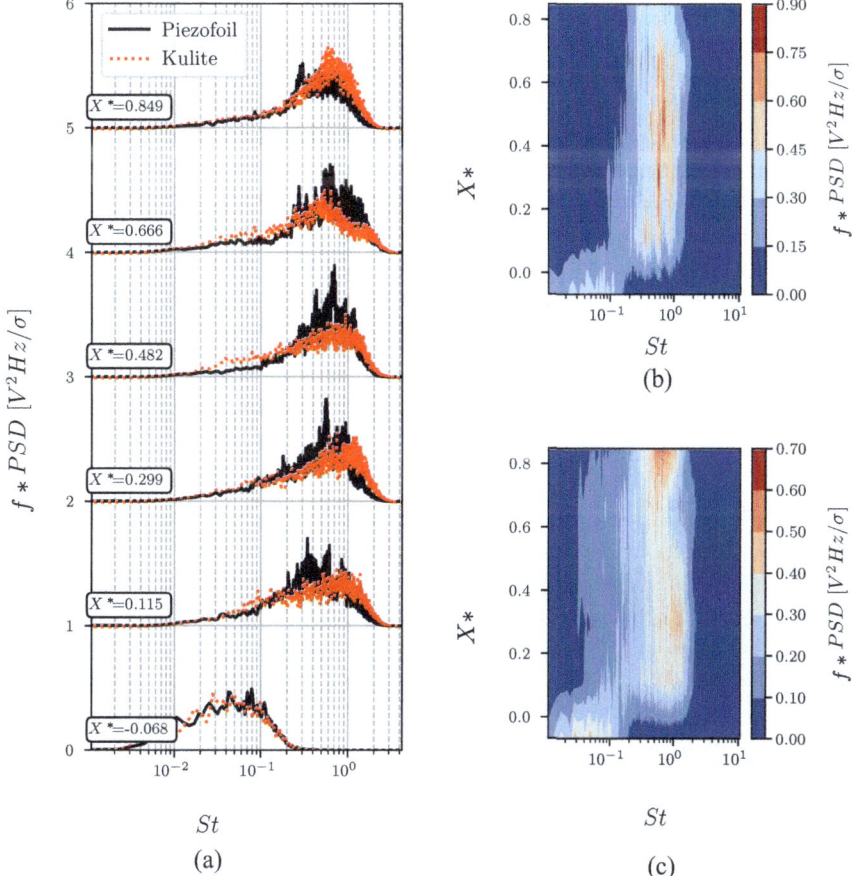

Fig. 8 Frequency domain results obtained in a Mach 2 turbulent SBLI obtained with a 10° shock generator: **a** comparison between Kulita (red) and Piezofoil (black) $f * PSD$ spectra of the acquired signal for every streamwise position across the SBLI; **b** $f * PSD$ map distribution obtained with the PVDF-array; **c** $f * PSD$ map distribution obtained with the Kulite sensor

frequency nature at $St_L \simeq 1$. The IFZ generally symbolizes the pressure footprint of large-scale coherent structures that are shed downstream of the separation bubble.

Figure 8 additionally presents a comparison between the $(f \cdot PSD)$ captured with the innovative PVDF sensors and those captured with a cutting-edge Kulite pressure transducer. The Kulite in use is the fast response piezoresistive transducer, XCQ-062, whose signal underwent an hardware low-pass filter with a cut-off frequency of $f_L = 100$ kHz. The Kulite was installed flush with the test section floor and the acquisitions were sequentially taken for each position corresponding to the PVDF array ones. The signal was recorded with the same DAQ as the piezofoil, and the data were processed with identical parameters. The findings suggest that the newly engineered piezofoil sensors can precisely match the Kulite data, and accurately record the unsteady features of the various flow regions. Specifically, both the LFZ and the IFZ are well recorded by the piezofoil sensors. Within the IFZ, numerous sharp peaks are observed solely on the PVDF spectra. These peaks are likely a result of acoustic interference caused by air leaks near the instrumented plug. The PVDF sensors, which possess higher sensitivity compared to the Kulite, are capable of capturing these acoustic interferences. Therefore, it is crucial to ensure that the wind tunnel is adequately sealed and there are no air leaks that could produce interferences. Despite these peaks, the results exhibit the high fidelity and reliability of the PVDF sensors in recording the dynamic behavior of the flow field.

4 Conclusion

A newly designed flexible unsteady wall-pressure sensor array, based on piezoelectric PVDF films, has been created and trialed in a turbulent SBLI setup at Mach 2. It's flexibility makes it possible to fit it on complicated surfaces, thereby facilitating pressure measurements in more sophisticated flow geometries. The array is composed of 18 round sensors with a diameter of 3 mm, fabricated using screen printing techniques from a PVDF film of $110\,\mu$m thickness. The array delivers outstanding spatial resolution and minimal flow intrusion at a cost significantly lower than traditional dynamic pressure transducers. The sensor was dynamically calibrated using a ball drop impact test device, demonstrating a uniform response across the range produced by the ball impulse up to $f = 25\,000$ Hz. The outcomes of the multi-point surface measurement aligned well with the benchmark measurements taken with a state-of-the-art Kulite pressure sensor, validating the proposed sensor array for unsteady wall pressure measurements. The acquired $f \cdot PSD$ distributions aligned well with the findings reported in the literature, with the low-frequency unsteadiness region beneath the separation shock foot ($X^* = 0$) characterized by a Strouhal range of $St = 0.03 - 0.05$. These results highlight the potential of the new sensor array for unsteadiness studies where high spatial resolution and low intrusion are required, such as in SBLI flows. For future research, spatial cross correlation investigations will be conducted across both streamwise and spanwise directions under the SBLI, to fully leverage the high spatial resolution and the simultaneous acquisition capabilities of the sensor array. In summary, the proposed PVDF flexible sensor presents a promising and cost-effective substitute to conventional dynamic pressure transducers, offering enhanced flexibility, robustness, and sensitivity for unsteady wall pressure measurements in complex flow fields.

Acknowledgements This project has received funding from the European Union's Horizon 2020 research and innovation program under grant agreement No. EC grant 860909.

References

1. Agostini, L., Larchevêque, L., Dupont, P.: Mechanism of shock unsteadiness in separated shock/boundary-layer interactions. Phys. Fluids **27**(12), 126103 (2015). https://doi.org/10.1063/1.4937350
2. Bartolome, E.: Signal conditioning for piezoelectric sensors. Technical report (2010)
3. Corsi, C., Rohlfs, L., Weiss, J., Huber, K., Röediger, T.: Fluctuating heat flux measurements in an incident shock/boundary-layer interaction. In: 57th 3AF International Conference AERO (2023)
4. Corsi, C., Rohlfs, L., Weiss, J., Wang, B., Kahf, M., Obloch, P., Ngo, H.D.: Development of a PVDF Piezo-film sensor array for unsteady wall-pressure measurements in a turbulent SBLI. In: AIAA AVIATION 2022 Forum, pp. 1–12 (2022). https://doi.org/10.2514/6.2022-4135
5. Dailey, C.L.: Supersonic diffuser instability. J. Aeronaut. Sci. **22**(11), 733–749 (1955). https://doi.org/10.2514/8.3452. https://arc.aiaa.org/doi/10.2514/8.3452
6. Dolling, D.S.: Fifty years of shock-wave/boundary-layer interaction research: What next? AIAA J. **39**(8), 1517–1531 (2001). https://doi.org/10.2514/2.1476
7. Dussauge, J.P., Piponniau, S.: Shock/boundary-layer interactions: possible sources of unsteadiness. J. Fluids Struct. **24**(8), 1166–1175 (2008). https://doi.org/10.1016/j.jfluidstructs.2008.06.003. https://linkinghub.elsevier.com/retrieve/pii/S0889974608000595
8. Ganapathisubramani, B., Clemens, N., Dolling, D.: Effects of upstream coherent structures on low-frequency motion of shock-induced turbulent separation. Amer. Inst. Aeronaut. Astronaut. (2007). https://doi.org/10.2514/6.2007-1141
9. Harvey, J., Babinsky, H.: Shock Wave-Boundary-Layer Interactions. Cambridge University Press (2013)
10. Kawai, H.: The piezoelectricity of poly (vinylidene Fluoride). Jpn. J. Appl. Phys. **8**(7), 975–976 (1969). https://doi.org/10.1143/JJAP.8.975. https://iopscience.iop.org/article/10.1143/JJAP.8.975
11. Lee, B.: Self-sustained shock oscillations on airfoils at transonic speeds. Prog. Aerospace Sci. **37**(2), 147–196 (2001). https://doi.org/10.1016/S0376-0421(01)00003-3. https://linkinghub.elsevier.com/retrieve/pii/S0376042101000033
12. Lee, I., Sung, H.J.: Development of an array of pressure sensors with PVDF film. Exper. Fluids **26**(1–2), 27–35 (1999). https://doi.org/10.1007/s003480050262. http://link.springer.com/10.1007/s003480050262
13. McLaskey, G.C., Glaser, S.D.: Acoustic emission sensor calibration for absolute source measurements. J. Nondestr. Eval. **31**, 157–168 (2012)
14. McLaskey, G.C., Lockner, D.A., Kilgore, B.D., Beeler, N.M.: A robust calibration technique for acoustic emission systems based on momentum transfer from a ball drop. Bull. Seismolog. Soc. Amer. **105**(1), 257–271 (2015). https://doi.org/10.1785/0120140170
15. Measurement Specialties, I.: Piezo film sensors technical manual (2008). https://doi.org/10.1148/100.2.415
16. Rohlfs, L., Stab, I., Weiss, J.: Experimental investigations of incident shockwave boundary layer interactions in a continuously operating supersonic wind tunnel. In: AIAA Aviation 2022 Forum. American Institute of Aeronautics and Astronautics (2022)
17. Threadgill, J.A.S., Bruce, P.J.K.: Unsteady flow features across different shock/boundary-layer interaction configurations. In: AIAA J. **58**(7), 3063–3075 (2020). https://doi.org/10.2514/1.J058918. https://arc.aiaa.org/doi/10.2514/1.J058918
18. Touber, E., Sandham, N.: Oblique shock impinging on a turbulent boundary layer: low-frequency mechanisms. In: 38th Fluid Dynamics Conference and Exhibit. American Institute of Aeronautics and Astronautics, Reston, Virigina (2008). https://doi.org/10.2514/6.2008-4170. https://arc.aiaa.org/doi/10.2514/6.2008-4170

19. Weiss, J., Little, J.C., Threadgill, J.A., Gross, A.: Low-frequency unsteadiness in pressure-induced separation bubbles. In: AIAA Scitech 2021 Forum, pp. 1–19. American Institute of Aeronautics and Astronautics, Reston, Virginia (2021). https://doi.org/10.2514/6.2021-1324. https://arc.aiaa.org/doi/10.2514/6.2021-1324

Open Access This chapter is licensed under the terms of the Creative Commons Attribution 4.0 International License (http://creativecommons.org/licenses/by/4.0/), which permits use, sharing, adaptation, distribution and reproduction in any medium or format, as long as you give appropriate credit to the original author(s) and the source, provide a link to the Creative Commons license and indicate if changes were made.

The images or other third party material in this chapter are included in the chapter's Creative Commons license, unless indicated otherwise in a credit line to the material. If material is not included in the chapter's Creative Commons license and your intended use is not permitted by statutory regulation or exceeds the permitted use, you will need to obtain permission directly from the copyright holder.

Transitional/Turbulent SBLI and Flow Control

Non-linearities in the Low-Frequency Dynamics of Transitional SBLI

Mariadebora Mauriello, Lionel Larchevêque, and Pierre Dupont

Abstract The need for a better understanding of the low-frequency unsteadiness observed in shock wave/boundary layer interactions (SBLI) has driven research in this area for several decades. While numerous studies have been conducted on interactions with a turbulent boundary layer, in the context of transitional SBLI, research is still in its early stages and the low-frequency unsteadiness and the mechanisms underlying its origin remain poorly defined. In this study, large eddy simulations (LES) are performed in a $M = 1.7$ transitional shock reflection with separation. The objective is to examine any unsteadiness and the underlying mechanism. Beginning with a thorough assessment of stability theory, which suggests that transition in low supersonic compressible flows arises from the breakdown of oblique unstable boundary layer modes, the following question is posed. Do non-linear couplings between these oblique unstable modes and low-frequency unsteadiness emerge, and to what extent? To investigate quadratic couplings, high-order diagnostic is required. The results demonstrate that the unstable modes of the boundary layer interact non-linearly. High-frequency modes cascade non-linearly towards higher frequencies, initiating the turbulent cascade process, and towards lower frequencies. The low-frequency quadratic coupling with the flow characteristics at the separation point is responsible for the unsteadiness.

Keywords Transonic flow · Shock wave—boundary layer interaction · Large Eddy Simulations · Shock induced separation · Low-frequency unsteadiness

M. Mauriello (✉) · L. Larchevêque · P. Dupont
CNRS, IUSTI, Aix Marseille University, Marseille, France
e-mail: mariadebora.mauriello@univ-amu.fr

L. Larchevêque
e-mail: lionel.larcheveque@univ-amu.fr

P. Dupont
e-mail: pierre.dupont@univ-amu.fr

1 Introduction

Shock wave/boundary layer interaction (SBLI) is a common phenomenon in high speed flights, and can significantly affect the aerothermodynamic loads and performances. Multiple shocks, flow separation, transition to turbulence, unsteadiness, and three-dimensionality occur near the interaction region, which can lead to loss of control or failure of aerospace components. One of the first experimental investigations of the influences of shock waves and boundary layers, both in laminar and turbulent regimes, at transonic and low supersonic Mach numbers was conducted by [2, 3]. Since then, numerous experiments and numerical simulations have been carried out to study SBLI in detail.

Much of the interest has been devoted to interactions between shock waves and turbulent boundary layers. A wide variety of geometric configurations have been covered such as normal shock interactions, incident reflecting interactions, compression corner, over-expanded nozzle, etc., and investigations have covered a wide range of Mach and Reynolds numbers. The global space and time organisation has been described [4, 5] and the qualitative mean organisation of the flow is currently well understood [6]. In the case of turbulent separated SBLI, evidence showed that the interaction is highly unsteady with very low-frequency motions of the separated region. The review paper [7] offers a detailed overview of all the investigations over the past few decades that have sought to understand the source of the low-frequency unsteadiness of shock-induced turbulent separation in canonical flow fields. Nevertheless, the origin of this phenomenon is still under debate and divides the scientific community into different strands of thought.

Less attention has been paid to unsteadiness in interaction between shock waves and an incoming laminar boundary layer. Only recently has the scientific community taken steps in this direction, and the European TFAST (Transitional Location Effect on Shock Wave Boundary Layer Interaction) project was one such initiative. Both numerical and experimental studies were carried out with the aim of understanding the mutual influence of the shock waves and the laminar boundary layer. These studies confirmed that unsteadiness occurs when transition to turbulence occurs. The work [8] revealed that as the breakdown to turbulence occurs, broadband disturbances travel upstream in the subsonic region of the boundary layer with a corresponding response near the separation point. Furthermore, temporal measurements showed that several frequencies are amplified along the interaction region [9]. The work [10] confirmed that the reattachment region could be the origin of the low-frequency unsteadiness and that its amplitude varies significantly with the size of the separated region and the location of the transition to turbulence. The work [11] suggested a feedback acting through the high sensitivity of the separation point as it is perturbed by density waves travelling backwards and growing in the separated flow region. However, all these studies agree that the low-frequency unsteadiness is associated with a feedback from the reattachment region, although many open questions remain.

The present work is part of the European TEAMAero (Towards Effective Flow Control and Mitigation of Shock Effects In Aeronautical Applications) project, which is an extension of its precursor TFAST. The supersonic group of research of Marseille contributed to the previous project, conducting experimental explorations on a laminar boundary layer interacting with an impinging-reflecting shock system in the supersonic wind tunnel at the IUSTI (Institut Universitaire des Systèmes Thermiques Industriels) laboratory. Concurrently, within the same group of research, numerical investigations started on the same flow configuration. The current work builds on these numerical investigations, complemented by extensive post-processing using high-order statistical tools. Numerical comparisons, together with tentative experimental comparisons, have been pursued to improve the overall understanding of the source of low-frequency unsteadiness in transitional SBLI. The starting point is based on a firmly assessment of the stability theory. Given that the transition in low-supersonic compressible flows is driven by the breakdown of oblique instability modes, the aim was to explore whether a connection exists between the unstable boundary layer modes and the observed low-frequency of the head shock.

This chapter focuses on a part of the results obtained during the Ph.D. programme at the IUSTI laboratory in Aix-Marseille. It presents the results of large eddy simulations (LES) carried out on an oblique shock impinging on a laminar boundary layer over a flat plate, together with a comprehensive analysis using advanced statistical techniques. For a more detailed explanation of the high-order statistical tools, as well as results related to the second geometry analysed (6° compression ramp) and findings from direct numerical simulations (DNS), readers are referred to the corresponding Ph.D. manuscript thesis [12].

2 Flow Conditions

The flow configuration under investigation is an impinging-reflecting shock interacting with an incoming laminar boundary layer. As the interaction progresses, transition occurs, leading to the development of a turbulent boundary layer by the end of the shock system. The flow conditions used to simulate this scenario are the same as those employed in an experimental campaign carried out at the IUSTI laboratory in Aix-Marseille and documented in [1]. The aerodynamic conditions are summarised in Table 1. Notably, the subscript "∞" denotes freestream properties.

The experimental campaign of [1] documented the flow using Hot Wire Anemometer (HWA) and high resolution Laser Doppler Anemometer (LDA). While with the former technique the extraction of time scales developed along the interaction was obtained, the complete characterisation of the incoming boundary layer was

Table 1 Aerodynamics flow conditions

Mach	Re_{unit} [m^{-1}]	θ [deg]	P_∞ [Pa]	T_∞ [K]
1.7	5.7×10^6	5	8351	296

possible with the latter technique allowing the extraction of mean quantities fundamental for the present numerical setup. Specifically the boundary layer thickness δ and the boundary layer displacement thickness δ^* at two meaningful locations. At the inlet of the computational domain, they are $\delta_{inlet} = 0.53$ [mm] and $\delta^*_{inlet} = 0.25$ [mm] respectively. The knowledge of δ_{inlet} was required to setup the inlet conditions for the simulation. The undisturbed boundary layer thickness and the boundary layer displacement thickness were extracted at the impingement point as well. Their values are $\delta_{imp} = 0.91$ [mm] and $\delta^*_{imp} = 0.4$ [mm]. The corresponding Reynolds numbers are $Re_{\delta_{imp}} = 5187$ and $Re_{\delta^*_{imp}} = 2280$, respectively.

3 Numerical Setup

The size of the domain in the three spatial dimensions is defined by different criteria, described as follows:

Length

The streamwise dimension L_x is such that the numerical inflow is located $70\delta_{inlet}$ upstream the separation point. The end of the domain is located $126\delta_{inlet}$ downstream of the impingement point allowing to capture the whole interaction region and to include the downstream relaxation zone.

Height

The wall-normal dimension L_y corresponds to $50\delta_{inlet}$. The choice of this value derives from a physical consideration. When the boundary layer separates due to the impinging oblique shock wave, a system of combined reflected shocks and an expansion fan is generated. Although non-reflective boundary conditions are applied to the top of the domain, a weak reflection occurs. The height of the domain is the minimal height that avoids the reflection of the wave system from the top boundary to impinge onto the downstream boundary layer.

Width

The span dimension L_z is set 7 times larger than the separation-bubble height. A similar choice for the width was adopted in the studies conducted during the TFAST project [13]. They proved to be sufficient to capture the low frequency dynamics.

The grid resolution is chosen such that to resolve transition and turbulence at the back of the bubble. For this reason, in the streamwise direction the grid is stretched using a two-side hyperbolic tangent distribution. The spatial step size $\Delta x = 0.12$ [mm] remains constant until $115\delta_{inlet}$, just before the interaction point. It then slowly decreases to $\Delta x = 0.075$ [mm] and remains constant at this value throughout the separated region. Once the reattachment point is reached, the grid resolution increases again in the relaxation zone where the maximum cells are 8% of the inlet cells. The grid distribution in the wall-normal direction is stretched with a stretching rate

Table 2 Numerical domain size normalised by δ^*_{inlet}, number of cells and grid resolutions in wall units

$L_x \times L_y \times L_z$	$N_x \times N_y \times N_z$	Δx^+	y^+_{wall}	Δz^+
$584 \times 108 \times 60$	$1254 \times 240 \times 200$	≤ 30	≤ 1.25	≤ 16

of 1.5% starting from the edge of the boundary layer thickness $y = 0.53$ [mm]. It clusters about 24% of the grid points within the boundary layer at the inlet being the total number of cells equal to 240. In the spanwise direction, the grid resolution is kept constant and equal to $\Delta z = 7.5 \times 10^{-5}$ [mm]. Periodic boundary conditions are applied. Although in the span direction the resolution value is outside the recommended range [14], at the point with the most stringent constraint, corresponding to the transitional-to-turbulent region, the resolution is satisfactory being $\Delta x = 1 \times 10^{-4}$ [mm], $\Delta y = 6 \times 10^{-6}$ [mm], and $\Delta z = 7.5 \times 10^{-5}$ [mm].

Table 2 details the simulation parameters. Note that the domain size is normalised with the displacement thickness at the inlet and grid resolutions are in wall units defined as $x_i^+ = (x_i\, u_\tau)/v$, with $u_\tau = \sqrt{\frac{\tau_{wall}}{\rho}}$ being the friction velocity and v the kinematic viscosity.

The boundary conditions applied to the computational domain are no-slip and adiabatic conditions at the wall, i.e. $u_i = 0$ and $dT/dy = 0$. The top (freestream) and outflow boundaries make use of non-reflective boundary conditions based on characteristic formulations in order to minimise unwanted reflections of waves from the computational box boundaries. To mimic the pressure jump $p_3/p_1 = 1.6$ of the reference experiments,[1] the Rankine-Hugoniot relation was set at the top wall. Non-reflecting boundary conditions are also set at the inflow and outflow. The inflow condition is the boundary condition to which particular attention has been paid in the early stages of this work and will be discussed in the following section.

All simulations were performed using the FLU3M code from ONERA. The code relies on a finite volume discretisation in space, and an implicit Gear scheme for the temporal discretisation, both being second-order accurate. To minimise the numerical dissipation, the space scheme is modified by adding the dissipative part of the Roe scheme to a centered scheme in regions where strong compressibility/low vorticity occurs, as identified by means of Ducros sensor. The time integration is performed with a maximum Courant–Friedrichs–Lewy number CFL of 11, and the non-linear system is solved through 7 sub-iterations resulting in a reduction of the residuals of more than three orders of magnitude in the laminar and transitional regions and about two orders of magnitude in the turbulent region. The LES modeling is built from an implicit grid filtering coupled with an explicit sub-grid modeling through the selective mixed-scale model.

---

[1] Note that p_3 indicates the pressure state after the reflected shock.

Simulations have been run for 1.2 million time steps providing data sampled over a physical duration of 0.12 s. The corresponding number of typical low-frequency cycles is about 70 and the total cost of the simulation was 250000 CPU hour. This run time ensured good statistical convergence for all spectral estimators.

4 Inflow Conditions

4.1 Delicate Choice

In this work the transition to turbulence has been stimulated using the synthetic eddy method (SEM). First evaluated for external flow by [15] and further extended for wall bounded flow by [16], the SEM method is based on the classical view of turbulence as a superposition of eddies. Each eddy is represented by specific shape functions of position and time that describe its spatial and temporal coherent properties, i.e. with correlation in space/time. SEM acts as a low-pass filter and the cut-off corresponds to the length scale of the turbulent structures supported by the boundary layer. In general, it is a broadband and stochastic forcing that is well suited to the purpose of this study. In fact, one of the objectives was to include oblique modes in the simulation and let the mean flow to select the most unstable family.

The procedure included to superimpose small velocity fluctuations to a compressible Blasius profile. The characterisation of the Blasius profile was available from the experimental measurements documented in [1], while the inflow perturbations were modulated in the wall-normal direction using the following ad hoc polynomial function:

$$A_0 \left[2 \left(\frac{x}{5\delta_{inlet}} \right)^3 - 3 \left(\frac{x}{5\delta_{inlet}} \right)^2 + 1 \right] \quad (1)$$

Using this function, disturbances remained primarily confined within the boundary layer thickness, with the highest disturbance intensity positioned near the generalised inflection point. The amplitude A_0 was applied identical to fluctuations in the streamwise, wall-normal, and spanwise directions. The integral scale of the perturbation was arbitrary selected to be $0.8\delta_{inlet}$ in the streamwise direction and $0.4\delta_{inlet}$ in both the wall-normal and spanwise directions. The streamwise and wall-normal velocity fluctuations were correlated using a correlation coefficient of -0.5. The next step was to select the amplitude A_0 of the inflow perturbations. This parameter plays a fundamental role in the extent of the separated region in a SBLI problem [10] and it required multiple iterations before obtaining the length of interaction as that provided by the experiments. The final value of the amplitude of the inflow perturbations is 0.0055% of U_∞.

4.2 Incoming Boundary Layer

The incoming boundary layer has been examined with the purpose of gathering information regarding the flow features enforced at the inlet by using the SEM. An a posteriori local linear stability analysis was performed to extract information about the most unstable family of boundary layer modes. The reader is referred to Chapter 4 of the corresponding manuscript [12] for details. It is important to bear in mind that such analysis is local. As long as the zone of influence of the shock system does not influence the evolution of the boundary layer, the hypothesis of self-similar flow can be considered valid.

Figure 1 shows the spectra extracted at $x = 40\delta_{inlet}$ from the inlet of the computational domain. It ensures that the boundary layer at that location is naturally developing. All the quantities are made dimensionless with the local boundary layer displacement thickness $\delta^*_{local} = 0.28$ [mm] and the local freestream velocity $U_{local_\infty} = 463$ [m s^{-1}]. They are indicated with the superscript "*". The spectrum of the real and imaginary part of the streamwise wavenumber $\alpha^* = \alpha \times \delta^*_{local}$ is shown in Fig. 1a. At that location, an amplified wave is observed as $-\alpha^*_i > 0$ for $\alpha^*_r > 0$ (see zoom insert in panel (a)). It is responsible of the growth of the unstable boundary layer modes. To identify the most unstable perturbation wave, the neutral maps show the growth rate α^*_i into the pulsation frequency-spanwise wavenumber

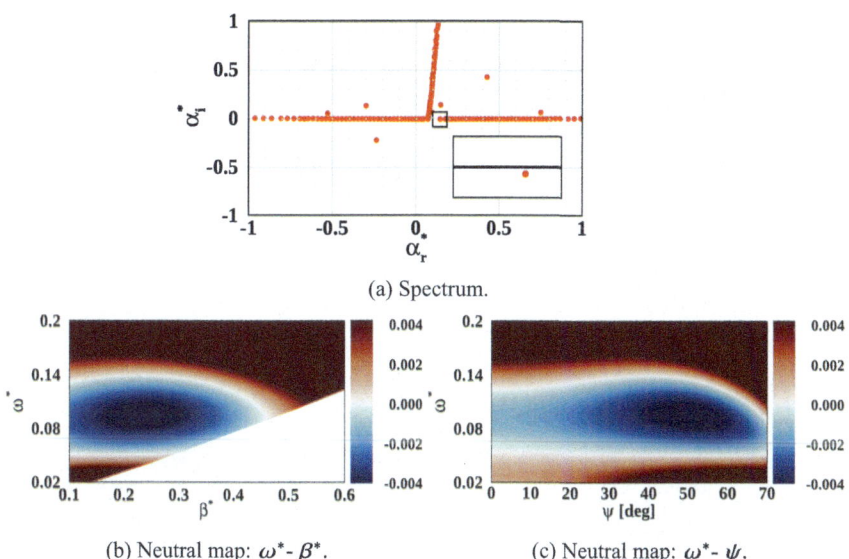

(a) Spectrum.

(b) Neutral map: ω^*-β^*.

(c) Neutral map: ω^*-ψ.

Fig. 1 Spectra of the incoming boundary layer at location $x = 40\delta_{inlet}$: spectrum (**a**) and neutral maps: ω^*- β^* (**b**), and ω^*- ψ (**c**). The neutral maps contours correspond with the growth rate α^*_i. All the quantities are made dimensionless with the local boundary layer displacement thickness δ^*_{local} and the local freestream velocity $U_{\infty local}$

(Fig. 1b) and pulsation frequency-wave angle (Fig. 1c) spaces. It is clear that the boundary layer entails a collection of amplified unstable modes (see blue contours) which are 3D in nature as the spanwise wavelengths β^* differ from zero. They are known as oblique modes. The most amplified family occurs at the pulsation frequency $\omega^* = (2\pi f \delta^*_{local})/U_{local_\infty} = 0.09$, which corresponds to the dimensional frequency of $f \simeq 24$ KHz. Figure 1b gives information about its spanwise wavelength being known $\beta^* = \beta \times \delta^*_{local} = 0.23$. It is $\lambda_z = 0.0075$ [mm] and it corresponds with half of the domain width $L_z = 0.015$ [mm]. It derives that the spanwise dimension of the computational domains restricted the most unstable oblique mode family to be at the most of $\lambda_z = L_z/2$. Figure 1c shows that the wave angle at which the oblique mode family is travelling is deviated of $\psi = 51°$ with respect the main flow direction x. For completeness, the streamwise wavelength corresponds to $\lambda_x = 0.0092$ [mm] (plot not shown).

Figure 2 shows the frequency-wavenumber spectra extracted at the same location where local stability analysis has been conducted ($x = 40\delta_{inlet}$). The variable investigated is the spanwise velocity component w. In this way, all possible influences that could result from pressure waves rising in the subsonic region of the boundary layer due to the presence of the shock system are removed. In these spectra, the frequency is normalised with the inlet boundary layer displacement thickness δ^*_{inlet} and the undisturbed velocity U_∞. It corresponds with the Strouhal number:

$$St_{\delta^*_{inlet}} = \frac{f \times \delta^*_{inlet}}{U_\infty} \qquad (2)$$

Through this normalisation, $St_{\delta^*_{inlet}}$ directly corresponds to the normalised frequency $w^*/2\pi$ previously defined for the stability analysis.

The wavenumber is normalised with the width of the domain L_z. In such a way, values equal to ± 1 correspond with the maximum extent of the computational domain, while all multiple integers are a fraction of it. Figure 2a shows that the flow features characterised by wavenumbers included in the range [± 1; ± 5] entail most of

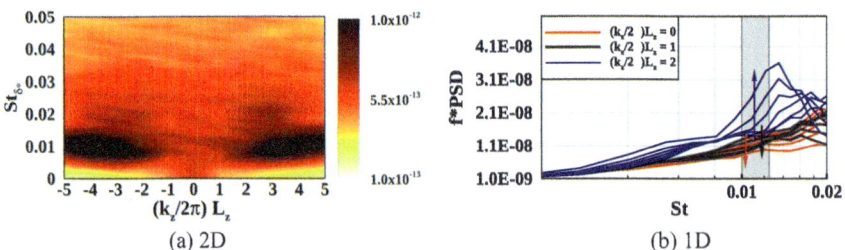

Fig. 2 2D frequency-wavenumber spectrum (**a**) and premultiplied power spectral density spectrum (**b**) of the spanwise velocity component w. In panel (**b**), the grey region indicates the limiting Strouhal numbers sustainable by the boundary layer instabilities. Starting from $x = 40\delta_{inlet}$, each group of lines is equispaced with step of $5\delta_{inlet}$. The arrows indicate the direction of the growth of the waves amplitude

the power spectral content. In particular, starting from $(k_z/2\pi)L_z = 2$ for all subsequent wavenumbers most of the energy content belongs to the range of Strouhal [0.002; 0.015]. Figure 2b restricts the attention to the 2D $((k_z/2\pi)L_z = 0)$ and 3D $((k_z/2\pi)L_z = 1$ and $(k_z/2\pi)L_z = 2)$ features emerging in that range of Strouhal. Starting from the $x = 40\delta_{inlet}$, their streamwise evolution is followed. Lines equally spaced with step of $5\delta_{inlet}$ trace back the growth rate of the amplitude of the evolving 2D and 3D boundary layer modes up to a point that is still upstream to the shock system influence. The grey region of the graph identifies the upper and lower limits of the Strouhal number obtained from the local stability analysis and appropriately scaled with the δ^*_{local}. The upper limit is defined by the local stability analysis and indicates the frequency that the most unstable perturbation can exhibit. It is $St_{\delta^*} = 0.0125$ and it falls within the range of the frequencies associated with the mostly emerging features at the inlet (see Fig. 2a). Conversely, the lower limit has been extracted from the last streamwise location before the perturbations enter the zone of interaction with the shock system. As the boundary layer naturally develops, the corresponding δ^*_{local} has increased at that location and the L_z/δ^*_{local} ratio has decreased consequently. Same considerations hold for the corresponding frequency, and the unstable mode, compatible with this normalised domain width, have Strouhal number $St_{\delta^*} = 0.01$ that represents the lower limit. In the range [0.1; 0.0125], the arrows indicate the trend of the amplitude of the 2D and 3D instability boundary modes. Within this interval, the group of least stable modes obtained by the stability analysis, aligned with the wavelength $\lambda_z = L_z/2$, exhibits a pronounced exponential increase in the amplitude of disturbances as they evolve. In contrast, 2D features and modes of $\lambda_z = L_z$ demonstrate a decrease in amplitude as they progress.

4.3 Length of Interaction

Various studies [10, 13] suggested that the length of the separated region is governed by the amplitude of the inlet perturbations, at least for a given shock strength. The strategy used in the present work to impose the inflow boundary conditions allowed such an amplitude to be easily tuned.

The target length provided by the experimental measurements of [1] was $L_{int} = 43.7$ [mm], being:

$$L_{int} = x_{imp} - x_{int} \qquad (3)$$

with $x_{int} = 63.6$ [mm] the interaction point defined as the inflection point of the pressure rise across the compression waves and $x_{imp} = 107.3$ [mm] the impingement location of the incident shock. In the present simulation, the impingement location has been extrapolated by prolonging the incident shock down to the wall, while the interaction point was extracted from the center of the head compression region again down to the wall.

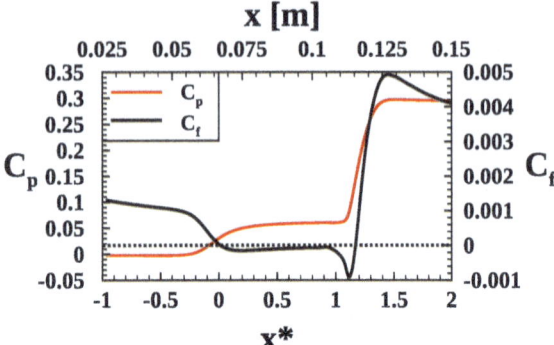

Fig. 3 Dimensional (bottom axis) and non-dimensional (top axis) streamwise evolution of the pressure and friction coefficients. The dot black line indicates the zero value of the friction coefficient

Figure 3 shows the streamwise evolution of the pressure and friction coefficients defined respectively as:

$$C_p = \frac{2(P - P_0)}{\gamma P_0 M_0^2}, \quad C_f = \frac{2\,\tau_{wall}}{\gamma P_0 M_0^2} \qquad (4)$$

In addition to the dimensional streamwise evolution, the streamwise coordinate normalised using the length of interaction is also presented in the same plot. The normalisation reads:

$$x^* = \frac{x - x_{int}}{L_{int}} \qquad (5)$$

In this way, x^* ranges from 0 and 1 within the interaction region. In dimensional unit, the range [0;1] corresponds to 43.7 [mm]. The strategy of adjusting the amplitude of the inflow perturbations led to a similar separation length observed by the experimental measurements.

The tuning of the inflow perturbations amplitude entails consequences on the location where transition occurs as it is governed by the nonlinear saturation of unstable modes that emerge within the mixing layer across the bubble. On the other hand, from a numerical point of view, the transition mechanism might be influenced by the mesh resolution. To verify that the transition process remains largely unaffected despite alterations in grid resolution, an additional refined mesh F was tested and compared with the standard mesh S. The finer mesh F was refined of 40, 20 and 25% in the x, y and z directions with respect to the standard mesh. Moreover, the inflow conditions amplitude was decreased by 10% to achieve an equivalent interaction length, considering that a finer mesh exhibits lower dissipation properties.

Figure 4 shows the Mach contours overlapped by the dividing streamline $h(x)$ (white line). It corresponds to the mean line of zero longitudinal mass flux computed as:

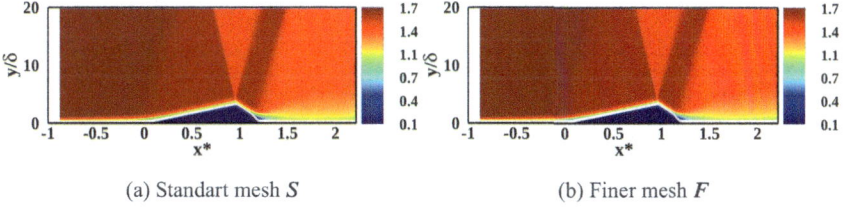

Fig. 4 Mach contours for simulations run with the standard mesh S (**a**) and finer mesh F (**b**). The white line denotes the separation streamline $h(x, t)$

$$h(x) \quad \text{such that} \quad \int_0^{h(x)} \left[\int_{-\frac{W}{2}}^{\frac{W}{2}} \overline{\rho(x, y, z) u(x, y, z)} dy \right] dz = 0 \qquad (6)$$

with W the width of the domain.

From Fig. 4 it can be concluded that although the optimal amplitude differs by less than 10% between the standard S and refined F computations for the reasons previously mentioned, the grid convergence study shows that the transition process is largely unaffected, as similar downstream mean flow characteristics are observed. It also shows that the previously described procedure for adjusting the inflow perturbation amplitude is independent of grid resolution.

5 Downstream Flow

The description of the downstream nature of the flow is presented in this section. In the rear part of the interaction region, the laminar boundary layer undergoes a transition. This process is influenced by the amplitude of the inflow perturbation fields, which in turn are chosen so as to obtain the same separated region length of the experiments.

Figure 5 shows the non-dimensional as well as the dimensional streamwise evolution of the turbulent kinetic energy normalised by the friction velocity k/u_τ^2 and the friction coefficient C_f. The turbulent kinetic energy profile is extracted from a plane located at $y/\delta = 1.2$, which corresponds to the distance from the wall where the maximum value of k is observed. Attention is focused on the downstream region of the flow from $x^* = 1$ onwards. It corresponds to the trace on the wall of the impinging shock. The separated region reattaches at $x^* = 1.18$. This value has been extracted from the streamwise evolution of the friction coefficient (see Fig. 3), and its location is highlighted in the plot Fig. 5 with the black dashed vertical line. Both peaks of k/u_τ^2 and C_f occur downstream of the reattachment point at $x^* = 1.25$ ($\simeq 2\delta_{inlet}$) and $x^* = 1.44$ ($\simeq 17\delta_{inlet}$) respectively. The latter is shifted by almost $15\delta_{inlet}$ with respect to the former. A possible explanation for this shift can be attributed to the convective nature by which the fluctuations generated by the turbulent kinetic energy

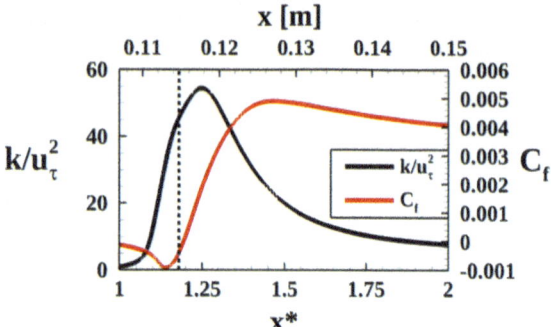

Fig. 5 Streamwise evolution of the normalised turbulent kinetic energy k/u_τ^2 at $y/\delta = 1.2$, and friction coefficient C_f. The black dashed vertical line indicates the point at which the separated region reattaches ($x^* = 1.18$ or $x = 0.115$ [m])

in the reattaching mixing layer are transferred to the wall, filling the velocity profile and thus increasing the wall shear stress τ_{wall}. The velocity at which they are convected requires not only time but also space, resulting in the different streamwise locations of the two peaks. After reaching the maximum, the friction coefficient begins to approach the classical trend of turbulent profiles.

Figure 6 helps to understand whether the boundary layer is fully turbulent or still in a transitional state. Figure 6a shows the streamwise mean velocity profiles at equally spaced locations. The entire streamwise evolution is spanned so that differences between the typical laminar and transitional and/or turbulent profiles are highlighted. At the inlet, a canonical laminar profile develops (see black curve in both panels), followed by a region where reverse flow occurs. This corresponds to the separated region. After the impinging point ($x^* = 1$), the boundary layer continues to recover the fuller profile as it flows downstream. The red lines describe the downstream flow, which appears to be turbulent in nature. This is supported by Fig. 6b, where the log-law region appears for velocity profiles downstream of the separated region.

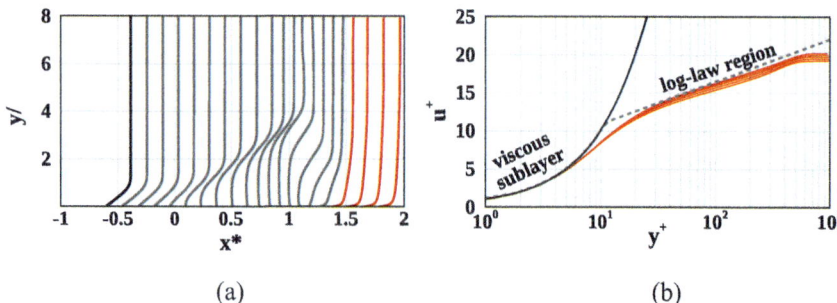

Fig. 6 Panel (**a**): streamwise mean velocity profiles at equally spaced x locations. The black profile is extracted at the inlet location, grey profiles in the middle region, and red profiles correspond with locations in the downstream zone. Panel (**b**): inlet (black) and downstream (red) profiles in wall units. The log-low is overlapped and indicated with the black dashed line

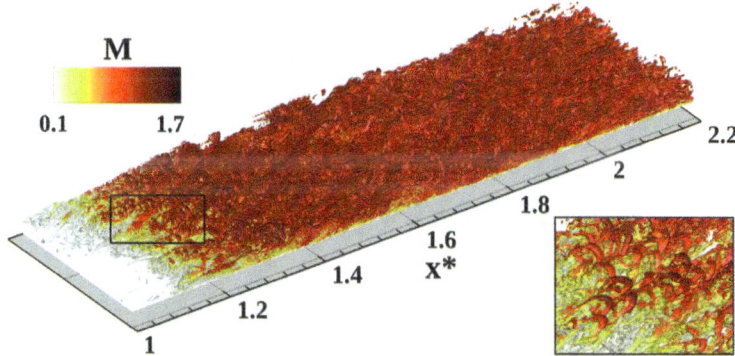

Fig. 7 Q-criterion visualisation in the downstream region colored with the values of the Mach number

They are compared with the standard log-law for wall-bounded flows, see the black dashed line in the same figure. It reads:

$$\frac{U}{u_\tau} = \frac{1}{k} ln\left(\frac{y}{\delta_v}\right) + A \tag{7}$$

with u_τ and δ_v being the viscous velocity and viscous length respectively, $k = 0.41$ is the von Kármán constant and $A = 5.2$ [17].

Figure 7 summarises the downstream flow and provides a complete picture of the structures that populate the rear zone. Vortical structures are extracted using the Q-criterion coloured with Mach isocontours. This highlights both density and velocity changes. Corresponding to the reattachment point, which occurs at $x^* = 1.18$, hairpin structures emerge, leading the boundary layer into the turbulent state (see zoom inset).

6 Separated Region: Unsteady Aspects

In the context of SBLI, the separated zone is a key region to investigate. The bubble is affected by the low-frequency *breathing*. Turbulent interactions consistently experience the influence of this phenomenon, and the source of the low-frequency unsteadiness remains a subject of ongoing debate. In the context of laminar and transitional interactions, [18] categorised laminar interactions as steady, while transitional interactions as highly unsteady. In this section, spectral analysis is considered to identify the key features that govern the reversal flow zone. The characteristic time scales are extracted by spectral decomposition of the wall pressure fluctuations.

Figure 8 shows the streamwise distributions of the power spectral density spectrum of the wall pressure fluctuation field p'. It is premultiplied by the frequency and

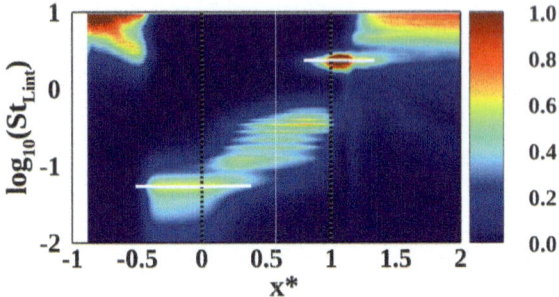

Fig. 8 Premultiplied and normalised wall pressure power spectral density spectrum. Two black dashed vertical lines indicate the interaction $x^* = 0$ and the impinging $x^* = 1$ points, whereas the white horizontal lines mark two significant Strouhal numbers, $St_{L_{int}} = 0.055$ and $St_{L_{int}} = 2.3$

normalised by the local value of the variance. The horizontal axis shows the streamwise evolution normalised by the interaction length L_{int} (see Eq. 3). The vertical axis shows the Strouhal number on a logarithmic scale. In the problem of low-frequency unsteadiness, the Strouhal number can describe the unsteady phenomena associated with the separated region if we select the length of interaction L_{int} as a length scale, which is an intrinsic scale of the phenomenon, while for the other scale, we use the external velocity U_∞. The Strouhal number thus reads:

$$St_{L_{int}} = \frac{f L_{int}}{U_\infty} \qquad (8)$$

The spectrum highlights several regions of activity: the inlet, the beginning of the interaction region near $x^* = 0$, downstream the impingement point $x^* = 1$, and the outlet zone. The inlet region stands out from the spectrum due to the normalisation and premultiplication used. The approach used to superimpose the fluctuation fields on the mean laminar boundary layer profile involved the use of broadband forcing with a cutoff corresponding to $St_{\delta_{inlet}} = 1$. Consequently, the cutoff retained low frequencies and excluded high frequencies, so the premultiplication resulted in an amplification of the higher frequency spectral content within this region. The second region of activity extends about $34\delta_{inlet}$ upstream of the interaction point (see white horizontal line). The pressure rise caused by the impinging shock is transmitted upstream through the subsonic part of the boundary layer and, due to its laminar nature, the upstream influence length is extended in this region [19]. This zone propagates at low frequencies with a Strouhal number of $St_{L_{int}} \simeq 0.055$. Such a value is close to that found in the DNS work of [8] and in the LES study of [10], but it is significantly lower than the $St_{L_{int}} \simeq 0.1$ seen in the reference experiments of [1]. In the recent experimental work of [20] on a Mach 4 compression ramp, a value of $St_{L_{int}} = 0.025$ was observed, and to the author's knowledge this is the only experimental Strouhal value that more closely matches the numerical results. However, the RANS and LES work of [11] gives an intermediate value of $St_{L_{int}} \simeq 0.08$. The discrepancies in the low frequency Strouhal numbers between the numerical and experimental studies are not clear. Nevertheless, next section is dedicated to a detailed discussion of the discrepancy between the present work and the reference experiments [1]. Moving in the streamwise direction, the typical frequency increases up to $St_{L_{int}} \simeq 0.6$ within the separated region. Once past $x^* = 1$, another region of activity appears. It is located in the region of the flow downstream of the impingement point at $x^* \simeq 1.1$ and the associated Strouhal number is $St_{L_{int}} \simeq 2.3$ (see white horizontal line). The presence of this intense activity of the wall pressure field has already been observed in [10] for transitional SBLI. The downstream region accommodates the transition mechanism of the boundary layer, whose footprint can appear in the spectrum. Indeed, the last region highlighted by the spectrum is located in the outlet zone, confirming that the boundary layer is approaching the turbulent state (see Figs. 6b and 7).

Figure 9 provides an instantaneous 3D visualisation of the structures surrounding the reversal flow region. The horizontal and vertical planes show the pressure gradient isocontours to help visualise the locations of the compression waves, while the Q-criterion, coloured with the values of the Mach number, highlights the vortex structures. The choice of Mach isocontours is explained in the description of Fig. 7. Note that the level of the Q-isosurface has been reduced compared to Fig. 7 in order to better visualise low amplitude flow structures. The interaction region is detected by the gradient of the pressure fluctuation field p' at the wall (see dark region in the plot). The separation point occurs at $x^* = 0$, and the reversal flow extends to $x^* = 1.18$, where the boundary layer reattaches through the reattachment shock. The spanned length corresponds to the separation length L_{sep}, which is defined as the difference between the two points above. Remembering that for low supersonic compressible boundary layers the transition to turbulence occurs due to the breakdown of the oblique mode instability waves [21–23], oblique modes are observed in this region starting from one third of L_{sep}. They start to develop much further upstream than this point. However, due to the low amplitude of the inlet oblique modes, they are partially visible in Fig. 9. From $x^* = 0.37$ the unstable modes continue to grow with

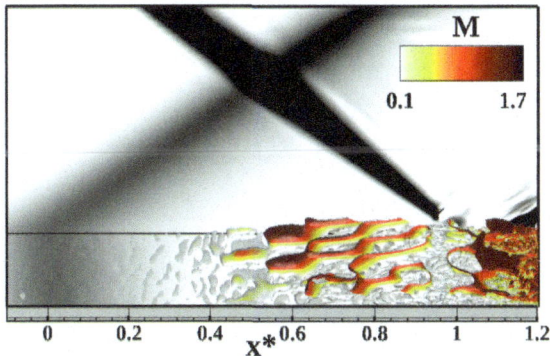

Fig. 9 Q-criterion visualisation in the separated region colored with the values of the Mach number. The horizontal and vertical planes show the gradient of the pressure isocontours

a linear growth rate to a high perturbation amplitude. In the vicinity of the reattachment point, they enter the nonlinear phase and their breakdown occurs. The boundary layer undergoes a transition to the turbulent state.

7 Deviations from Experiments

As stated from the beginning, the setup of the present simulation is based on the experimental data provided by the work [1]. In this section, we point out the discrepancies that occur as well as an attempt is made to explain them.

Figure 10a shows the longitudinal mean velocity profiles U at several sections, including locations before and after the interaction region L_{int} (see grey zone in the plot). The numerical results are compared with experimental measurements obtained using the LDV technique (the reader is referred to [1] for technical details). The data show excellent agreement in the upstream region as well as in the bubble, where the velocity profiles gradually evolve from laminar to inflectional profiles. From the impingement point, the curves diverge. Although the seeding problem of the LDV technique affects the near-wall region, if the experimental mean velocity profile at the last location is extrapolated to the wall, it appears that the boundary layer recovers the turbulent state faster. To confirm this hypothesis, Fig. 10b plots the longitudinal fluctuation field u' at the same locations as in panel (a). For the mean flow, the numerical simulation agrees with the experimental measurements in the upstream region. In the reversal zone, the agreement is partially observed. In the first half there is no discrepancy, whereas from the second half onwards the fluctuation peaks start to deviate. From the point $x^* = 0.5$ the difference is remarkable. This point corresponds to half of the region where the shear layer develops, and if we look again at the Fig. 9, we can see that new structures start to appear. Perhaps the growth rate of the unstable modes populating the boundary layer is slowly increasing. The peaks of the fluctuations are less pronounced in the case of the numerical data, and

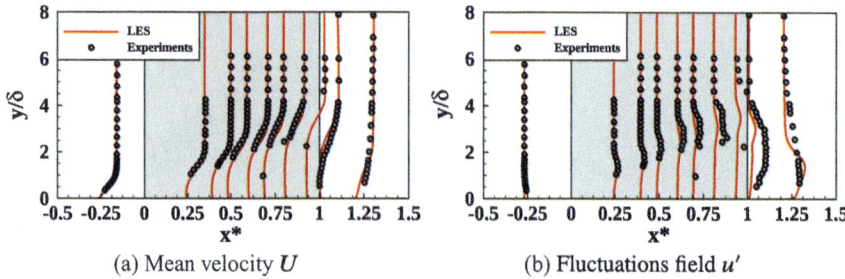

Fig. 10 Validation with experiments: streamwise mean velocity U (**a**) and the corresponding fluctuations field u' (**b**) profiles at several streamwise locations. The grey zone in both panels indicates the interaction region L_{int}

they gradually start to enter the nonlinear regime. On the other hand, the peaks of the experimental measurements increase dramatically, possibly fed by the receptivity process of the external fluctuation field and the wind tunnel noise, although very low (0.1% of the external velocity). The experimental flow is less organised as the profiles are more distorted, possibly influenced by competing dynamics, including vortex structures shed by the shear layer, the increase in velocity due to the expansion fan and the development of turbulent fluctuations due to the transition process. The nature of the flow downstream of the reattachment point $x^* = 1.18$ is different. Figure 6b showed a well defined logarithmic region, indicating that the boundary layer has reached the turbulent state. Such a state is not observed in the experiments. The work [1] documented the downstream flow and showed that the boundary layer did not reach the classical turbulent state as a non well-defined logarithmic region was observed (see Fig. 12a of their paper). This result seems to be incompatible with what was previously observed from the mean velocity field U extracted from experimental data (see Fig. 10a) and remains a question to be clarified. Nevertheless, an attempt has been made to address the discrepancy between the numerical data and the experimental measurements. In general, when performing a numerical simulation based on experimental data, one aim is to reproduce identical input conditions as those observed in the wind tunnel. In addition to the problem of receptivity, this involves matching various aspects of the boundary layer, including its displacement thickness, the amplitude of the inflow perturbations, and possibly even the specific oblique mode family. If the initial conditions are indeed identical and the LES model can accurately replicate the real flow, the resulting conditions should theoretically be the same. However, in this particular study, the full characterisation of the boundary layer was incomplete (only the boundary layer thickness and displacement thickness data were available from those required and listed above), along with the length of the separation zone, which was consequently used as an input parameter. This discrepancy could potentially explain the observed differences in downstream flow between the numerical simulation and the experimental measurements.

Another aspect to consider is the different time scales extracted at the separation point by the two studies, although the order of magnitude is the same, i.e. $\mathcal{O}(10^{-2})$.

Figure 11 compares the premultiplied and normalised power spectral density of the streamwise momentum ρu extracted at the separation point. x_{sep} has been translated along the characteristic line up to $z = 0.005$ [m]. In fact, for a reasonable comparison, the spectral content for both numerical data and experimental measurements is extracted at the same height $z = 0.005$ [m]. A broadband low-frequency spectral content, attributed to the head shock, emerges from the LES data, with a main peak at $St_{L_{int}} = 0.055$ (see black dashed vertical line). Conversely, hot wire measurements show a distinct frequency distribution characterised by a narrower band with a sharp peak at $St_{L_{int}} = 0.11$. It is unclear why the experimental measurements fail to extract the same numerical time scale.

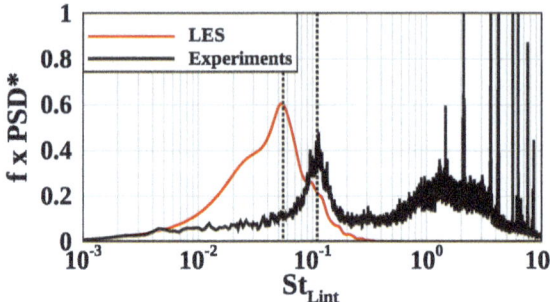

Fig. 11 Validation with experiments. Premultiplied and normalised power spectral density spectrum of the streamwise momentum extracted at the separation point at height $z = 0.005$ [m]. The two black dashed vertical lines corresponds with $St_{L_{int}} = 0.055$ and $St_{L_{int}} = 0.11$, respectively

8 Multiple Mechanisms and Frequencies

In the previous sections, the global organisation in terms of space and time dynamics has been presented and the qualitative mean organisation of the flow has been given. However, the problem of boundary layer interaction with a shock system hides complex features that are often difficult to distinguish.

Figure 12 helps the reader to immediately visualise different regions of interest where sensors are located for the investigation. S_i correspond to the end of the ascending mixing layer (S_1), the centre of the descending mixing layer (S_2) and three locations distributed over a horizontal plane fairly close to the wall, either upstream, at or downstream of the reattachment point (S_3, S_4 and S_5 respectively).

First, attention is focused on the near-wall region (horizontal plane in Fig. 12) to see if there is a linear relationship between the slow motion of the head shock and the downstream flow. Phase information is extracted for the three characteristic Strouhal numbers observed in Fig. 8, i.e. $St_{L_{int}} = 0.055$ at the interaction point, $0.11 < St_{L_{int}} < 0.33$ in the separated region, and $St_{L_{int}} = 2.3$ at $x^* = 1.1$.

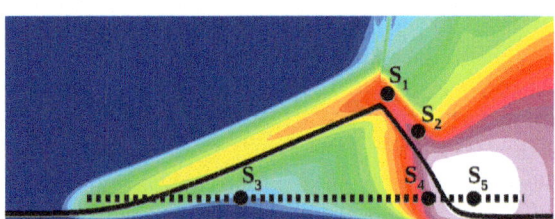

Fig. 12 Locations of the sensors overlapping the turbulent kinetic energy map. The black solid line indicates the dividing streamline, whereas the dashed black line corresponds with the horizontal plane passing through sensors S_3, S_4 and S_5 and where data were sampled

Fig. 13 Streamwise evolution of the normalised phase associated with ρu at fixed Strouhal numbers. The reference sensor is located at $x^* = 0$ and the black dashed vertical line indicates $x^* = 0.95$. It corresponds with the projection of sensor S_1 (see Fig. 12) down to the wall

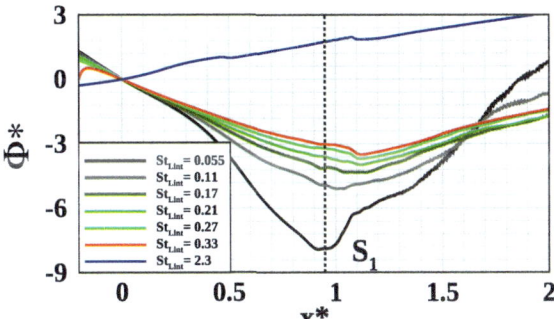

Figure 13 plots the streamwise evolution of the phase associated with the streamwise momentum ρu at the fixed aforementioned Strouhal numbers. The phase data are then normalised to the Strouhal number and unwrapped so that jumps of 2π are removed. The inverse of the slope gives information about the phase velocity between the reference ρu at the interaction point $x^* = 0$ and the streamwise momentum signals over the plane. It reads:

$$v_\Phi = \frac{(2\pi f \Delta x)}{\Delta \Phi} = \frac{\Delta x^*}{\Delta \Phi^*} \qquad (9)$$

The sign of v_Φ indicates the direction of motion. With respect to the reference point, a positive sign indicates downstream motion, while a negative sign indicates upstream motion. It is important to note that the phase velocity is associated with a fixed frequency, but is contaminated by all the streamwise wavenumbers k_{x_i}. Since no particular wavenumber has been chosen, it can be assumed that $v_\Phi \sim v_g \sim U_c$, where v_g is the group velocity and U_c is the convection velocity. In other words, Fig. 13 provides the following information.

For $St_{L_{int}} = 2.3$ the phase increases linearly along the flow direction with a convection velocity of $U_c \simeq +0.64 U_\infty$. The experimental work of [24] on a supersonic laminar boundary layer along a flat plate found a similar value for the convection velocity, i.e. $U_c \simeq +0.62 U_\infty$. They attributed this motion to three-dimensional perturbations, more likely the oblique modes. Following their results, it can be postulated (it is confirmed in the following section) that the downstream travelling waves we observe are the oblique modes. Back to Fig. 13, for lower Strouhal numbers an abrupt phase change is observed at $x^* = 0.95$ (see black dashed vertical line in the plot). This corresponds to the projection of the sensor S_1 onto the investigated plane (see Fig. 12). Up to $x^* = 0.95$ the phase decreases almost linearly, while everywhere else it increases linearly. Before the jump, upstream propagation velocities are found which correspond to $U_c \simeq -0.09 U_\infty$ and $U_c \simeq -0.25 U_\infty$ for the low and medium frequencies respectively. These velocities were compared with the motion within the reversal region. Taking into account the maximum reversal velocity $U_{rev} = -0.04 U_\infty$ and the local speed of sound $c = 0.72 U_\infty$, the propagation velocity $U_p = |U_{rev} - c| = 0.68 U_\infty$ is obtained in the range $0 < x^* < 1$. Therefore, the low

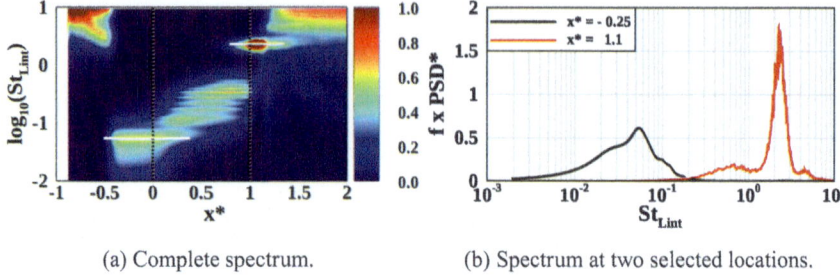

(a) Complete spectrum.   (b) Spectrum at two selected locations.

Fig. 14 Premultiplied and normalised wall pressure fluctuations power spectral density spectrum: **a** complete, and **b** at the two energy prevailing locations corresponding to the two white horizontal lines in (**a**)

and medium frequency waves travel at lower speeds than this propagation velocity U_p and can be considered fluidic rather than acoustic. After the jump, downstream propagation speeds are observed with $U_c \simeq 0.11 U_\infty$ and $U_c \simeq 0.3 U_\infty$ for the low and medium frequencies respectively. These results are partially consistent with [8]. They found upstream travelling disturbances in the separated region of the boundary layer with a strong link to the separation point. However, they found that the pressure waves travel with a constant phase velocity equal to $v_\Phi = -0.6 U_\infty$, and suggested the acoustic nature of the waves responsible for the appearance of the low frequency unsteadiness.

Together with the phase information, Fig. 14 summarises the spatial and frequency scale organisation of the flow. The reattachment region is characterised by high-frequency scales $St_{L_{int}} = 2.3$, more likely associated with the breakdown of the oblique mode waves. Indeed, we have previously pointed out that a transition mechanism occurs in this region. Conversely, the separation point is dominated by low frequency scales $St_{L_{int}} = 0.055$ associated with the unsteadiness of the head shock. It appears that two main physical mechanisms are involved: the low frequency driven processes and the oblique modes. However, the clear separation in frequency scales suggests that the two physics can only be coupled non-linearly, with the simplest case being quadratic interactions resulting in flow features driven by low frequencies $(f_1 - f_2)$ and/or harmonics $(f_1 + f_2)$.

9 Splitting in Two Physics

Of all the physical mechanisms that contribute to the problem of low-frequency transients, two have emerged. This section attempts to characterise them fully. The spectral POD (SPOD) method has been used to highlight the 2D and/or 3D nature of the coherent structures in both spatial and frequency domains [25, 26]. The mode selection allowed a global view of the spectral content organisation and only the

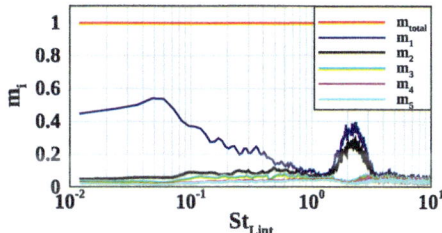

Fig. 15 Spectral POD relative energy spectrum of the streamwise momentum ρu for modes 1–5

most energetically prevalent modes were considered for analysis. For a detailed explanation of the method, the reader is referred to [12].

Figure 15 shows the SPOD relative energy spectrum of the streamwise momentum ρu. From the full data set, 60 modes m_i were extracted, which are able to reproduce the dynamics of the flow in the whole space with a frequency resolution of $\Delta St_{L_{int}} = 0.01$. The graph shows only the first five modes. The mode m_{total} corresponds to the sum of all modes (the total energy is 1) and is shown for reference. In accordance with the approach recommended by [27] for normalising the inner product, the modes are intrinsically classified according to their decreasing energy content. This implies that the first modes have the largest fraction of the total energy on a frequency by frequency basis. The first mode m_1 captures the largest fraction of the total energy, which is confined to the low frequency range. Up to Strouhal values $St_{L_{int}} \simeq 0.1$, 40% of the total energy belongs to flow features characterised by low frequency dynamics. Thereafter, the energy content decreases with increasing Strouhal and then peaks again in the high frequency range $1 < St_{L_{int}} < 3$. In the same range, m_2 also shows a peak: both peaks are centred around $St_{L_{int}} = 2.3$. Such a value has already been pointed out in the previous section as a typical frequency characterising the region of activity downstream of the impingement point. It is more likely to be associated with the unstable boundary layer modes. The other modes do not reach 10% of m_{total}. For this reason, only modes m_1 and m_2 are considered in the following analysis.

9.1 Oblique Modes

The unstable boundary layer modes have been partially characterised in the previous section: 3D in nature, the most unstable family for the current simulation has a spanwise wavelength equal to half the domain width, corresponding to the wavenumber $(k_z/2\pi)/L_z = 2$. For simplicity, we will refer to $(k_z/2\pi)/L_z$ as k_z^* in the rest of the discussion. Here we try to extract their spatial organisation, which may result from quadratic interactions $(f_1 + f_2)$, possibly between all 3D waves characterised by $k_z^* = \pm 2$. Furthermore, since the SPOD allows to determine the energy content for each frequency scale, the Strouhal number $St_{L_{int}} = 2.3$ is chosen. This corresponds

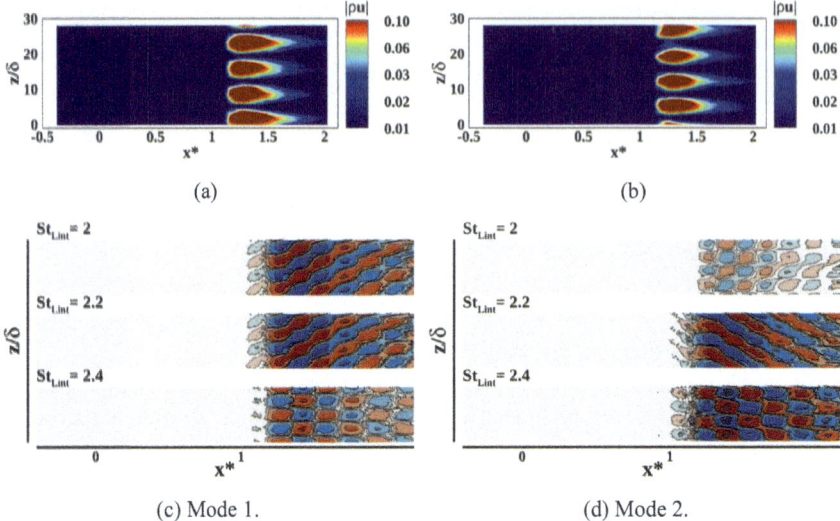

Fig. 16 Dominant SPOD modes at $St_{L_{int}} = 2.3$ and the corresponding reconstructed flow fields over a horizontal plane located at $y/\delta = 1.2$ from the wall. Left column: mode 1; right column: mode 2

to the leading frequency at the reattachment location as seen in Fig. 14. Instead, three Strouhal values around $St_{L_{int}} = 2.3$ were chosen for the flow field reconstruction.

Figure 16 shows the energy organisation of the norm of the streamwise momentum (first row) as well as the corresponding reconstructed flow field for an arbitrary time (second row) for the modes m_1 (left column) and m_2 (right column) over a horizontal plane located at $y/\delta = 1.2$ from the wall. Both modes indicate that 3D waves are dominant in the reattachment region. Their spanwise dimension is $k_z^* = 2$: both panels (a) and (b) show four peaks of concentrated energy. As the norm of the energy is shown in the plots, two of these peaks belong to the positive fluctuations, while the other two belong to the negative fluctuations, giving a total of four peaks. This result is not surprising as it confirms the predicted wavenumber associated with the most unstable boundary layer mode ($k_z^* = 2$) extracted from the stability analysis. In addition, mode m_2 exhibits a quadrature relationship in the spanwise direction with respect to mode m_1, meaning that the peaks associated with m_2 are shifted by 1/4 of the spanwise wavelength of the peaks associated with m_1. The corresponding flow fields clearly show that the downstream region is populated by high frequency 3D coherent structures with a spanwise dimension of $k_z^* = \pm 2$.

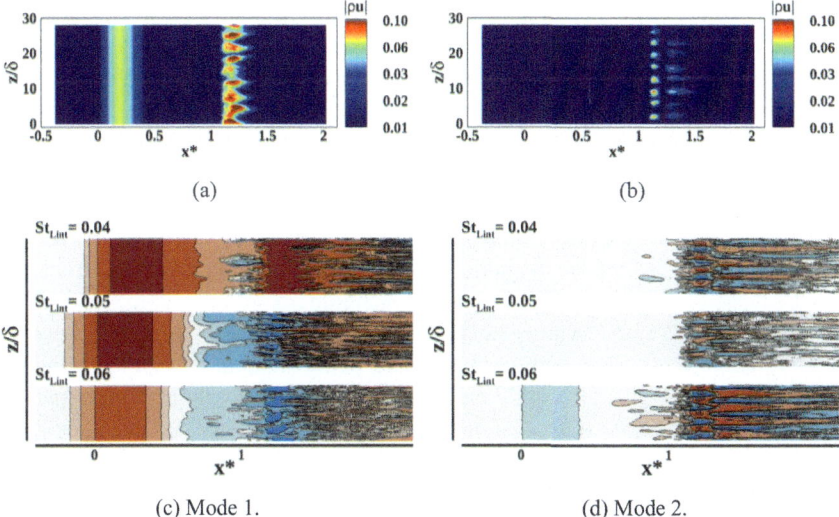

Fig. 17 Dominant SPOD modes at $St_{L_{int}} = 0.055$ and the corresponding reconstructed flow field over a horizontal located at $y/\delta = 1.2$ from the wall. Left column: mode 1; right column: mode 2

9.2 Low-Frequency Driven

The flow features driven by low frequencies are here characterised.

Figure 17 shows the spatial energy organisation of waves acting at $St_{L_{int}} = 0.055$ as well as the reconstructed flow field for three Strouhal values around $St_{L_{int}} = 0.055$. Note that the same organisation of the plots as in Fig. 16 applies here. In contrast to the oblique mode interaction seen earlier, the low frequency couplings introduce waves at both the separation and reattachment points. However, the spanwise periodicity is different: $k_z^* = 0$ for the former and $k_z^* = 4$ for the latter (see 8 peaks in Fig. 17a and b). The 3D waves persist at the reattachment point at low frequency, but they are no longer oblique modes. They resemble streaks (see the reconstructed flow fields in Fig. 17d). By definition, streaks are stable in time, and they can result from low-frequency ($f_1 - f_2$) nonlinearities that are close to zero [28].

10 Scale-by-Scale Frequency Energy Transfer: Local or Global Process?

The region of the flow around the reattachment point is dominated by small scale motions resulting from the breakdown of oblique mode waves whose frequencies are much higher than the low frequency range associated with the upstream fluidic feedback ($St_{L_{int}} = 2.3 \gg St_{L_{int}} = 0.055$). The question that arises is whether these two phenomena have only local characteristics in terms of space and frequency, or whether there is a broader, global connection between them. One suggestion comes from the different frequency scales that separate them. Because of this frequency scale separation, both phenomena may be related by quadratic interactions. Such interactions can be highlighted in spectral space from the bispectrum, which is a measure of quadratic nonlinearities at the bifrequency (f_1, f_2) [29, 30].

Before highlighting any non-linear interactions between the unstable modes of the boundary layer and the separation point, we will focus our attention on the reattachment zone. Our previous results indicate that this is the region where the transition process takes place. The nonlinear spectral technique is well suited to highlighting this phenomenon. Figure 18 plots the (squared) bicoherence maps $Bic^2_{g_1,g_2,g_3}$ between the time series $g_1(t)$, $g_2(t)$ and $g_3(t)$ extracted at the locations indicated in Fig. 12. A value of $Bic^2_{g_1,g_2,g_3}$ significantly different from zero is associated with a (partial) phase relationship between $g_1 \times g_2$ and g_3 that is stable over time and indicates a possible non-linear energy exchange between the frequencies considered. The variable analysed is the streamwise momentum ρu, since it undergoes quadratic interactions through the convective term of the Navier-Stokes momentum equation. Figure 18a shows the bicoherence map between sensors $S_1 \times S_1$ and S_2. The linearly unstable modes develop in the ascending mixing layer and enter the nonlinear regime near the apex of the bubble. Indeed, the positive quadrant of the bicoherence map (a) is typical of a transition mixing layer. It shows an emerging direct turbulent cascade with triadic interactions between frequency domains that are multiple integers of the initial unstable wave at location S_1, centred around $St_{L_{int}} = 2.3$ (see white circle in panel (a)). On the contrary, the quadrant defined by $St_{L_{int_1}} < 0$ and $St_{L_{int_2}} > 0$ shows a quadratic relationship with $\left|St_{L_{int_1}}\right| \simeq \left|St_{L_{int_2}}\right|$ towards lower frequencies. The large low frequency band along the diagonal corresponds to $\Delta St_{L_{int}} = 0.35$ and the white straight lines in panel (a) help to visualise the band. Visible nonlinearities occurring between several integers of the harmonics associated with the oblique modes are observed from the near diagonal region.

Let's turn our attention to the specific area of interest. As a reminder, the goal is to investigate whether there is a non-linear correlation between the oblique modes

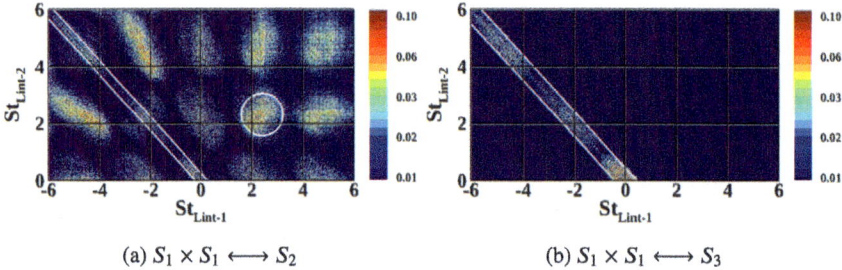

(a) $S_1 \times S_1 \longleftrightarrow S_2$ (b) $S_1 \times S_1 \longleftrightarrow S_3$

Fig. 18 (Squared) bicoherence maps: transitioning mixing layer (**a**) and *breathing* of the bubble (**b**)

and the 2D features located within the boundary layer in the separation region. Figure 18b shows the bicoherence map between sensors $S_1 \times S_1$ and S_3 (see Fig. 12). The positive quadrant of the map lacks significant values of bicoherence because the turbulent energy cascade does not occur within the reversal region, but begins to develop between the impingement and reattachment points. Focusing on the negative quadrant, there is a visible quadratic coupling for a band of low frequency, again spread along the diagonal with a larger bandwidth belonging to the medium frequency range of $\Delta St_{L_{int}} = 0.6$. $St_{L_{int}} = 0.055$ belongs to this region: it can be speculated that the quadratic phase coupling (QPC) between small-scale high frequency structures results in the large-scale low frequency flow features located in the (upstream part of the) separation region. They are responsible for the unsteadiness of the head shock.

However, for triadic interactions between band-limited signals, bicoherence is relatively easy to analyse. The oblique modes are not associated with a single frequency, but with a large frequency band as observed from the neutral map Fig. 1. Consequently, the quadratic interactions associated with the resulting single frequency can be multiple. For this reason, such an analysis becomes more difficult to interpret for more broadband signals, as a single frequency may be involved in multiple triadic interactions. One way to overcome this difficulty is to use summed bicoherence, which includes all potential quadratic couplings associated with any pairs of frequencies of constant sum [29].

Figure 19 shows the summed bicoherence between the "source" signal $S_1 \times S_1$ and "target" sensors located within the descending shear layer (S_2) and near the wall (S_4), both upstream (S_3) and downstream (S_5) of the reattachment point. The variable studied is the streamwise momentum ρu for the reasons explained earlier. Note that the low values associated with the summed bicoherence indicated on the y-axis are a consequence of the normalisation applied. The energy transfer towards higher harmonics of the original instability wave at $St_{L_{int}} = 2.3$ is mostly achieved when approaching the vicinity of the reattachment point (see sensor S_2). When the reattachment point is reached, the turbulence is almost fully developed. The energy cascade appears to be almost saturated as only two peaks resulting from triadic interactions are observed (see sensor S_4). However, within the separation bubble

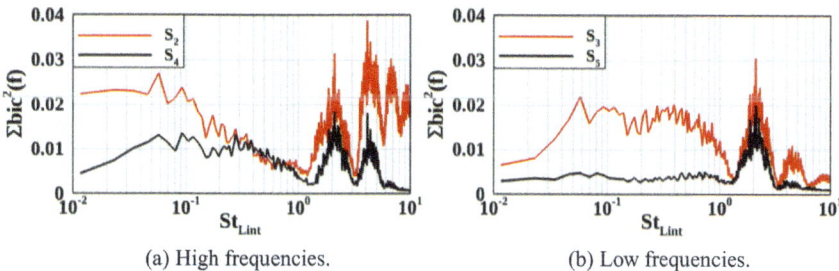

Fig. 19 Summed bicoherence: high frequencies destinations S_2 and S_4 (**a**) and low frequencies destinations S_3 and S_5 (see Fig. 12 for the location of the sensors)

there is a transfer of energy towards the low-medium frequency range, as shown by sensor S_3 in figure (b). No such process is observed downstream of the reattachment point (see sensor S_5). It is worth noting that the relative energy content at low-medium frequency increases as one moves upstream within the bubble (see sensor S_4 in the low-medium frequency range compared to sensor S_3). Based on these results, it can be postulated that the upstream low-frequency feedback is driven by quadratic interactions between high-frequency oblique modes found in the mixing layer developing over the separated region.

Additional information about the convective or stationary nature of the square link is extracted from the biphase spectrum (see [12] for details). The streamwise evolution of the biphase between the $S_1 \times S_1$ and sensors distributed over the horizontal line crossing sensors S_3 to S_5 is plotted in Fig. 20. Note that the biphase is normalised using the same procedure as for the phase evolution. The biphase shows several frequency pairs resulting from quadratic couplings. High frequency pairs have been selected whose sums correspond either to the various typical low frequency Strouhal fluctuations identified in the separated region ($St_{L_{int}} < 1$) or to the oblique modes and their harmonics ($St_{L_{int}} \simeq n \times 2.3, n \in \mathbf{N}$). Among the different frequency pairs ($f_1 + f_2$) of a given sum f_3 that have been computed, the three that give the highest bicoherence levels have been kept. In this sense, the streamwise biphase evolution plotted is representative of the strongest quadratic contributions in the low frequency range. The comparison of Figs. 13 and 20 confirms that in the separated region the linear and quadratic phase evolutions are very similar for the frequency range considered. However, while for the linear case the jump in propagation velocity is observed from $x^* = 0.95$, the quadratic coupling restricts the upstream motion to a shorter region (from $x^* = 0.75$ towards the separation point), at least in the case of the lowest frequency range. Despite this, the group velocity associated with the motion of the fluctuations quadratically coupled to the oblique mode is roughly the same as that found from the linear phase analysis. A possible explanation is as follows. Non-linear interactions arise from the interaction of oblique modes in the vicinity of the reattachment point. As a result, low frequency structures are formed which

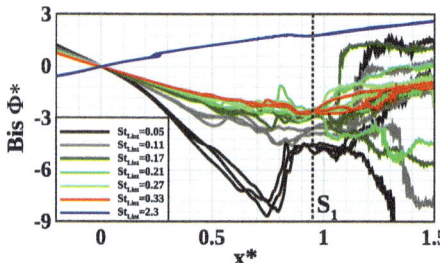

Fig. 20 Streamwise evolution of the normalised biphase along the horizontal line crossing sensors S_3 to S_5. For each $St_{L_{int}}$ spanning from low to high values, the three combinations resulting in the highest values of bicoherence are plotted. The black dashed vertical lines indicates $x^* = 0.95$ and corresponds with the projection of sensor S_1 down to the wall

undergo linear convection towards the separation point. Despite this linear convection, these structures persist in the nonlinear phase. This phenomenon occurs because when a quadratic process undergoes a linear transformation, the resulting transformation remains quadratic with respect to the original (in this case linear) process. This, together with the rather large values of bicoherence found in this region, confirms that the upstream convected streamwise momentum fluctuations are largely induced by nonlinear coupling with the oblique modes of the mixing layer. Since such upstream motion is associated with the low-frequency breathing of the separated region, it is strongly suggested that the breathing is induced by nonlinear beating near the reattachment point of the high-frequency oblique modes that have developed in the mixing layer. Such a scheme could explain the lack of low-frequency breathing in the case of fully laminar interaction, for which quadratic interactions of modes are poorly energetic and do not result in transition.

Overall, the upstream convected streamwise momentum fluctuations are for a large part induced by non-linear coupling with the mixing layer oblique modes. The interaction is non-linear and involves different regions of the flow. The oblique mode at the top of the bubble not only communicates quadratically with the downstream region where the turbulent cascade energy transfer occurs, but also results in 2D flow features emerging at the separation point and governed by the low frequency $St_{L_{int}} = 0.055$.

11 Conclusions

This chapter explored the interaction between an incoming laminar boundary layer and an impinging-reflecting shock, with the aim of improving the understanding of the physical origins and mechanisms of the breathing phenomenon. The initial efforts were aimed at reproducing this flow configuration in a numerical context using large eddy simulation, starting from the available experimental measurements carried out during the experimental campaign at the IUSTI laboratory in Aix-Marseille and documented in [1]. Beside the limit of LES in resolving transitional flows, the initial challenge involved selecting input conditions for the simulation. The experimental data were insufficient to fully characterise the incoming laminar boundary layer. While numerous methods, available in the literature, were suitable for turbulent boundary layers, choosing for a naturally evolving boundary layer required careful considerations, ultimately leading to the adoption of synthetic eddy method as the final decision. The SEM technique achieved a twofold purpose. It allowed to incorporate multiple three-dimensional unstable modes into the incoming laminar boundary layer, fundamental for the transitioning mechanism in low supersonic flows. The most unstable oblique mode family was fully characterised by means of a posteriori local linear stability analysis, being $\lambda_z = 0.75 \times 10^2$ [m^{-2}], $\lambda_x = 0.92 \times 10^2$ [m^{-2}], and $\psi = 51°$ its space organisation. On the other side, the SEM allowed to easily tune the amplitude of the inflow perturbation fields, that affects the length of the separated region, as suggested by studies of [10, 13]. In this way, the target length provided by the experimental measurements was matched.

However, the tuning of the inflow perturbations amplitude entailed consequences on the location where transition of the laminar boundary layer occurs as it is governed by the non-linear saturations of unstable modes that emerge within the mixing layer across the bubble. Deviations from the reference experiments were observed in the downstream region of the shock system, evidencing routes for the boundary layer transition that differ. Probably, the mismatch of incoming boundary layer characterisation represents the key to understanding this discrepancy.

Beside the different space arrangement in the downstream region, different temporal scales were extracted at the separation point. Broadband low-frequency spectral content with a peak at $St_{L_{int}} = 0.055$ was extracted from LES data, compatible with the recent literature [8, 10], whereas narrower band with a sharp peak at $St_{L_{int}} = 0.11$ was observed by the reference experiments. The discrepancy in time scales has long perplexed the scientific community, with an explanation remaining elusive thus far.

Nevertheless, linear and non-linear statistical analyses made it possible not only to confirm results already observed in the literature, but to extract new information, useful in clarifying the nature of the interactions between the shock system and the breathing phenomenon. In particular, the linear spectral diagnostic confirmed an upstream motion from the reattachment region toward the separation point. It has been shown to be slow and convective in nature, being $U_c \simeq -0.09 U_\infty$ the convection velocity associated with low-frequency scale deputed to the slow motion of the head shock. These results are partially in agreement with [8]. They found the value $-0.6 U_\infty$ for the upstream travelling pressure waves, suggesting an acoustic nature of the motion. Furthermore, high-order statistical techniques permitted the identification of quadratic coupling, observed for the mixing layer oblique modes in the low-frequency dynamics, and resulting in two-dimensional $k_z = 0$ flow features beneath the head shock. Indeed, it has been shown that the arrangement of the flow, which emerges when it is taking into account the quadratic phase coupling between selected flow structures and the associated frequency range in which they are active, sees in the low frequency dynamics 2D flow features at the separation point, and three-dimensional arrangement downstream the reattachment point. The sequence of events involves the breakdown of the oblique modes after the apex of the separated region, followed by the development of quadratic couplings of 3D oblique waves. They enter the separated region with a convection velocity that decreases linearly with distance, eventually reaching the separation point, which is then populated by 2D flow features.

During the Ph.D. program at the IUSTI laboratory in Aix-Marseille, another flow geometry was examined with the purpose of demonstrate that non-linearities in low-frequency dynamics are a prevalent occurrence in all transitional SBLI. A purely numerical comparison between an impinging-reflecting shock and a 6° compression ramp was carried out, and similarities and discrepancies were pointed out. Irrespective of the flow geometry, whether it is a the simple compression of the flow through a ramp or a flat plate interacting with a shock system, and of the transitional state of the boundary at the reattachment point, it was observed that the quadratic coupling between the oblique waves and the separation point persists. It was concluded that the non-linear feedback involving the unstable waves and the shock system remains consistent regardless of the geometry and the transitional path of the laminar boundary layer. The full analysis is available in the corresponding Ph.D. manuscript thesis [12].

References

1. Diop, M., Piponniau, S., Dupont, P.: High resolution LDA measurements in transitional oblique shock wave boundary layer interaction. Exp. Fluids (2019)
2. Liepmann, H.W.: The interaction between boundary layer and shock waves in transonic flow. J. Aeronaut. Sci. (1946)
3. Ackeret, J., Feldmann, F., Rott, N.: Investigations of compression shocks and boundary layers in gases moving at high speed. Technical Report (1947)
4. Dolling, D.S.: Fifty years of shock-wave/boundary-layer interaction research: what next? AIAA J. (2001)
5. Dupont, P., Haddad, C., Debiève, J.F.: Space and time organization in a shock-induced separated boundary layer. J. Fluid Mech. (2006)
6. Agostini, L., Larchevêque, L., Dupont, P., Debiève, J.F., Dussauge, J.P.: Zones of influence and shock motion in a shock/boundary-layer interaction. AIAA J. (2012)
7. Clemens, N.T., Narayanaswamy, V.: Low-frequency unsteadiness of shock wave/turbulent boundary layer interactions. Annu. Rev. Fluid Mech. (2014)
8. Sansica, A., Sandham, N.D., Hu, Z.: Instability and low-frequency unsteadiness in a shock-induced laminar separation bubble. J. Fluid Mech. (2016)
9. Diop, M., Piponniau, S., Dupont, P.: On the length and time scales of a laminar shock wave boundary layer interaction. In: 54th AIAA Aerospace Sciences Meeting (2016)
10. Larchevêque, L.: Low-and medium-frequency unsteadinesses in a transitional shock–boundary reflection with separation. In: 54th AIAA Aerospace Sciences Meeting (2016)
11. Bonne, N., Brion, V., Garnier, E., Bur, R., Molton, P., Sipp, D., Jacquin, L.: Analysis of the two-dimensional dynamics of a mach 1.6 shock wave/transitional boundary layer interaction using a RANS based resolvent approach. J. Fluid Mech. (2019)
12. Mauriello, M.: Non-linearities in the low-frequency dynamics of transitonal shock wave-boundary layer interactions. Ph.D. thesis, Aix-Marseille University (2024)
13. Doerffer, P., Flaszynski, P., Dussauge, J.P., Babinsky, H., Grothe, P., Petersen, A., Billard, F.: Transition Location Effect on Shock Wave Boundary Layer Interaction: Experimental and Numerical Findings from the TFAST Project. Springer Nature (2020)
14. Garnier, E., Adams, N., Sagaut, P.: Large Eddy Simulation for Compressible Flows. Springer Science & Business Media (2009)
15. Jarrin, N., Benhamadouche, S., Laurence, D., Prosser, R.: A synthetic-eddy-method for generating inflow conditions for large-eddy simulations. Int. J. Heat Fluid Flow (2006)
16. Pamiès, M., Weiss, P.E., Garnier, E., Deck, S., Sagaut, P.: Generation of synthetic turbulent inflow data for large eddy simulation of spatially evolving wall-bounded flows. Phys. Fluids (2009)
17. Pope, S.B.: Turbulent flows. Measurement Science and Technology (2001)
18. Chapman, D.R., Kuehn, D.M., Larson, H.K.: Investigation of separated flows in supersonic and subsonic streams with emphasis on the effect of transition. Technical Report (1958)
19. Babinsky, H., Harvey, J.K.: Shock Wave-Boundary-Layer Interactions. Cambridge University Press (2011)
20. Threadgill, J.A.S., Little, J.C., Jesse, Wernz, S.H.: Transitional Shock Boundary Layer Interactions on a Compression ramp at Mach 4. American Institute of Aeronautics and Astronautics (2021)

21. Fasel, H., Thumm, A., Bestek, H., Kral, L.D., Zang, T.A.: Direct numerical simulation of transition in supersonic boundary layers: oblique breakdown. ASME-PUBLICATIONS-FED (1993)
22. Sandham, N.D., Adams, N.A.: Numerical simulation of boundary-layer transition at Mach two. Appl. Sci. Res. (1993)
23. Sandham, N.D., Adams, N.A., Kleiser, L.: Direct simulation of breakdown to turbulence following oblique instability waves in a supersonic boundary layer. Appl. Sci. Res. (1995)
24. Laufer, J., Vrebalovich, T.: Stability and transition of a supersonic laminar boundary layer on an insulated flat plate. J. Fluid Mech. (1960)
25. Nekkanti, A., Schmidt, O.T.: Frequency-time analysis, low-rank reconstruction and denoising of turbulent flows using SPOD. J. Fluid Mech. (2021)
26. Schmidt, O.T., Colonius, T.: Guide to spectral proper orthogonal decomposition. AIAA J. (2020)
27. Chu, B.T.: On the energy transfer to small disturbances in fluid flow (Part I). Acta Mech. (1965)
28. Schmid, P.J., Henningson, D.S.: Stability and Transition in Shear Flows. Springer, New York, NY (2012)
29. Tynan, G.R., Moyer, R.A., Burin, M.J., Holland, C.: On the nonlinear turbulent dynamics of shear-flow decorrelation and zonal flow generation. Phys. Plasmas (2001)
30. Cui, G., Jacobi, I.: Biphase as a diagnostic for scale interactions in wall-bounded turbulence. Phys. Rev. Fluids (2021)

Open Access This chapter is licensed under the terms of the Creative Commons Attribution 4.0 International License (http://creativecommons.org/licenses/by/4.0/), which permits use, sharing, adaptation, distribution and reproduction in any medium or format, as long as you give appropriate credit to the original author(s) and the source, provide a link to the Creative Commons license and indicate if changes were made.

The images or other third party material in this chapter are included in the chapter's Creative Commons license, unless indicated otherwise in a credit line to the material. If material is not included in the chapter's Creative Commons license and your intended use is not permitted by statutory regulation or exceeds the permitted use, you will need to obtain permission directly from the copyright holder.

The Length and Time Scales of Transitional SBLIs

Nikhil Mahalingesh, Sébastien Piponniau, and Pierre Dupont

Abstract Shock-wave boundary layer interaction is a commonly found flow phenomenon in transonic and supersonic aerodynamics. However, interactions involving laminar boundary layers have received relatively less interest, due to the complexity of transition. As a result, experiments were performed to study laminar boundary layers and their interaction with shock-waves. The Mach number was 1.65 and the unit Reynolds number was varied between 5.6 and 11 million m^{-1}. Pitot probes and hot-wire anemometry were employed for flow measurements. Experiments of the transition process of a natural laminar boundary layer captured the modal growth mechanisms of the primary instability, and a new time scale was found in the latter stages of the transition process. A new multi-sensor hot-wire probe was developed to study this new time scale, which revealed strange physical properties. Experiments of transitional SBLIs were performed on a 6° and a 10° compression ramp. A new non-dimensional parameter was developed for scaling the strength of the imposed shock, that was able to reconcile the large scatter in a diverse collection of length scales of transitional interactions. Measurements of the boundary layer transitional mechanisms over the interaction showed an accelerated growth over the separated shear layer, but surprisingly the growth of sub-harmonic instabilities was bypassed at reattachment. Finally, low-frequency unsteadiness at separation was found at Strouhal number of 0.05, similar to other studies on transitional interactions. A possible link between the presence of non-linearities over the separated shear layer and the low-frequency unsteadiness was found.

Keywords Transonic flow · Shock wave—boundary layer interaction · Boundary layer transition · Low-frequency unsteadiness · Measurements

---

N. Mahalingesh · S. Piponniau (✉) · P. Dupont
CNRS, IUSTI, Aix Marseille University, Marseille, France
e-mail: sebastien.piponniau@univ-amu.fr

N. Mahalingesh
e-mail: nikhilmahalingesh@gmail.com

P. Dupont
e-mail: pierre.dupont@univ-amu.fr

1 Introduction

Shock wave Boundary Layer Interaction (SBLI) is a classical flow phenomenon of high speed aerodynamics, which are encountered in various practical situations such as engine inlets, transonic wings, rocket nozzles, compressors, etc. SBLIs generally occur when a boundary layer confronts an adverse pressure gradient, which can result in a number of unfavourable effects including separation of the boundary layer, increased viscous dissipation, intense thermal loading, structural fatigue, and also unsteadiness of the associated shock-waves. These issues can lead to dangerous loss of control or even catastrophic failure of aerospace components, thus restricting the design and envelope of air-borne vehicles. Consequently, SBLIs have been the subject of research since the 1940s.

Experiments and numerical simulations have been performed for various geometrical configurations (such as oblique shock impingement, compression ramps, aerofoils, compressor and/or turbine blades, etc.) and for a range of flow parameters, including various Mach numbers, Reynolds numbers, and flow deflections. These studies have made great strides in the understanding of the length and time scales of SBLIs. However, limitations from an experimental and computational point of view, had restricted much of the focus on the interaction between shock waves and turbulent boundary layers. And hence, attention to the interaction between shock waves and laminar boundary layers had been relatively lower. Recent advancements in experimental techniques and computational resources, together with emerging focus on sustainability and increased efficiency, have renewed interest on laminar interactions.

The European TFAST (Transition Location Effect on Shock Wave Boundary Layer Interaction) project was one such initiative, with the objective to understand the interaction between shock waves and laminar boundary layers in more detail [38]. As part of this project, experiments and numerical simulations were performed at various academic and industrial institutions across Europe for a period of five years between 2012 and 2017. Although significant progress was made as a result of this project, several open questions were left unanswered, including the complex effects of boundary layer transitional mechanisms on the spatial and temporal scales of such interactions [11].

The current work is an extension of those experiments, being performed as part of the subsequent European project, called TEAMAero (Towards Effective Flow Control and Mitigation of Shock Effects In Aeronautical Applications). During the previous TFAST project, the interaction between an externally generated oblique shock wave and a laminar boundary layer was studied experimentally in the current experimental facilities of the IUSTI laboratory. To complement these experiments, the focus was shifted towards the interaction between a laminar boundary layer and a compression ramp in the current work. The main objective was to further develop the general understanding of transitional SBLIs, building on the findings of the TFAST project.

This chapter is intended to be a brief summary of the work carried out as part of the Ph.D. program at the IUSTI laboratory of Aix-Marseille University. Hence, the reader is referred to the Ph.D. thesis manuscript for a more detailed description of the literature review, experimental methodology, verification, validation, and detailed analyses of the results [37].

2 Experimental Methodology

2.1 Experimental Setup

The experiments were performed at the IUSTI laboratory of Aix-Marseille University and CNRS, France. The supersonic wind tunnel was a closed-loop system which could be continuously operated for several hours without significant drift in free-stream properties (an increase of 1K/hr and a maximum change in total pressure of ±0.5 mbar). Experiments were performed in the S8 test section, where a symmetric converging-diverging nozzle accelerated the flow to a Mach number of 1.65, corresponding to a free-stream velocity (u_∞) of 464 m/s (Fig. 1).

The total temperature was maintained at ambient conditions (approximately 295 K, depending on weather conditions), while the total pressure (p_t) of the free-stream could be varied. The current experiments were performed for total pressures of 0.4, 0.6 and 0.8 atm (Table 1).

Downstream of the nozzle, the test section was 105 mm in height and 170 mm in span-wise width. The geometric models of the flat plate and the two compression ramps were similar in construction with a sharp leading edge, total length of 280 mm

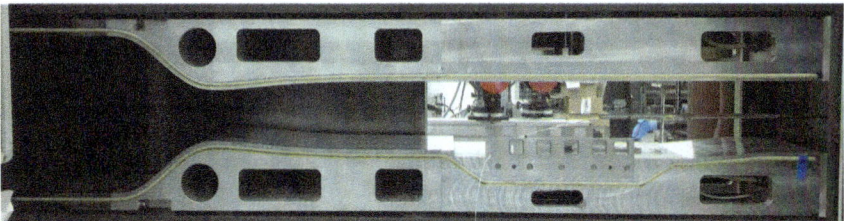

Fig. 1 S8 Nozzle and test section

Table 1 Operating conditions of the experiments

p_t [atm]	M	Re_u ($\times 10^6$) [m^{-1}]	Re_c ($\times 10^6$)	$\sigma_{\rho u}$ [%]	σ_u [%]	σ_p [%]
0.4	1.64	5.61	0.65	0.07	0.04	0.16
0.6	1.64	8.37	0.96	0.06	0.03	0.13
0.8	1.65	11.01	1.27	0.05	0.03	0.11

and spanning the entire width of the test section. The location of the corner of the ramp (x_c) was 115 mm from the leading edge for both ramps. Reynolds number based on the location of the corner is shown in Table 1 for different total pressures of the free-stream. The two models were placed at a height of 25 mm from the floor, using supports near the span-wise edges of the wind tunnel. The floor of the test section was modified to have an additional depth of 10 mm to alleviate chocking of the secondary flow underneath the models.

Pictures and schematics of the models are shown in Figs. 2, 3 and 4. It is to be noted here that the adiabatic wall temperature for these models nearly reached ambient conditions (total temperature of the freestream), considering a recovery factor: $r \approx Pr^{1/2} \approx 0.84$ for a fully laminar boundary layer and $r \approx Pr^{1/3} \approx 0.89$ for a fully turbulent boundary layer [3], where Pr is the Prandtl number.

The mean flow in the test section was found to be quite uniform, with static pressure varying within 1% over the entire test section, and the Mach number evolution showed a similar behaviour, with a trivial deceleration from 1.64 to 1.63 over the test section. Hence, it was concluded that the mean flow was uniform within reasonable limits of pressure and velocity.

The free-stream noise at the exit of the nozzle was measured using the classical single sensor hot-wire anemometer. Table 1 shows the measured turbulence intensities

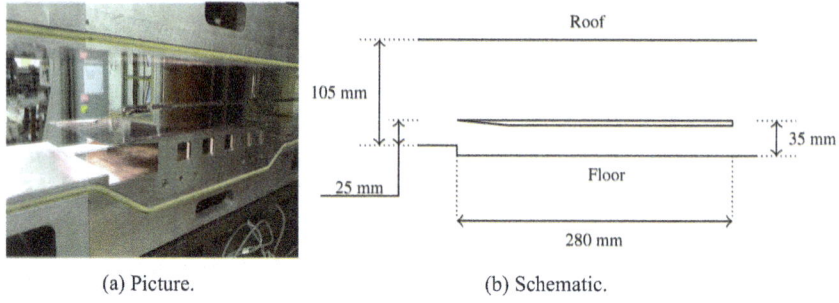

Fig. 2 Flat plate geometrical configuration

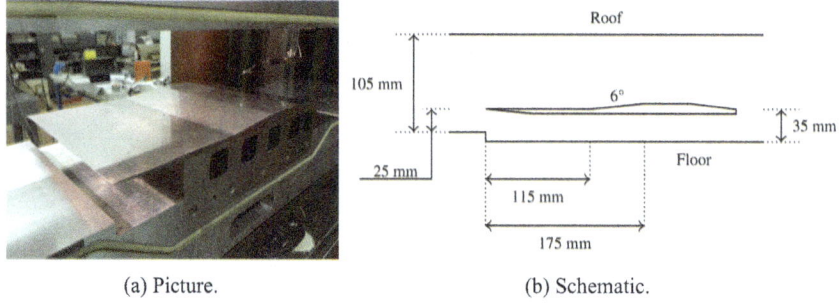

Fig. 3 6° compression ramp geometrical configuration

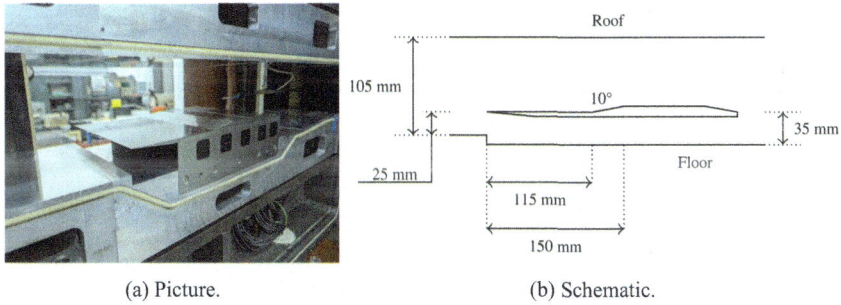

Fig. 4 10° compression ramp geometrical configuration

in terms of non-dimensional fluctuations (root mean square) of mass-flux, velocity and pressure. The free-stream noise was found to be hypo-turbulent in nature, with low turbulence intensities for all operating conditions. Thus, ensuring that the laminar boundary layer would not undergo bypass transition [5, 12].

The flow deflection (φ) of the two compression ramps were chosen such that direct comparisons could be drawn with the oblique shock reflection experiments from the previous European project TFAST [11].

2.2 Experimental Techniques

A classical Pitot probe was used to make measurements of the mean flow field. The tip of the probe was 0.3 mm in height with a opening of 0.15 mm in height, which measured the mean stagnation pressure of the flow. Flow properties such as Mach number, pressure, temperature, velocity and density were determined using standard equations.

To address temporal properties of the flow, hot-wire anemometry was used. The wire was made of platinum and tungsten, with a diameter of 2.5 μm and length of 0.5 mm. The Streamline amplifier from Dantec Dynamics was used in constant temperature mode and the bridge was operated in a symmetrical configuration. An overheat ratio of 0.8 was used and the anemometer was sensitive to voltage fluctuations up to 150 kHz. The acquisition frequency was 400 kHz and the signal was acquired for 10 s. The National Instruments NI6133 analog-to-digital signal acquisition system was used to save the data with 14 bits of resolution, with a low pass filter of 300 kHz before acquisition.

Figure 5 shows the two types of hot-wire probes that were used for this study. The standard straight probe, which was suitable for measurements made in the external flow (Fig. 5a). In the potential region, the probe received fluctuations radiated by the boundary layer [8–10] and the general noise of the wind tunnel [7] (Fig. 6).

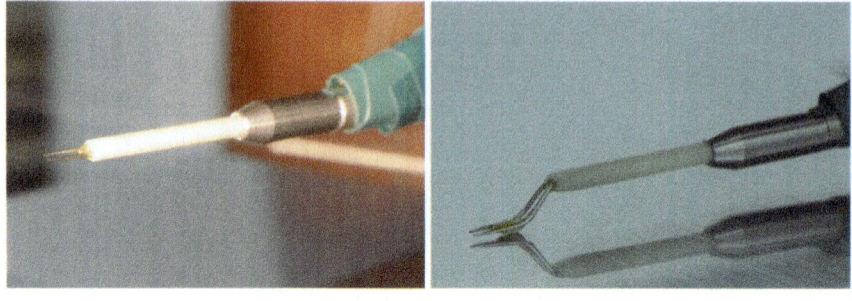

(a) External probe. (b) Boundary layer probe.

Fig. 5 Types of hot-wire probes

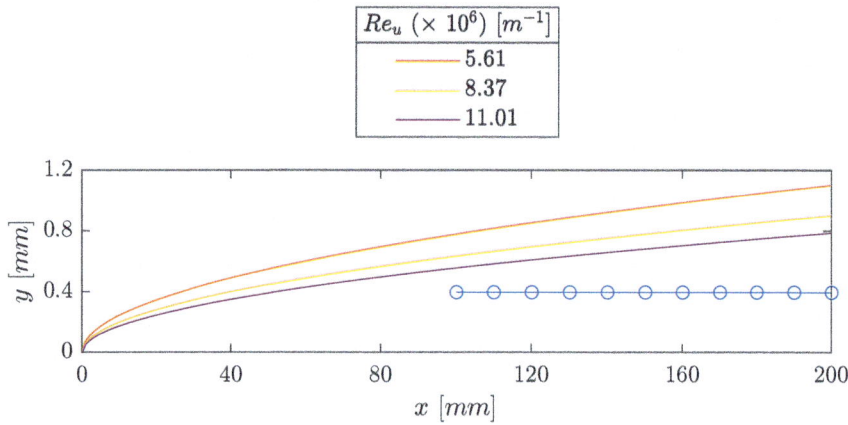

Fig. 6 Illustration of the hot-wire measurements (blue circles) made inside the boundary layer

Additionally, a boundary layer hot-wire probe was also used (Fig. 5b). Boundary layer probes have historically been used for boundary layer research in both low-speed [1, 2] and high-speed [4, 6] experimental facilities, without significant effect on the flow field.

The non-dimensional fluctuations of hot-wire voltage was directly related to the non-dimensional fluctuations of mass-flux through the King's law coefficient [13]. Hence, the data processing and analyses were directly performed on the measured voltage of the hot-wire. The measured time signals were de-trended, and the mean was subtracted before being transformed into Fourier space, and Power Spectral Density (PSD) was estimated using Welch's method. The time signals was broken into segments of 2^{16} points, with an overlap of 50%. The Hamming window was applied to the overlapping segments and the periodogram was calculated by computing the discrete Fourier transform. The individual periodograms were averaged to reduce the variance of the measurements, and estimates of the power spectral density were obtained. Fully converged spectral statistics were obtained using a total of 2^{22}

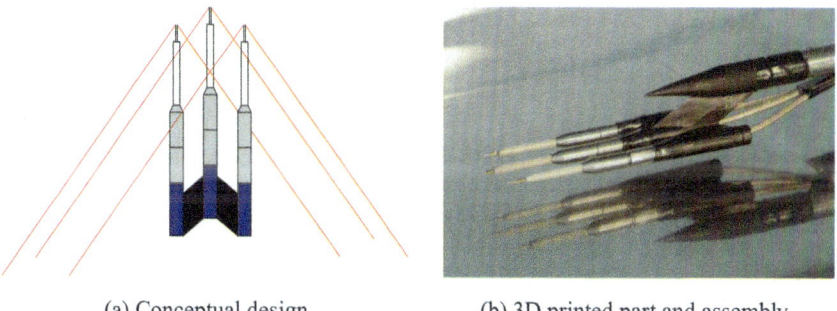

(a) Conceptual design.   (b) 3D printed part and assembly

Fig. 7 Trishula probe

sampled points. The PSD was normalized by the square of the mean voltage at each location.

A new probe was developed using three hot-wire sensors (named as Trishula). The three sensors were placed on the vertices of an isosceles triangle on the stream-wise span-wise plane (Fig. 7a). The spacing between the sensors were chosen be as close as possible without mutual interference from each other (Fig. 7b). This probe was developed to resolve characteristics of 3D plane waves in the flow such as spatial orientation with respect to the stream-wise direction, span-wise wavelengths, phase, and group velocities. The reader is referred to the Ph.D. thesis manuscript for more details about the design, construction and working principle of this new probe [37].

3 Results: Boundary Layer Transition at Zero Pressure Gradient

3.1 Evolution of Time Scales Along the Boundary Layer

Time signals of the laminar boundary layer were measured for different total pressures and at various stream-wise locations on the flat plate using the single-sensor boundary layer probe hot-wire probe. Figure 8 shows the estimates of the PSD represented in pre-multiplied and normalized format to highlight the dominant frequencies at each stream-wise location. Frequency is shown on the vertical axis, represented in non-dimensional angular frequency ($\omega^* = 2\pi f/(Re_u u_\infty)$). Reynolds number is shown on the horizontal axis, based on the local Blasius reference length scale ($R = \sqrt{Re_x}$).

Inside the boundary layer, a coherent time scale was observed in the High Frequency (HF) range, over the first part of the flat plate. These time scales were not observed outside the boundary layer (some spots of HF scales corresponded to the general background noise of the wind tunnel). Hence, these HF scales did not radiate any fluctuations outside the boundary layer.

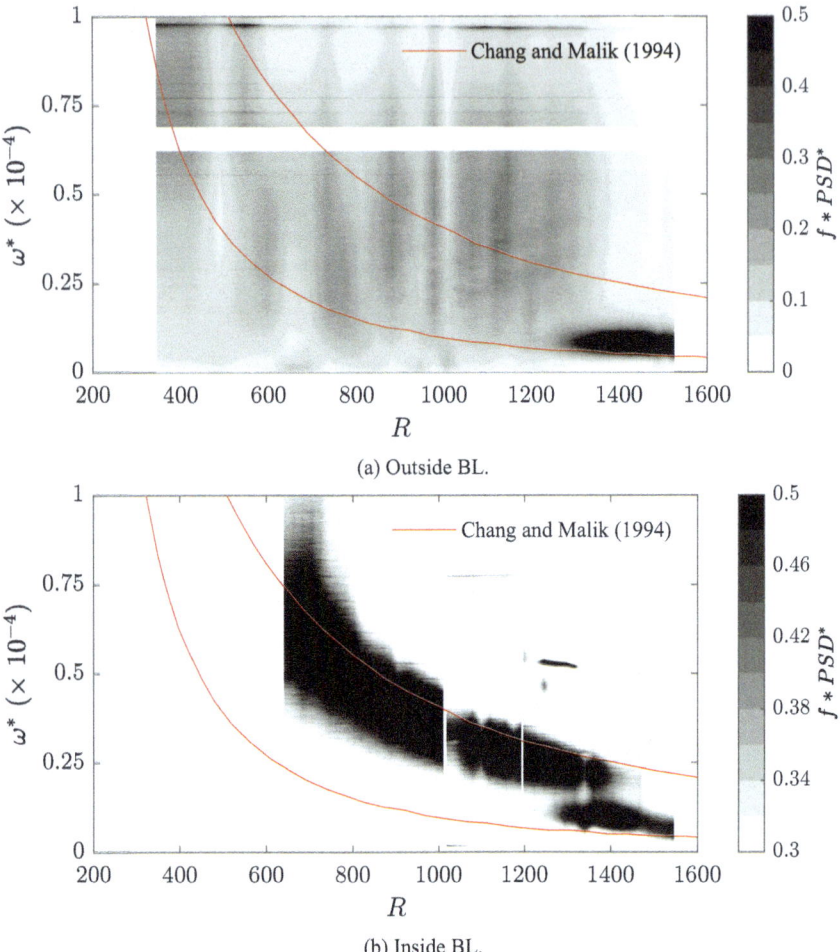

Fig. 8 Space-time evolution of PSD represented in terms of ω^*

Figure 8 also compares the contours of PSD from the current hot-wire measurements with the neutral map of the most unstable frequencies found by [14].

Frequencies inside the red curve correspond to the most amplified modes according to stability theory and frequencies outside the red curve were expected to be damped. Accordingly, the non-dimensional angular frequency of the most unstable mode decreases with increasing Reynolds number. This trend was well captured by the hot-wire measurements inside the boundary layer, and there was reasonable agreement between theory and experiments to suggest that these HF are consistent with the compressible TS waves. At low supersonic speeds, these self-excited modes of the laminar boundary layer are three-dimensional in nature i.e. they are oriented at an angle with respect to the free-stream flow direction and are referred to as Oblique

Modes (OM) [15]. The laminar boundary layer is receptive to small disturbances in the free-stream, which trigger these naturally unstable eigen modes. OMs have been studied extensively by experimental [4, 16, 17] and numerical [14, 18, 19] studies for a range of Mach and Reynolds numbers.

Apart from the HF scales, a second time scale was observed in the range of Medium Frequencies (MF) near the end of the flat plate at $R \approx 1300$, from both inside and outside the boundary layer. In fact, Fig. 8 highlighted that these MF scales were approximately equal to half of the HF scales. This suggested that these MF time scales were possibly the sub-harmonics of the HF OMs from the first part of the laminar boundary layer.

3.2 Evolution of Amplitudes Along the Boundary Layer

Figure 9 shows the evolution of the band-passed non-dimensional amplitude of hot-wire fluctuations for different Reynolds numbers, where $A^* = 1$ corresponded to the background noise of the wind tunnel. Inside the boundary layer, both time scales exhibited exponential growth over the entire flat plate. It was observed that the HF scales had a higher amplitude than the MF for the first part of the measurements. Nevertheless, both scales showed the same exponential growth rate up to $R \approx 1200$. This growth rate in the first part of the flat plate was compared with predictions from stability theory, and a reasonable agreement was found [37].

In the second part of the flat plate, it seems that the growth rate of the MF was higher than the HF and subsequently, the two curves meet at $R \approx 1300$. From this point on, both time scales exhibited a much higher growth rate than the first part of the flat plate, reaching amplitudes of nearly 200 times the free-stream noise of the

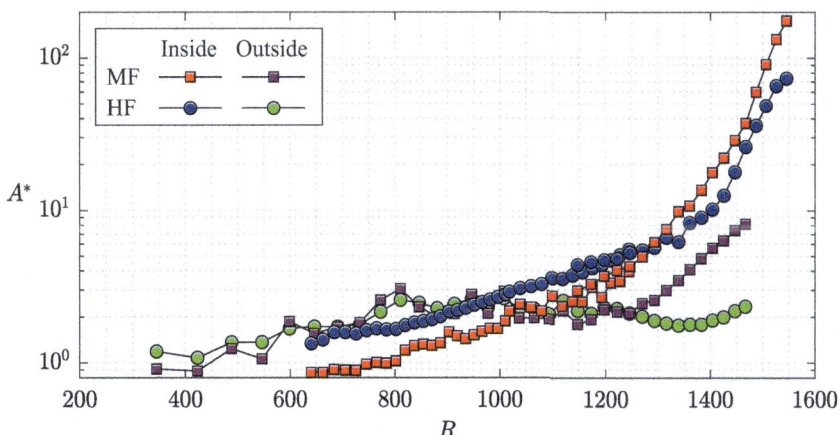

Fig. 9 Stream-wise evolution of the non-dimensional amplitude of the hot-wire measurements

wind tunnel. This high growth rate in the second part of the flat plate, dominated by MFs, suggested that non-linear interactions were taking place inside the boundary layer.

3.3 Dispersion Relationship of MF

In order to uncover the nature of these mysterious MF time scales, measurements were made using the three sensor hot-wire probe over the flat plate in the external flow. Using the phase difference between the three sensors, the wavenumber (k) and angle of orientation (ψ) were obtained for the MFs. Given the large spacing (with respect to the boundary layer thickness) between the sensors of this probe, such measurements could be susceptible to spatial aliasing (Fig. 10).

An anti-aliasing analysis was performed using synthetic signals with different combinations of wavenumbers and angular orientation. The results of this anti-aliasing analysis is shown in Fig. 10, which highlighted that the MFs could have multiple possible input combinations of (k, ψ) (shown in grey contour lines). A summary of the different possibilities is shown in Table 2. Family I corresponds to actual (or real) measurements by the Trishula probe, whereas families II, III and IV correspond to results from the (synthetic) anti-aliasing analysis.

It was observed that the families III and IV obtained from the anti-aliasing analysis of the current experiments were very similar to previous experiments from literature [20], where it was shown that a phase synchronization between the HF OMs and its sub-harmonics resulted in a resonance phenomenon. The synchronization of the wavenumbers in the current experiments (from families III and IV) and in the experiments of [20] are demonstrated in Table 3. Further, the wavenumber and angular orientation of OM (for equivalent Mach and Reynolds numbers) from LST analy-

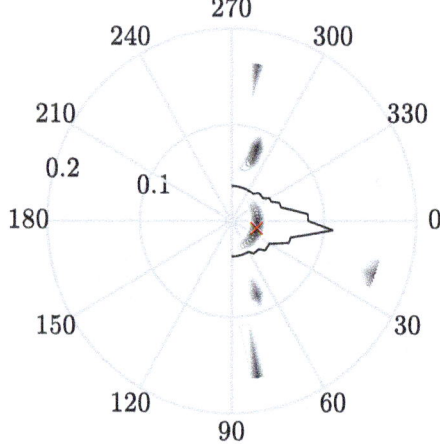

Fig. 10 Synthetic data analysis of the multi-sensor probe. (legend: ×: Measured Point, solid black line: Aliasing limit, grey contour lines: low residuals)

Table 2 Possible properties of MF obtained from synthetic analysis

	k_x^*	k_z^*	ψ	u_p/u_∞
I	0.0273	0.0067	14°	0.25
II	0.1553	0.0565	20°	0.04
III	0.0283	±0.0776	±70°	0.09
IV	0.0287	±0.1628	±80°	0.04

Table 3 Comparison of phase synchronization between sub-harmonics

	$k_x^1 + k_x^2 = k_x^3$	$k_z^1 + k_z^2 = k_z^3$	ψ^3
Current experiments	$0.0283 + 0.0287 = 0.0570$	$-0.0776 + 0.1628 = 0.0852$	56°
Kosinov et al. [20]	$0.0368 + 0.0308 = 0.0676$	$-0.0849 + 0.1980 = 0.1131$	59°
Mack [15]	0.05	0.0714	55°

sis of [15] are also shown, and very good agreement was found between the three results. These measurements confirmed that the MF observed here corresponded to sub-harmonic modes of the OM, and phase synchronization was possible between the sub-harmonics and the OM, leading to non-linear interactions.

DNS studies have shown that this sub-harmonic resonance phenomenon can accelerate or "enrich" the oblique breakdown transition scenario [21, 22]. The non-linear regime of the transition process had received relatively less attention from an experimental point of view. And, the current experiments used novel measurement techniques which provided a detailed description of this phenomenon involving the sub-harmonic modes of the boundary layer, including their frequency, amplitude, wave-numbers, angular orientation and phase velocity.

4 Results: Length Scales of Transitional SBLIs

The interaction between a laminar boundary layer and a compression ramp has been well documented for a number of flow deflections, Mach, and Reynolds numbers [23]. Pitot measurements were used to verify and validate the two-step pressure rise over the separated SBLI. Comparisons were made with the inviscid pressure rise due to the compression ramp, and also the pressure rise at separation was verified with free-interaction theory [24]. These verifications and validations are described in more detail in the Ph.D. thesis manuscript [37].

Additionally, Pitot probe measurements provided quantitative insight into the upstream influence of the SBLI through the length of interaction, which was defined as the stream-wise distance between the corner of the ramp and the mean location of boundary layer separation (measured at the wall). It was observed that the length of interaction decreased when Reynolds number was increased, for both the ramps. And comparing between the two ramps, the length of interaction was larger for higher ramp angles.

An attempt was made to compile a number of experimental measurements of the lengths of interaction for transitional SBLIs. Particularly, experiments of oblique shock reflections and compression ramps were collected. This compilation had a collection of independent experiments performed over the past 70 years, in different wind tunnel facilities, and with a wide range of operating conditions. It is important to note that such a compilation was made for the first time for laminar and transitional SBLIs. The reader is referred to the Ph.D. thesis manuscript regarding the details of this compilation [37].

For comparisons to be made between different geometries (e.g. oblique shock reflections and compression ramps), there was a need to utilise a common length scaling for both geometries. A common length scaling was developed for turbulent SBLIs, based on the mass flow deficit between the outgoing and incoming boundary layer [25]. Although this scaling was developed for turbulent SBLIs, it should also be applicable to transitional SBLIs, given that this formulation was developed on a mass conservation basis. Additionally, a non-dimensional separation criterion was also proposed [25], comparing the overall pressure rise across the interaction with the dynamic pressure of the freestream.

Figure 11 shows the compilation of the transitional SBLI experiments represented using the scaling proposed above [25]. The reader is referred to the Ph.D. thesis manuscript for the legend of this compilation [37]. It was observed that this particular set of scaling parameters resulted in different trends in the data, and this variety in the trends was greater than the uncertainty associated with the measurement techniques.

The reason why such a scaling worked for the compilation of turbulent SBLIs [25], was because Reynolds number had a weak effect on the separation criterion of turbulent SBLIs. However, it was clear that from the current experiments that Reynolds

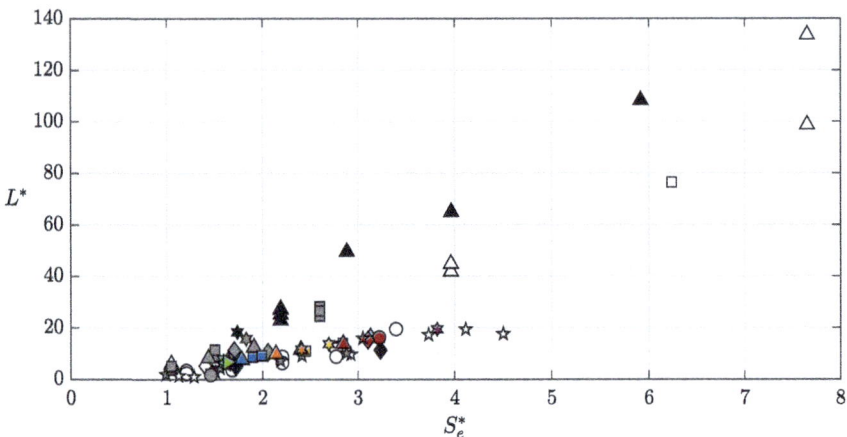

Fig. 11 Length of interaction based on the mass-balance scaling [25] with $k = 9$ (legend: [37])

number had a significant effect on the pressure required to separate a laminar boundary layer. And the inability of the shock strength scaling to take into account this Reynolds number effect, possibly led to this scatter observed in transitional SBLIs.

In order to introduce Reynolds number in the shock strength scaling, the concept of free-interaction theory [24] was revisited. The main idea behind this formulation was that the non-dimensional pressure rise at separation showed universal behaviour for all SBLIs (apart from the constant being different for turbulent and laminar SBLIs).

The original idea for the non-dimensional shock strength scaling [25], was to compare the overall increase in pressure across the interaction with the pressure rise across separation $(\Delta p/(\Delta p)_{sep})$. The current compilation showed that Reynolds number had a significant effect on the pressure rise at separation for laminar and transitional SBLIs, and this behaviour could be universally described by free-interaction theory.

Hence, a new scaling for the shock strength was proposed, based on the the ratio of pressures, as opposed to comparing the pressure differences:

$$S_r^* = \frac{p_3/p_1}{(p/p_1)_{sep}} = \frac{p_3}{p_1}\left[1 + \frac{\gamma M^2}{2} F_p \sqrt{\frac{2 c_f}{(M^2 - 1)^{1/2}}}\right]^{-1} \quad (1)$$

Where p_3/p_1 is the overall pressure rise across the interaction, $(p/p_1)_{sep}$ is obtained from the free-interaction theory [24], $F_p = 1.5$ is the coefficient of free-interaction at plateau and c_f is the skin-friction coefficient at the mean location of separation.

This new scaling seemed to collapse most of the data set, as shown in Fig. 12. The effect of Reynolds number appeared to be well captured by this new scaling as the data points did not fall on a vertical line as the previous scaling (Fig. 11) and instead of the different trends observed in the previous figure, the same linear relationship

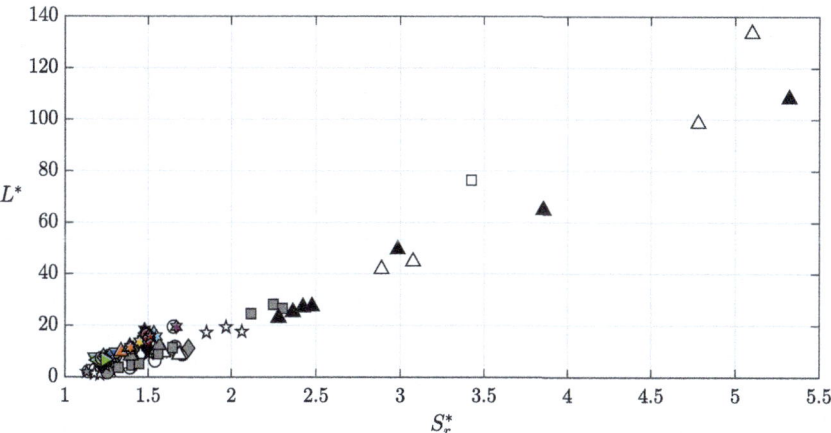

Fig. 12 Compilation of length scales for different non-dimensional shock strengths, S_r^*

(i.e. same slope) was obtained for most of the data points from the compilation in Fig. 12.

The non-dimensional nature of this scaling automatically adjusted the shock strength; for low shock strengths with no separation, the imposed adverse pressure ratio is lower than the pressure ratio required to separate the boundary layer and hence, $S_r^* < 1$. For incipient interactions involving intermittent separation, the imposed pressure ratio is close to the pressure ratio required to separate the boundary layer ($S_r^* \approx 1$). And finally $S_r^* > 1$ corresponded to a typical separated SBLI.

The reader is referred to the Ph.D. thesis manuscript for a more detailed description of this separation criterion, including the effect of Reynolds number and freestream noise of the wind tunnel.

5 Results: Boundary Layer Transition at Adverse Pressure Gradient

Figure 13 compares the hot-wire spectra measured over the separated region between the two ramps ($0 < X^* < 1$). The PSD at different stream-wise locations are shown in absolute pre-multiplied format, to highlight the difference in amplitude between different measurements.

The bottom horizontal axis shows the non-dimensional angular frequency (ω^*) according to stability theory of laminar boundary layers, and the top horizontal axis shows the non-dimensional Strouhal number according to SBLI normalization (based on the length of interaction and the free-stream velocity). Each colour represents the spectra at a given stream-wise location, and hence, several spectra are shown corresponding to different locations over the bubble.

The spectra for the 6° ramp remained approximately the same over the interaction, with no growth of amplitudes for any frequency. However, on the 10° compression ramp, the growth of MF time scales was observed. These MF time scales were similar in frequency to the ones found in the non-linear regime of the flat plate measurements.

Based on these measurements, it was presumed that these MF time scales were the same type of structures found in the flat plate boundary layer, that arose due to non-linear interactions of the upstream OMs. The wavelength, convection velocities and direction of propagation of such time scales have been documented in Table 2. The reason as to why they are only observed on the 10° ramp and not on the 6° ramp might be related to the size of the separated region. As the length was longer for the 10° ramp, the height of the shear layer was also higher. The higher height might have moved the generalized inflection point higher, resulting in the destabilization of the KH instability [26]. It has been shown that boundary layer modes are convectively amplified through the KH mechanism [27, 28].

Figure 14 shows the non-dimensional amplitude of MF over the separated region for the both ramps, compared to the amplitude evolution of MF over the flat plate. The amplitudes of the 6° ramps did not show a large change over the separated region,

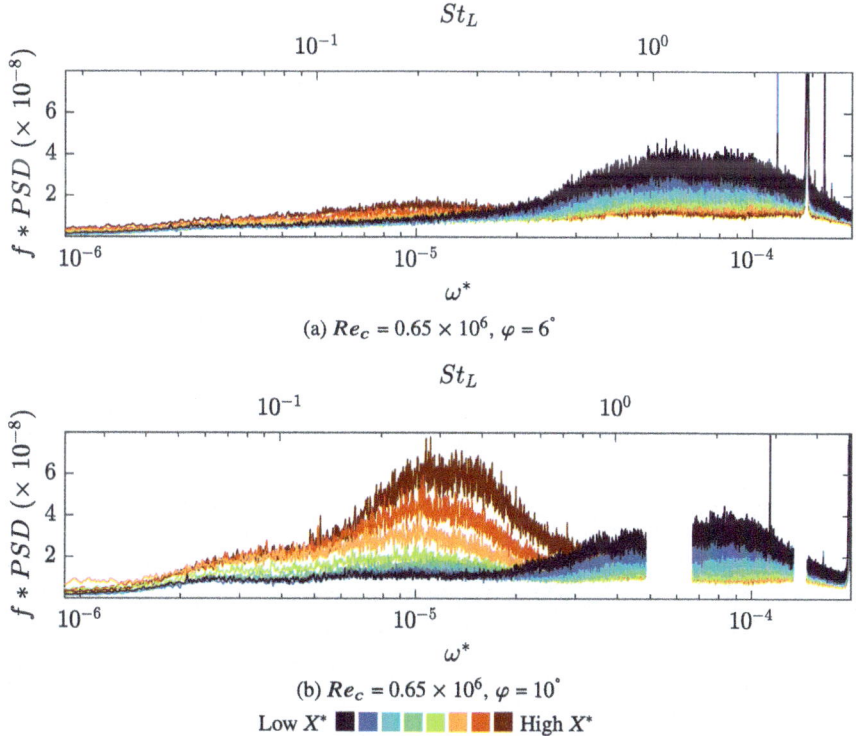

Fig. 13 Evolution of hot-wire spectra over the separation bubble

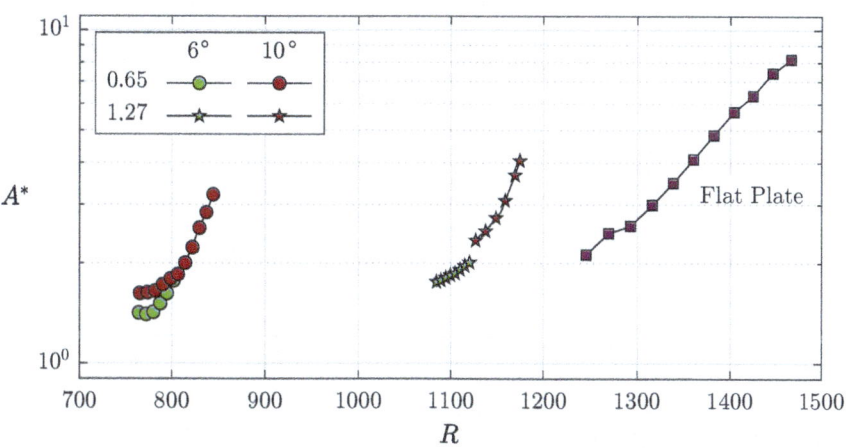

Fig. 14 Comparison of MF growth rate between the flat plate and the compression ramps

while an exponential growth was observed for the 10° ramp. This suggested that the growth of these time scales was strongly dependent on the size of the separated region, possibly due to the higher location of the generalized inflection point.

Additionally, the growth rate of the MF over the SBLI was nearly three times higher compared to their evolution on the flat plate. These results suggested that the separated region accelerated the growth of these MF scales.

Figure 15 compares the spectra of the hot-wire measurements made inside the boundary layer for the flat plate, 6° and 10° ramps. The last measurement point with the highest amplitude was chosen from the flat plate to be compared with the highest amplitude of the inside measurements on both ramps. This corresponded to the first point downstream of reattachment for both ramps. This comparison clearly highlighted the differences in amplitudes and frequencies of the boundary layer between the three geometries.

Surprisingly, in both ramp configurations, there were no evidences of MF inside the boundary layer, which exhibited strong growth over the separated region for the 10° ramp. Two dominant time scales were identified from the measurements made inside the boundary layer for both the ramps; HF, typically associated with OM of the laminar boundary layer were identified, in the same range of frequencies as observed from the flat plate measurements. Additionally, Very High Frequencies (VHF) were also identified, which were not found inside the natural boundary layer measurements on the flat plate. It was believed that the VHF were most likely the evidence of breakdown of coherent structures in the boundary layer, indicating that the boundary layer was undergoing transition to turbulence.

What was surprising to note was that the MF, which were found to be most dominant from the external measurements over the separated region (Figs. 13 and 14), was not the most dominant time scale inside the reattached boundary layer of the SBLI. It seemed that the reattached boundary layer was preferentially selective to

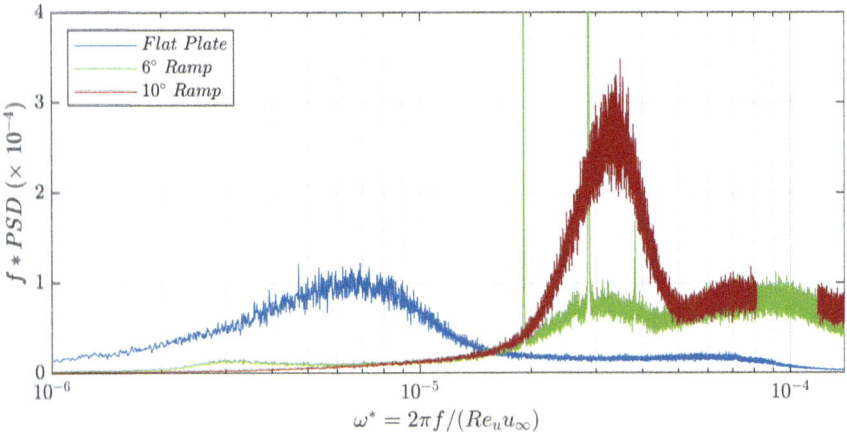

Fig. 15 Comparison of the hot-wire spectra between the flat plate, 6° and 10° ramp

higher frequency modes. Given that the growth of these MF was linked to a possible phase synchronization between the OM and the sub-harmonic MF (Table 3), perhaps the reattachment compression possibly ruined this synchronization and resonance condition could have been broken. Despite these large amplitudes inside the boundary layer, these hot-wire spectra were not typical of a boundary layer that was fully turbulent, where most of the energy would be associated with the turbulent boundary layer time scale (VHF). These results suggested that the boundary layer was not fully turbulent at reattachment.

6 Results: LF Unsteadiness of Compression ramp SBLIs

Figure 16 compares the absolute pre-multiplied PSD of the compression waves at separation for the 6° and 10° ramps. The spectra at separation in the current experiments were characterized by LF time scales ($0.04 \leq St_L \leq 0.08$), which were typically associated with the large-scale "breathing" motion of the recirculating region [29–32]. While the presence of LF was clearly identified for the 10° ramp, it was barely distinguishable for the 6° ramp. In fact, just looking at the spectra of the 6° ramp alone, without comparing with the spectra from the 10° ramp, it could be said that there was no discernable LF unsteadiness found in the 6° ramp. Hence, the imposed flow deflection had a significant influence on the relative amplitude of LF unsteadiness in the current experiments.

In addition to LF, a broad range of HF ($0.4 \leq St_L \leq 2$) was observed. These HF were in the same range of frequencies typically associated with OM from the laminar boundary layer. Although it was shown that these time scales do not radiate fluctuations outside the boundary layer, the compression waves at separation might

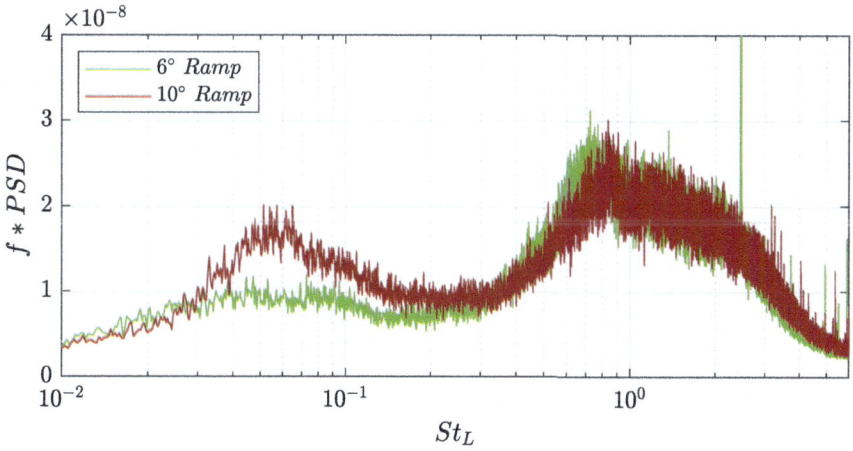

Fig. 16 Comparison of PSD over the mean separation point for the two compression ramps

be receptive to these time scales of the laminar boundary layer. Moreover, the background wind tunnel noise could also be locally amplified through the compression waves, and the hot-wire could have picked up such fluctuations. Therefore, it was difficult to determine the true nature of these frequencies without further experiments and analyses.

Several hypotheses and models have been proposed for the origin of LF unsteadiness. Several studies have reported the growth of convective instabilities (with characteristic MF) over the shear layer arising due to non-linear interactions of the laminar boundary layer modes [28, 30, 32, 33]. Additionally, a correlation was found between the presence of non-linear interaction over the shear layer to the LF unsteadiness of the transitional SBLI [34–36]. Hence, it was concluded that the growth of new MF time scales over the separated shear layer of the 10° ramp (Fig. 13b), which were possibly triggered by non-linear interactions of OM, might have played a role in the LF unsteadiness observed at separation. The absence of non-linear interactions on the 6° ramp did not result in the growth of MF time scales over the shear layer, and coincidentally no discernible LF unsteadiness was observed for the 6° ramp (Fig. 13a).

Unfortunately, a consolidated theory on the origin and mechanism of LF unsteadiness associated with transitional SBLI could not be developed with the current measurements. Nevertheless, the presence of new MF time scales and LF unsteadiness pointed to a speculative link between these flow features. Further investigations were necessary to understand this phenomenon in more detail.

7 Conclusions

Experiments were performed in the supersonic wind tunnel of the IUSTI laboratory, to study the length and time scales of transitional SBLIs. Given that such SBLIs often accelerated the transitional mechanisms of laminar boundary layers, another important objective was to study this phenomenon in more detail. The Mach number of the experiments was fixed to 1.65 and measurements were performed at relatively "low" unit Reynolds numbers ($5.61 \leq Re_u (\times 10^6) [m^{-1}] \leq 11.01$). A variety of experimental techniques were employed including Schlieren, Shadowgraph, Pitot probes and hot-wire anemometry. A new multi-sensor hot-wire probe was developed over the course of this research, to reveal the spatio-temporal nature of laminar boundary layer instability waves.

Measurements were first made on a simple flat plate geometry, to understand the transitional mechanisms of an undisturbed laminar boundary layer with zero pressure gradient. The transition mechanism was found to be divided into two regimes. In the first regime, modal mechanisms corresponding to the primary instability of the laminar boundary layer was found. These linear time scales agreed very well with the predictions of the most amplified frequencies (or OMs) from stability theory. In the second regime, a new MF time scale was found to be the most dominant, with higher growth rates compared to the previous regime. It was believed that this new time scale arose due to non-linear interactions between the self-excited eigen modes of the first regime. Measurements of this time scale were made using the newly developed multi-sensor hot-wire probe, which revealed large stream-wise wavelengths and low phase velocities compared to OM. The span-wise nature of these modes was uncertain due to possible spatial aliasing of the wave. Nevertheless, evidence was found to indicate that this MF could be related to the sub-harmonic resonance scenario. However, it was found in literature that these sub-harmonics do not drive the boundary layer to transition, but rather accelerate the oblique breakdown transition process involving stationary stream-wise vortex modes.

Following the measurements on the flat plate, experiments of transitional SBLI were made on nominally 2D compression ramps. The canonical nature of the current transitional compression ramps were verified with comparisons with free-interaction theory. A compilation of lengths of interaction reported from various experiments in literature was made, and a large scatter was found in the relationship between the length of interaction and the imposed shock strength. A new non-dimensional scaling was developed for the shock strength that included the effects of Reynolds number through the free-interaction theory. In particular, it was found that the non-dimensional length of interaction scaled with the ratio of pressure as opposed to the difference in pressure across the interaction. In particular, the pressure ratio across the interaction normalized with the pressure ratio needed to separate the boundary layer was developed. Such a scaling dramatically reduced the scatter in the data set, and all the experiments exhibited nearly the same linear relationship between this new shock strength parameter and length of interaction.

The hot-wire measurements over the compression ramp SBLI shed light on the transitional mechanisms of the separated laminar boundary layer. MF, in a similar range as that on the flat plate, were found to be amplified over the separated shear layer of the 10° ramp. Such an amplification was not found over the 6° ramp. The growth rate of these MF over the 10° ramp was found to be nearly three times higher compared to their growth rate on the flat plate (i.e. laminar boundary layer without interaction). Such MF were also found by other studies on transitional SBLIs, which reported that these frequencies were characteristic of convectively amplified modes of the boundary layer. Surprisingly, downstream of the interaction, these MFs were not the most dominant, but most of the energy was associated with HF OMs of the laminar boundary layer. The strong compression at reattachment was found to be preferentially selective towards higher frequencies, which eventually drove the boundary layer to turbulence at some point downstream of reattachment. This also suggested that the boundary layer was not fully turbulent after reattachment, contrary to other studies on transitional SBLIs.

The LF dynamics of such transitional SBLIs was also studied using hot-wire anemometry. It was found that while the 10° ramp was clearly characterized by LFs ($St_L \approx 0.05$) at the compression waves at separation, such LF unsteadiness was not found on the 6° ramp. While a conclusive theory on the origin and mechanism of LF unsteadiness was not determined, a "coincidental" link between the amplification of MF time scales over the separated shear layer and the presence of LF unsteadiness was noted, indicating that non-linearities (the possible cause for the growth of MF over the separated shear layer) possibly played a role in the origin and mechanism of LF unsteadiness in such transitional SBLIs.

Competing Interests The authors have no conflicts of interest to declare that are relevant to the content of this chapter.

Acknowledgements The authors would like to acknowledge the technical support of Pierre Lantoine while performing the experiments in the wind tunnel.

References

1. Schubauer, G.B., Skramstad, H.K.: J. Res. Natl. Bur. Stand. **38**, 251 (1947)
2. Dhawan, S., Narasimha, R.: J. Fluid Mech. **3**(4), 418 (1958)
3. Mack, L.: Jet Propulsion Lab. Report, pp. 20–80 (1954)
4. Laufer, J., Vrebalovich, T.: J. Fluid Mech. **9**(2), 257 (1960)
5. Laufer, J.: J. Aerospace Sci. **28**(9), 685 (1961)
6. Kendall, J.M.: AIAA J. **13**(3), 290 (1975)
7. Smits, A.J., Dussauge, J.P.: Turbulent Shear Layers in Supersonic Flow. Springer Science & Business Media (2006)
8. Agostini, L., Larchevêque, L., Dupont, P., Debiève, J.F., Dussauge, J.P.: AIAA J. **50**(6), 1377 (2012)
9. Jaunet, V., Debieve, J., Dupont, P.: AIAA J. **52**(11), 2524 (2014)
10. Diop, M., Piponniau, S., Dupont, P.: In: Tenth International Symposium on Turbulence and Shear Flow Phenomena. Begel House Inc. (2017)
11. Diop, M.: Transition à la turbulence en écoulements compressibles décollés. Ph.D. thesis, Aix-Marseille (2017)
12. Morkovin, M.V.: Fluctuations and Hot-wire Anemometry in Compressible Flows. North Atlantic Treaty Organization Advisory Group for Aeronautical Research (1956)
13. Dupont, P.: Etude expérimentale des champs turbulents dans une couche limite supersonique fortement chauffée. Ph.D. thesis, Aix-Marseille 2 (1990)
14. Chang, C.L., Malik, M.R.: J. Fluid Mech. **273**, 323 (1994)
15. Mack, L.M.: Boundary-layer linear stability theory. Technical report, California Inst of Tech Pasadena Jet Propulsion Lab (1984)
16. Demetriades, A.: Phys. Fluids A **1**(2), 312 (1989)
17. Graziosi, P., Brown, G.L.: J. Fluid Mech. **472**, 83 (2002)
18. Fasel, H., Thumm, A., Bestek, H.: In: Fluids Engineering Conference, pp. 77–92. Publ by ASME (1993)
19. Mayer, C.S., Von Terzi, D.A., Fasel, H.F.: J. Fluid Mech. **674**, 5 (2011)
20. Kosinov, A., Semionov, N., Shevel'kov, S., Zinin, O.: In: Nonlinear Instability of Nonparallel Flows: IUTAM Symposium Potsdam, NY, USA July 26–31, 1993, pp. 196–205. Springer (1994)
21. Fezer, A., Kloker, M.: In: Laminar-Turbulent Transition: IUTAM Symposium, Sedona/AZ September 13–17, 1999, pp. 415–420. Springer (2000)
22. Mayer, C.S., Wernz, S., Fasel, H.F.: J. Fluid Mech. **668**, 113 (2011)
23. Babinsky, H., Harvey, J.K.: Shock Wave-Boundary-Layer Interactions, vol. 32. Cambridge University Press (2011)
24. Chapman, D.R., Kuehn, D.M., Larson, H.K.: Investigation of separated flows in supersonic and subsonic streams with emphasis on the effect of transition. Technical report, Ames Aeronautical Laboratory (1958)
25. Souverein, L., Bakker, P., Dupont, P.: J. Fluid Mech. **714**, 505 (2013)

26. Diwan, S.S., Ramesh, O.N.: J. Fluid Mech. **629**, 263 (2009)
27. Pagella, A., Babucke, A., Rist, U.: Phys. Fluids **16**(7), 2272 (2004)
28. Lugrin, M., Beneddine, S., Leclercq, C., Garnier, E., Bur, R.: J. Fluid Mech. **907**, A6 (2021)
29. Robinet, J.C.: J. Fluid Mech. **579**, 85 (2007)
30. Sansica, A., Sandham, N.D., Hu, Z.: J. Fluid Mech. **798**, 5 (2016)
31. Larchevêque, L.: In: 54th AIAA Aerospace Sciences Meeting, p. 1833 (2016)
32. Threadgill, J.A., Little, J.C., Wernz, S.H.: AIAA J. **59**(12), 4824 (2021)
33. Niessen, S.E., Groot, K.J., Hickel, S., Terrapon, V.E.: Phys. Fluids **35**(2) (2023)
34. Sansica, A., Sandham, N., Hu, Z.: Phys. Fluids **26**(9) (2014)
35. Guiho, F., Alizard, F., Robinet, J.C.: J. Fluid Mech. **789**, 1 (2016)
36. Bonne, N., Brion, V., Garnier, E., Bur, R., Molton, P., Sipp, D., Jacquin, L.: J. Fluid Mech. **862**, 1166 (2019)
37. Mahalingesh, N.: Boundary layer transitional mechanisms and shock induced separation. Ph.D. thesis, Aix-Marseille University (2024)
38. Doerffer, P., Flaszynski, P., Dussauge, J.P., Babinsky, H., Grothe, P., Petersen, A., Billard, F.: Transition Location Effect on Shock Wave Boundary Layer Interaction: Experimental and Numerical Findings from the TFAST Project, vol. 144. Springer Nature (2020)

Open Access This chapter is licensed under the terms of the Creative Commons Attribution 4.0 International License (http://creativecommons.org/licenses/by/4.0/), which permits use, sharing, adaptation, distribution and reproduction in any medium or format, as long as you give appropriate credit to the original author(s) and the source, provide a link to the Creative Commons license and indicate if changes were made.

The images or other third party material in this chapter are included in the chapter's Creative Commons license, unless indicated otherwise in a credit line to the material. If material is not included in the chapter's Creative Commons license and your intended use is not permitted by statutory regulation or exceeds the permitted use, you will need to obtain permission directly from the copyright holder.

Parameter Influence on Porous Bleed Performance for Shock-Wave/Boundary-Layer Interaction Control

Julian Giehler, Pierre Grenson, and Reynald Bur

Abstract This chapter examines the influence of hole diameter, porosity, thickness-to-diameter ratio, and stagger angle on the performance of porous bleed control in mitigating the negative effects of shock-wave/boundary-layer interactions. A detailed numerical study focuses on the control of an irregular shock reflection, or Mach reflection, where a separation bubble below the shock foot is present in the uncontrolled case. Implementing bleed control modifies the flow field significantly, with variations in bleed rates upstream and downstream of the shock because of the external flow characteristics. The findings indicate that smaller hole diameters enhance bleed efficiency and control effectiveness, while porosity levels and thickness-to-diameter ratios exhibit complex trends, with a medium thickness-to-diameter ratio and a stagger angle of 45° emerging as optimal configurations for effective shock-wave/boundary-layer control.

Keywords Supersonic flow · Shock wave—boundary layer interaction · Flow control · Numerical simulations

1 Introduction

The deceleration of the airflow to subsonic conditions forces the presence of shock-wave/boundary-layer interactions in supersonic air intakes [1]. Thus, the boundary-layer separation may occur, accompanied by performance losses. Porous bleed systems are a common technique to control shock-wave/boundary-layer interactions in supersonic air intakes [2, 3]. These systems remove the low-momentum portion of the boundary layer through perforated plates, leading to a non-separating boundary layer due to control of its size and characteristics.

While the working principle is simple, the control effect of bleed systems is still challenging to forecast. Typical bleed systems consist of hundreds of holes, resulting

---

J. Giehler · P. Grenson (✉) · R. Bur
DAAA, ONERA, Institut Polytechnique de Paris, Meudon, France
e-mail: pierre.grenson@onera.fr

in the impracticality of three-dimensional simulations of the entire system in the pre-design phase because of the excessive numerical costs. Thus, bleed models, which can be applied as a boundary condition in RANS solvers, are desired. However, state-of-the-art bleed models are not able to capture important geometrical parameters of the system [4].

Our previous numerical study revealed parameter influences on *bleed efficiency* and *control effectiveness* when a bleed is applied to control a supersonic boundary layer [5]. The bleed efficiency describes the capability of removing the highest possible bleed rate for a given pressure drop from the external wall to the bleed plenum. On the contrary, the control effectiveness is unlinked to the efficiency and describes the ability to achieve the maximum effect on the boundary layer, i.e., the increase of the momentum in the near-wall region, for a given bleed rate. In the following, a parametric study on the control of a shock-wave/boundary-layer interaction using porous bleed is presented, showing the influences of four geometrical parameters: hole diameter, porosity level, thickness-to-diameter ratio, and stagger angle.

2 Methodology

2.1 Governing Fluid Equations and Flow Solver

The flow is modeled by the compressible Reynolds-averaged Navier-Stokes equations, which describe the behavior of an ideal gas in motion, where **u** denotes the velocity vector $[u, v, w]$:

Continuity:
$$\frac{\partial \rho}{\partial t} + \nabla \cdot (\rho \mathbf{u}) = 0 \quad (1)$$

Momentum:
$$\frac{\partial (\rho u)}{\partial t} + \nabla \cdot (\rho \mathbf{u} \otimes \mathbf{u}) = -\nabla p + \nabla \cdot \tau_\nu + \nabla \cdot \tau_t \quad (2)$$

Total energy:
$$\frac{\partial \rho E}{\partial t} + \nabla \cdot [(\rho E + p)\mathbf{u}] = \nabla \cdot [(\tau_\nu + \tau_t) \cdot \mathbf{u}] - \nabla \cdot \varphi - \nabla \cdot \varphi_t \quad (3)$$

The total energy E combines internal energy e and kinetic energy $\frac{\|\mathbf{u}\|^2}{2}$. For a calorically perfect gas, internal energy e is given by $c_v T$, where $c_v = 717.63\,\text{J}\,\text{kg}^{-1}\,\text{K}^{-1}$ is the specific heat at constant volume, and T is the temperature. For Newtonian fluids, the stress tensor τ_ν is determined by the velocity gradient $\nabla \mathbf{u}$ and dynamic viscosity μ:

$$\tau_\nu = \mu \left[\nabla \mathbf{u} + \nabla \mathbf{u}^T - 2/3(\nabla \cdot \mathbf{u}) \mathbb{I} \right] \quad (4)$$

Dynamic viscosity is calculated through Sutherland's law [6] from the static temperature obtained via the ideal gas law $T = p/(\mathbb{R}\rho)$, where $\mathbb{R} = 287.058\,\text{J}\,\text{kg}^{-1}\,\text{K}^{-1}$.

The heat flux vector φ follows Fourier's law as $\varphi = -\lambda_f \nabla T$, where λ_f represents the thermal conductivity coefficient.

Reynolds tensor τ_t and turbulent heat flux φ_t are determined using the Spalart-Allmaras turbulence model with quadratic constitutive relation [7], based on the Boussinesq hypothesis with the turbulent Prandtl number $Pr_t = 0.9$ and modeled turbulent viscosity coefficient μ_t. The turbulence model sensitivity was verified in our previous study [5].

To numerically solve the compressible Navier-Stokes equations, the ONERA-Safran finite-volume solver *elsA* [8] is employed, utilizing a second-order-accurate Roe upwind scheme with the minmod limiter and Harten entropic correction for spatial derivatives, alongside a backward-Euler implicit local time-stepping scheme.

2.2 Geometry and Mesh

Three-dimensional simulations of the S8Ch wind tunnel at the ONERA in Meudon, where experiments were performed [9], were conducted. A convergent-divergent nozzle upstream of the working section generates a $M = 1.62$ flow. A shock generator with a wedge angle of $\alpha = 10.5°$ is placed in the nozzle exit to provoke an oblique shock wave. The wedge angle was selected to obtain a so-called Mach reflection—an irregular shock reflection at the wall. Thus, the presence of a normal shock wave in the vicinity of the wall is guaranteed, enabling the investigation of the porous bleed under both supersonic and subsonic flow conditions. The numerical domain is illustrated in Fig. 1.

The structured mesh is generated using the in-house pre-processing tool and mesh-generator *Cassiopee* [10]. A fully parameterized mesh allows the variation of the bleed parameters. Therefore, each hole is modeled out of five blocks using a butterfly mesh. The four other blocks are part of a C-grid, including the wall boundary layer on the plate and plenum sides (see Fig. 2). For a proper resolution of the boundary layer, the minimum wall-normal cell size is $y^+ \approx 1 (0.2 \times 10^{-3}$ mm). Inside the holes, the wall is equally meshed. The cell-to-cell growth ratio is 1.1, the maximum value to

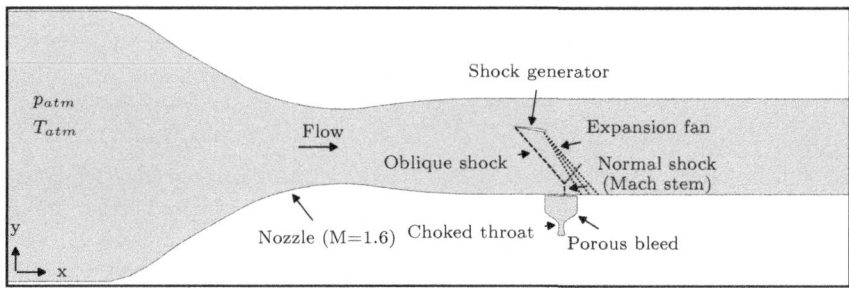

Fig. 1 Computational domain with shock generator [4]

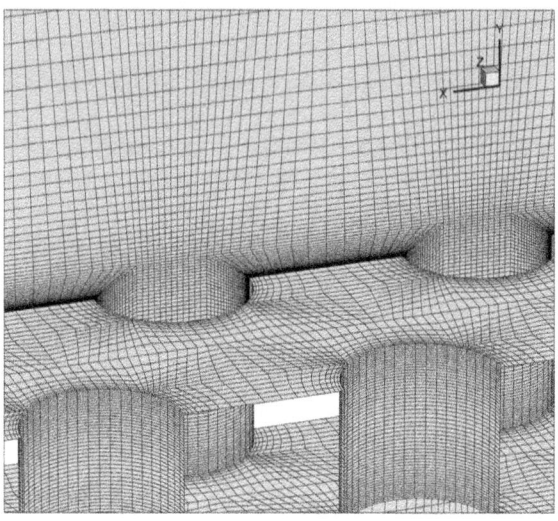

Fig. 2 View on mesh around the holes [5]

accurately predict the local mass flow rate, thanks to a preliminary mesh sensitivity study [5]. More details on the reference simulations can be found in [5, 11].

2.3 Evaluation of Bleed Efficiency and Control Effectiveness

The efficiency of porous bleed systems is typically quantified by the sonic flow coefficient. In this study, the scaling of Slater [12] is used to compute the surface sonic flow coefficient

$$Q_{sonic,w} = \frac{\dot{m}_{bl}}{\dot{m}_{sonic,w}}, \quad (5)$$

where the surface sonic mass flow rate $\dot{m}_{sonic,w}$ normalizes the bleed mass flow rate $\dot{m}_{bl}$. The surface sonic mass flow rate can be computed using the static flow quantities at the wall as follows:

$$\dot{m}_{sonic,w} = p_w A_{bl} \left(\frac{\gamma}{RT_w}\right)^{1/2} \left(\frac{\gamma+1}{2}\right)^{-\frac{\gamma+1}{2(\gamma-1)}} \quad (6)$$

With the aim to analyze the local mass flow rate for each hole, the local extraction of the hole mass flow rate $\dot{m}_{bl,i}$ and the local wall pressure is required. In the current study, the mass flow rate is extracted locally in each hole instead of using the bleed mass flow rate extracted at the plenum exit. Therefore the flow momentum in the plate normal direction is integrated

$$\dot{m}_{bl,i} = \int_S \rho v \, dS \tag{7}$$

at the hole center corresponding to half the plate thickness, as illustrated in Fig. 3. The (global) bleed mass flow rate can be easily computed by summing all hole mass flow rates:

$$\dot{m}_{bl} = \sum_{i=1}^{N} \dot{m}_{bl,i} \tag{8}$$

The static wall pressure and temperature cannot be extracted at the hole position since its value is disturbed by the local velocity caused by the flow into the holes. Therefore, a circular patch around the hole is defined with its size A_o fitting the porosity ϕ concerning the hole area A_h:

$$\phi = \frac{A_{bl,i}}{A_o} \tag{9}$$

The external wall quantities are then extracted by averaging their values above the patch size (see Fig. 3). Cells that partially lie within the circular patch are weighted using the areal fraction inside the circle. A similar method is applied to the plenum pressure, where the patch is equal in size but placed three hole diameters below the wall instead of using the wall pressure, which can be highly affected by the under-expanded jet at the hole exit. Higher patch distances lead to local information losses as the plenum pressure becomes more uniform with further distance from the holes.

The bleed effectiveness describes the ability of the bleed system to control the flow. In this investigation, the aim of the porous bleed is to increase the momentum in the wall vicinity. Therefore, characteristic quantities of the boundary layer are required, which are challenging to extract even in this relatively simple case of a flat plate because shock waves and expansion fans induced by the suction holes profoundly modify the boundary layer downstream of the region. As a result, we selected the wall shear stress to evaluate the effectiveness of the bleed system. The fuller and/or thinner the boundary layer, the higher the wall shear stress. Thus, comparing the wall shear stress upstream and downstream of the bleed regions enables the evaluation of the effectiveness. Therefore, the wall shear stress along the span is extracted on a line 10 mm upstream of the leading edge of the first hole and 10 mm downstream of the rear edge of the last hole. Using these quantities, we define the rise in the wall shear stress as

$$\varepsilon_\tau = \frac{\tau_{w,d} - \tau_{w,u}}{\tau_{w,u}}, \tag{10}$$

quantifying the relative increase of the wall shear stress along the bleed region. However, it is essential to note that this value is case-sensitive and dependent on the inflow conditions, which are kept constant in this study.

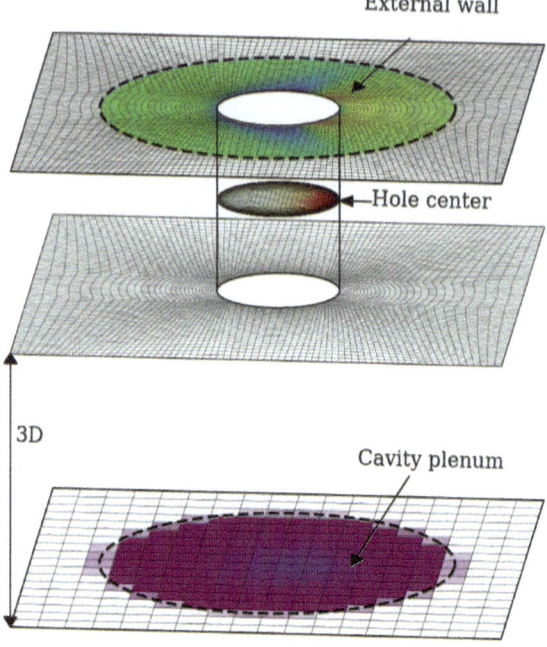

Fig. 3 Positions for the local extractions [5]

Moreover, the bleed mass flow rate must be constant relative to the inflow to evaluate the effectiveness. Hence, the displacement mass flow rate

$$\dot{m}_{\delta_{1,c}} = \int \delta_{1,c} \rho_\infty u_\infty dz, \quad (11)$$

which describes the theoretical missing mass flow rate required to obtain an inviscid flow, is computed. The ratio of bleed mass flow rate and displacement mass flow rate is kept constant ($\dot{m}_{bl}/\dot{m}_{\delta_{1,c}} = const.$).

3 Flow Topology in the Uncontrolled Case

Before controlling a shock-wave/boundary-layer interaction, the base flow without any control is presented for a wedge angle of the shock generator of $\alpha = 10.5°$. Figure 4 shows the flow field in the area. Three different shock waves are apparent in the outer inviscid flow: the oblique shock provoked by the shock generator, the reflected shock, and a normal shock—the so-called Mach stem—close to the wall. All three shocks meet in the so-called triple point. Downstream of the Mach stem, the flow is subsonic and has a higher entropy because of the higher total pressure losses caused by the normal shock. In contrast, the flow remains supersonic in the

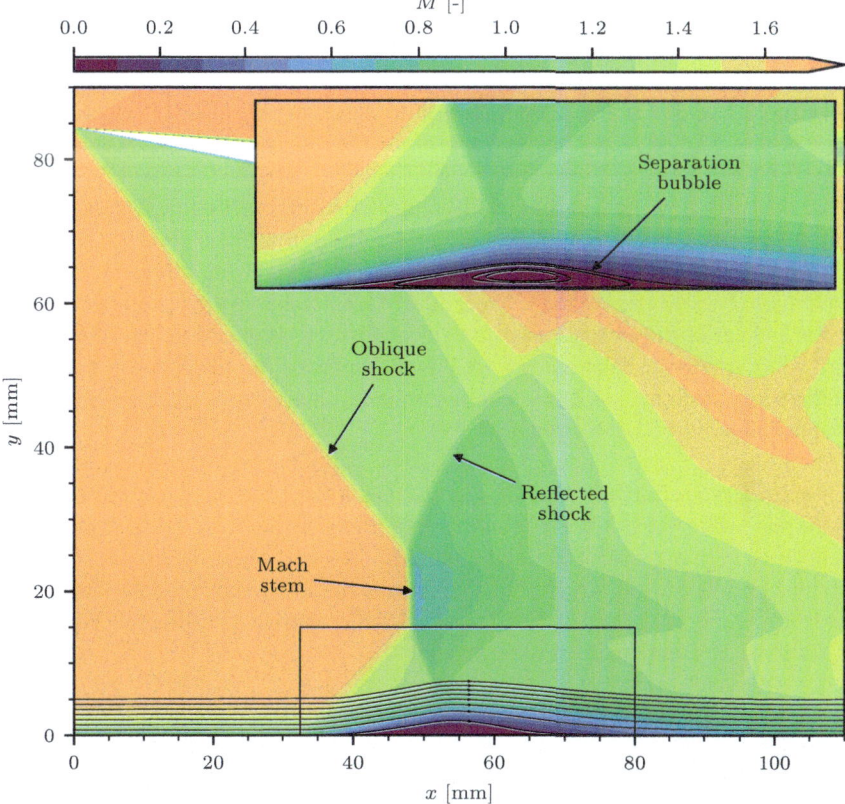

Fig. 4 Mach number field for the uncontrolled case with SBLI

upper part. As a result, a slip line is present between the supersonic and subsonic regions, separating the areas of different entropy.

Closer to the wall, a typical Lambda shock foot is observed. Again, an oblique shock is apparent, induced by the boundary-layer growth resulting from the adverse pressure gradient propagating upstream in the subsonic part of the boundary layer. Further downstream, we observe another normal shock. Again, the two shock waves meet with the Mach stem in a triple point.

A closer look at the flow field in the vicinity of the wall (see zoom-in view in Fig. 4) reveals the presence of a separation bubble. This separation bubble is limited in size because of an expansion fan generated by the rear edge of the shock generator, which accelerates the flow and leads to its reattachment. The aim of the porous bleed system is the mitigation of the separation bubble.

4 Geometry Influence on the Control

The flow field for a controlled case is presented in Fig. 5 for a bleed configuration with a hole diameter of $D = 2$ mm, a length-to-diameter ratio of $L/D = 1$, a porosity level of $\phi = 22.67\%$, and a stagger angle of $\beta = 60°$. Figure 5a shows the flow on the hole-cutting plane. The wedge angle is equal to the uncontrolled case with $\alpha = 10.5°$.

A different flow field is observed compared to the uncontrolled case. Again, the oblique shock is irregularly reflected, resulting in a Mach stem, as shown in Fig. 5b. However, no lambda shock is apparent, resulting from mitigating the flow separation. However, close to the shock, the Mach number in the vicinity of the wall is low, indicating the presence of a small local flow separation. Moreover, the suction through the perforated plate causes additional compressible effects. At the beginning of the perforated plate, an expansion fan is caused, and the so-called trailing shock is apparent at the end of the plate.

Following the streamlines in Fig. 5a, a slight thinning of the supersonic boundary layer is observed upstream of the shock, resulting in the expansion fan at the beginning of the plate. After passing the shocks, the streamlines are bent towards the wall, which indicates a high mass removal downstream of the shock where the flow is subsonic. No shock-induced flow separation is notable as the low-momentum flow is sucked into the plenum. At the end of the plate, a flow deflection in the wall-parallel direction is observed, resulting in the trailing shock.

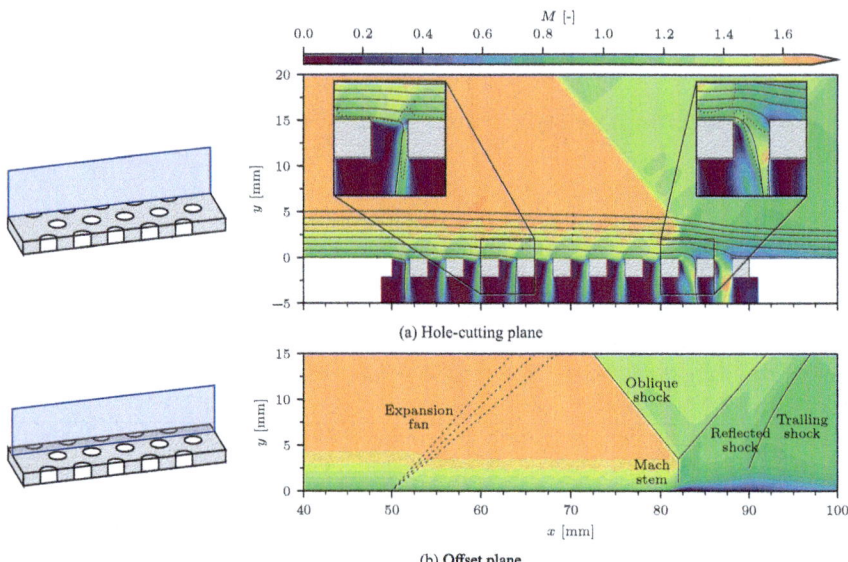

Fig. 5 Mach number field for an SBLI control; gray patches illustrate the out-of-plane holes ($TR = 0.7$)

The zoom-in view on the left-hand side shows the flow field inside a bleed hole upstream of the shock where the external flow is supersonic. The suction rate in this region is relatively low, the holes are unchoked, and the flow inside the hole becomes supersonic only in a small area as illustrated by the sonic line (dotted). Moreover, a sizeable separated region is observed. At the front of the hole, an expansion fan is induced by the suction, while further downstream, the so-called barrier shock is present, which redirects the flow either into the hole or in the wall-parallel direction.

A completely different flow field is apparent on the zoom-in view on the right, presenting the flow inside the hole in the subsonic region downstream of the shock. Here, the external flow is subsonic. Thus, no barrier shock is evident, and the expansion fan is limited to a small area close to the leading edge, where the flow is accelerated to supersonic conditions again. The flow inside the hole is choked, as visualized by the sonic line. The separated region inside the holes in the subsonic region is significantly lower. Moreover, the streamlines directed toward the plate indicate a substantial thinning of the boundary layer. Altogether, the flow inside the holes is found to behave like in the case of supersonic boundary-layer bleeding upstream of the shock, and equally to subsonic boundary-layer bleeding downstream of the shock [11].

The flow on the hole-cutting plane between the holes is detailed in Fig. 5b, where the pattern of the shock reflection, as well as the expansion fan and the trailing shock induced by the porous bleed, are illustrated. In the supersonic region upstream of the shock, small out-of-plane effects caused by the hole located outside the plane are evident in the form of irregularities in the Mach number inside the boundary layer. Downstream of the shock, where the external flow is subsonic, no out-of-plane effects are apparent. Close to the shock, the Mach number in the vicinity of the wall is low, which indicates the presence of a small local flow separation.

CFD simulations have been performed for different hole diameters, porosity levels, thickness-to-diameter ratios, and stagger angles over a range of plenum exit throat ratios. Here, the effect of the parameters on the control of the SBLI is investigated.

4.1 Hole Diameter

The flow fields for the different hole diameters are shown in Fig. 6. On the left, the contours are extracted on the hole-cutting plane so that the flow inside the holes is visible. The flow on the offset plane between the two rows is shown on the right side. From top to bottom, the hole diameter is increased. The gray patches illustrate the position of the second row of bleed holes.

A view of Fig. 6a reveals significant differences between the flow fields. For the smallest diameter $D = 0.25$ mm, the flow field upstream of the shock is very homogeneous. There is no apparent penetration of the barrier shocks in the boundary layer. The shock is located at $x = 83$ mm (dotted vertical lines), and no lambda foot is found. This is a consequence of the strong thinning of the upstream boundary layer. Thus, the interaction of the shock with the boundary layer is effectively mitigated with

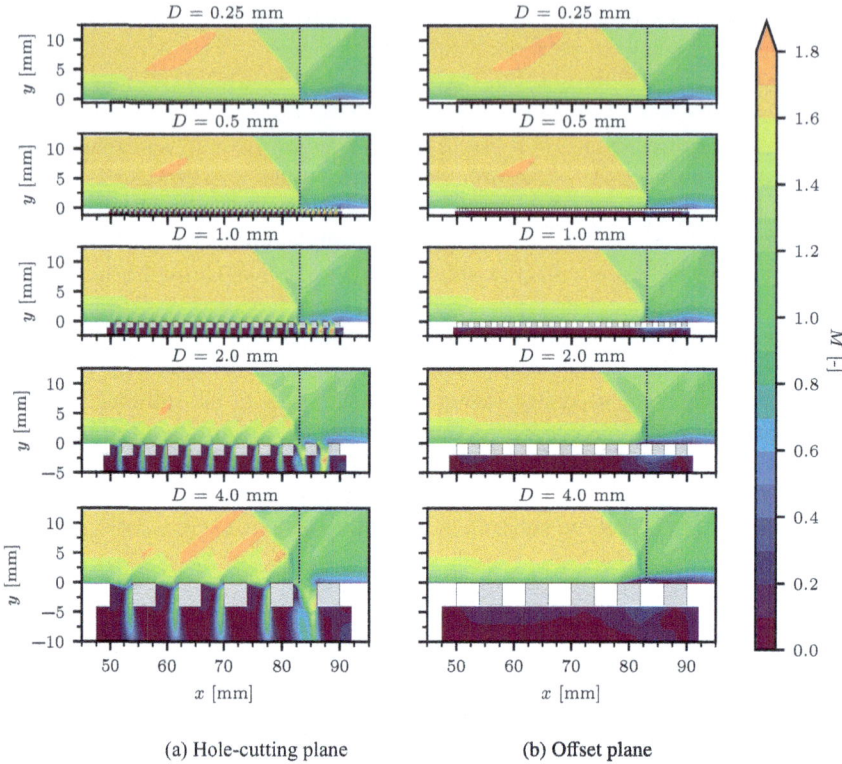

Fig. 6 Mach number contours for different hole diameters; gray patches illustrate the second row hole positions; dotted line highlights shock location for $D = 0.25$ mm

no apparent flow separation. Downstream of the shock, a strong transpiration flow is evident with the high-velocity jets in the plenum. As a result, no flow separation is apparent. At the end of the plate, the trailing shock, which leads to a weak thickening of the boundary layer, is visible.

With larger hole diameters, a slight upstream displacement of the shock foot is apparent. Moreover, for hole diameters $D \geq 1$. mm, the flow field is less homogeneous, and penetration of the barrier shocks through the boundary layer is found. This also affects the shape of the shock wave, which is strongly bent in the vicinity of the holes. For the largest diameter of $D = 4$ mm, a strong interaction of the holes with the boundary layer can be observed. Every hole creates an expansion fan and a barrier shock that penetrates the boundary layer and the incoming oblique shock wave. The shock wave is curved, and the shock foot is approximately 3 mm further upstream than for the smallest holes. Also, downstream of the shock, differences can be noted: The boundary layer is thicker, and the trailing shock is less prominent. These findings indicate a lower bleed efficiency.

In Fig. 6b, the effect of the hole diameter is more significant. For small hole diameters $D \leq 1$ mm, the flow field looks identical to the left-hand side. This means that the flow field is very homogeneous in the spanwise direction. For larger holes $D \geq 2$ mm, the impact of the barrier shock and the expansion waves is evident. Moreover, the Mach number in the vicinity of the wall below the shock foot is very low. Here, a lambda shock foot is observable, which indicates the presence of local flow separation.

Figure 7 shows the curves of the external wall pressure and the plenum pressure, which are scaled by the stagnation pressure p_t, and the surface sonic flow coefficient along the bleed region. The shock position is located between $\hat{x}/L_w = 75\%$ and 80%, as seen by the pressure rise in Fig. 7a. Upstream of the shock, The larger the hole, the higher the wall pressure. Moreover, the shock is more smeared for larger hole diameters. Hence, better control is achieved for small holes. Downstream of the shock, the differences are smaller. Only at the end of the plate, a larger pressure rise for tiny holes is found, leading to the assumption of a stronger trailing shock as the pressure information can spread upstream inside the boundary layer.

The surface sonic flow coefficient (Fig. 7c) behaves similarly to the wall pressure. Upstream of the shock, the flow coefficient is the highest for the smallest holes. Passing the shock foot, the sonic flow coefficient sharply increases as the pressure

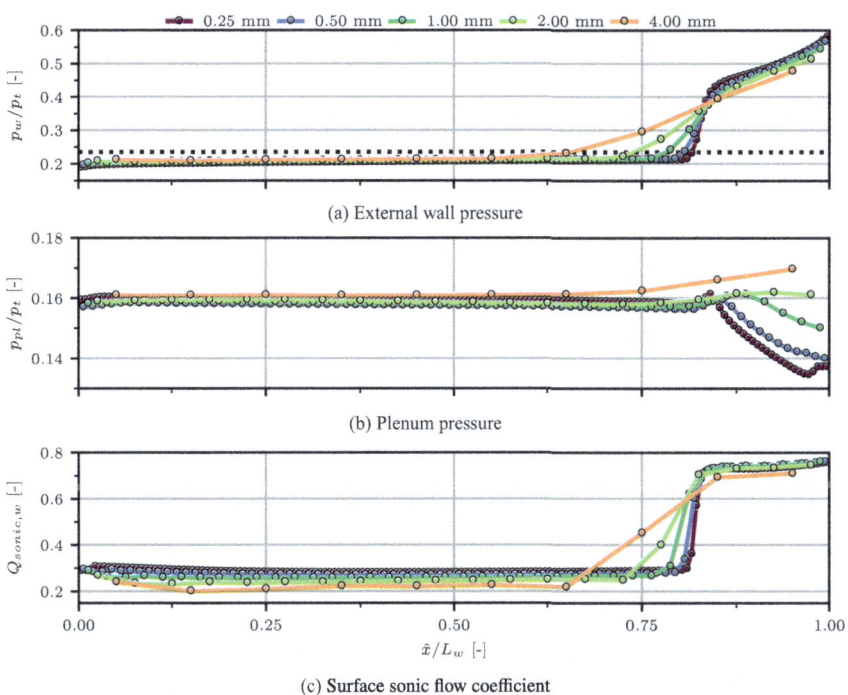

Fig. 7 Comparison of the flow quantities for different hole diameters

difference from the external wall to the plenum increases. Downstream of the shock, a lower efficiency is found for the largest diameter of $D = 4$ mm. The differences between the other hole diameters are small.

Also, the pressure inside the cavity slightly varies downstream of the incident shock, as shown in Fig. 7b. Since the bleed rate is higher in this area, the transpiration flow is stronger, leading to secondary flows inside the cavity plenum. Also, losses induced by the under-expanded jet are increased. However, the pressure difference inside the plenum is low compared to the change in the external wall pressure. Thus, the assumption of constant pressure inside the cavity, as used in the bleed modeling, is valid.

The surface sonic flow coefficient as a function of the pressure ratio is visualized in Fig. 8. Additionally, the bleed models of Slater [12] and Grzelak et al. [13] are shown. The model of Slater [12] is validated for supersonic external flows while the model of Grzelak et al. [13] is derived from a numerical study without external flow. In our previous study [4], we found a low prediction error of the first model upstream of the shock, while the second one predicts the bleed effect downstream of the shock better.

The trend for the sonic flow coefficient is challenging to grasp in one view since an extensive range of working regimes is covered by one throat ratio. For low throat ratios, the plenum pressure can exceed the external wall pressure upstream of the shock. Thus, the air is not removed but added to the boundary layer (blowing). Consequently, the momentum inside the boundary layer is decreasing instead of increasing as desired.

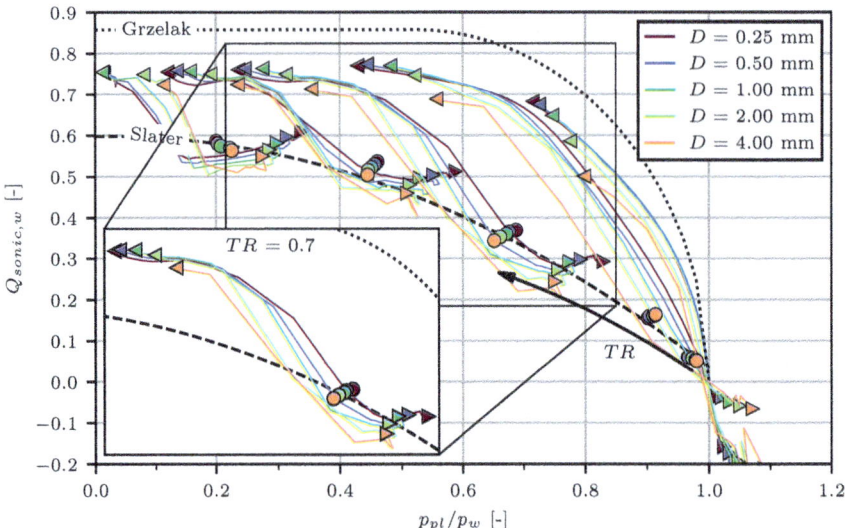

Fig. 8 Variation of the surface sonic flow coefficient for different hole diameters for the SBLI

Downstream of the shock, the pressure ratio is higher regardless of the throat ratio. Thus, the sonic flow coefficient sharply increases. Since the flow momentum in the vicinity of the wall is low, the size of the separated region decreases, resulting in a larger vena contracta area and hence a higher flow coefficient than for the supersonic bleeding. The bleed operates here in subsonic flow conditions, and the trend is similar to that of the model of Grzelak et al. [13].

A diameter dependence on the flow coefficient is apparent, similar as for supersonic and subsonic flows. The higher the hole diameter, the lower the sonic flow coefficient. However, this trend is smaller for the holes downstream of the shock, where the hole flow is choked, and the external flow is subsonic.

Interestingly, by observing the mean plate values, large holes are found to be slightly more efficient for low throat ratios. The sonic flow coefficient is equal for all cases, but the pressure ratio is higher. The reason seems to be the low "blowing efficiency" for large holes.

A more detailed view of the control of the shock-induced flow separation is shown in Fig. 9, which displays the streamwise wall shear stress below the shock foot. The absolute values for the coordinates are given to grasp the size of the separated region. For the largest holes of $D = 4$ mm, a large region of reversed flow is apparent between the rows starting from $x = 78$ mm. This region is induced by the adverse pressure gradient of the incident shock and ends with the end of the plate ($x = 90$ mm), where the flow reattaches due to the strong bleed rate, but also because of the expansion fan caused by the rear edge of the shock generator.

For smaller holes, the upstream influence of the incident shock inside the boundary layer is smaller, as indicated by the smaller extent of the region of reversed flow. With regard to Fig. 6b, it can be stated that the smaller hole sizes lead to a higher flow momentum inside the boundary layer, resulting in a shorter interaction length. The size of the lambda shock foot is smaller, and the separation bubble below decreases in size. Directly downstream of the bleed holes, there is no flow separation apparent, as the momentum in the wall vicinity is too high to separate. Thus, the flow separation is limited to the spanwise area between the holes, which is a function of the porosity level and the hole diameter. Hence, smaller diameters have a better control effect as the solid area between two holes is reduced.

The boundary-layer profiles downstream of the bleed region are illustrated in Fig. 10. The observed effects are similar to the investigations for the boundary-layer bleeding but even stronger. The use of tiny holes guarantees a higher momentum of the flow in the wall vicinity downstream of the impinging shock.

The application of large holes to control SBLIs results in strong variations of the boundary-layer profiles along the span. As observed in Giehler et al. [5], large hole diameters lead to a less homogeneous flow for both supersonic and subsonic conditions. The appearance of the adverse pressure gradient caused by the shock wave enforces this effect. Fluctuations in the sonic height result in variations of the interaction length along the span. Thus, the flow momentum between the holes is further decreased and negative effects sum up. Consequently, the variations along the span increase.

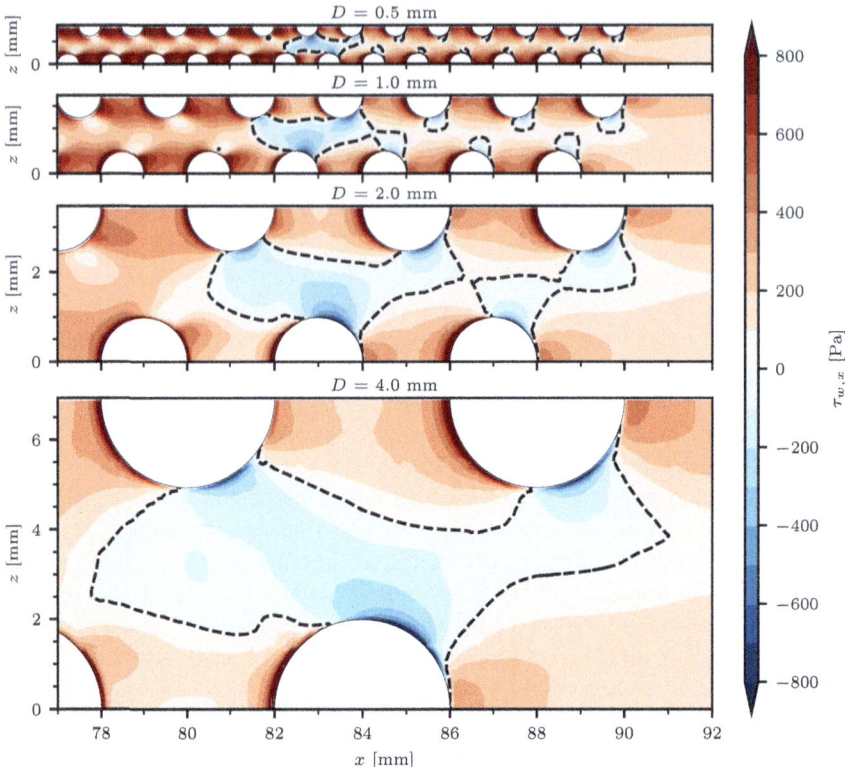

Fig. 9 Streamwise wall shear stress below the shock foot for different hole diameters; dashed lines mark $\tau_{w,x} = 0$ contour line

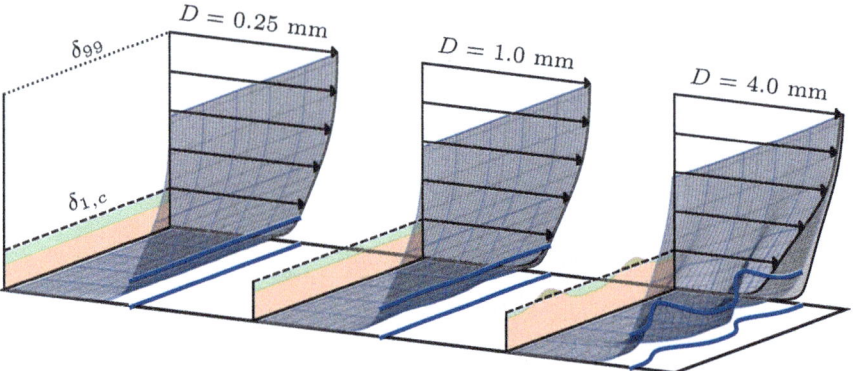

Fig. 10 Boundary layer 10 mm downstream of the bleed region for $D = 0.25$ mm, 1.00 mm and 4.00 mm (left to right); gray patch illustrates the envelope of the boundary-layer profiles along the spanwise direction; red area details the compressible displacement thickness and green area the difference to the inflow

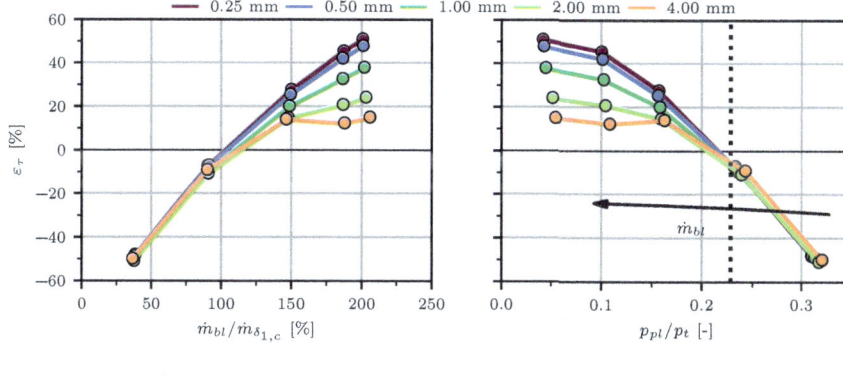

(a) Rise in wall shear stress as a function of the bleed rate

(b) Rise in wall shear stress as a function of the pressure ratio

Fig. 11 Rise in the wall shear stress as a function of the bleed rate $\dot{m}_{bl}$ and the pressure ratio p_{pl}/p_t for varying hole diameters; dotted line in (**b**) highlights isentropic pressure ratio for $M = 1.62$

The rise of the wall shear stress as a function of the bleed rate and pressure ratio p_{pl}/p_t is shown in Fig. 11. The stagnation pressure p_t is used to scale the plenum pressure since the wall pressure p_w varies along the bleed because of the incident shock. A view at Fig. 11a reveals significant differences in the boundary-layer bleeding: for low bleed rates, a degradation of the wall shear stress is apparent. This is a result of a lower Mach number downstream of the impinging shock, but also of the presence of blowing for lower bleed rates as highlighted in Fig. 8. As long as blowing occurs, no differences between the different hole diameters are found. With reaching a plenum pressure below the external wall pressure, the bleed operates in full-suction mode, resulting in differences in the effectiveness between the different hole diameters.

Figure 11b confirms that a pressure lower than the isentropic static pressure upstream of the shock is necessary to obtain a higher wall shear stress. If the plenum pressure is higher, the blowing occurs upstream of the shock, which eliminates diameter influences. The lower the plenum pressure, the higher the effectiveness. However, tiny holes are advantageous, which are linked to the lower extent of the reversed flow region below the shock foot.

4.2 Porosity Level

The influence of the porosity level on the SBLI is shown in Fig. 12. Again, the sonic flow coefficients and the models of Slater [12] and Grzelak et al. [13] are illustrated. Generally, it must be noted that comparing different porosity levels is challenging as the global bleed mass flux strongly differs. The higher the bleed mass flux, the better

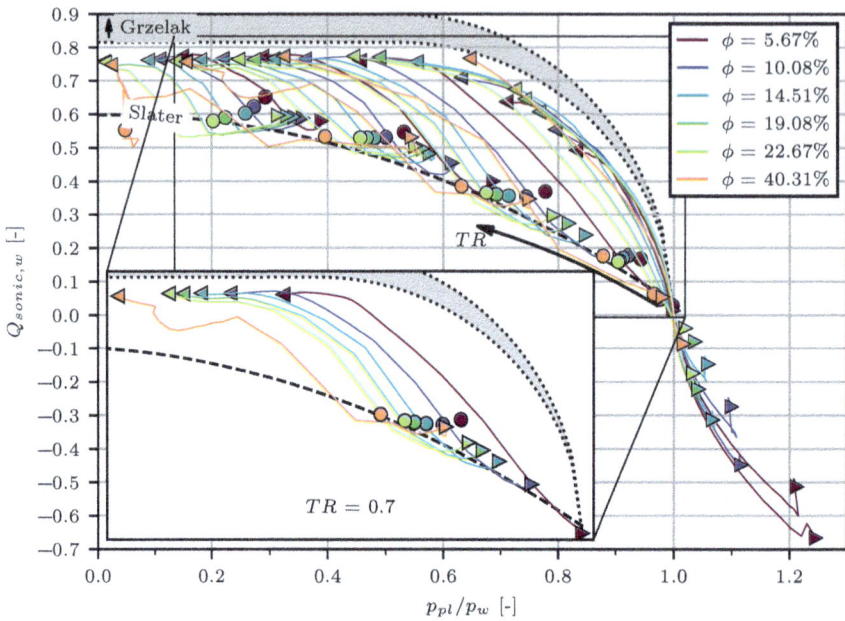

Fig. 12 Variation of the surface sonic flow coefficient for different porosity levels for the SBLI

the control of the SBLI because of the higher momentum close to the wall. Figure 12 still allows the comparison of the efficiency of different porosity levels.

For low throat ratios, blowing takes place upstream of the shock for all porosity levels with higher intensity for low-porosity plates. The reason is the weaker control of the shock. As a result, the location of the shock foot moves upstream, and the plenum pressure and the blowing rate increase.

Downstream of the shock, the differences between the different porosity levels are minor. All points of the last holes show a similar trend. For unchoked conditions, high porosity levels seem to be slightly more efficient, whereas the contrary is the case for choked holes. Also, the findings of Grzelak et al. [13] cannot be confirmed as they found a higher efficiency for larger porosity levels. However, their study was limited to small porosity levels and no tangential flow.

Since the porosity level is strongly linked to the achievable bleed rate, this parameter has an essential impact on the effectiveness of the bleed. The higher the bleed rate, the higher the effect. Figure 13 demonstrates the higher rise in the wall shear stress for higher porosity levels. As already shown for the hole diameter, a certain amount of bleed rate is required to obtain an equal wall shear stress downstream of the bleed compared to upstream. However, the maximum bleed rate for the lowest porosity level of $\phi = 5.67\%$ is not high enough to reach this point.

Nevertheless, a difference between the porosity levels in terms of effectiveness is observed in Fig. 13a: the lower the porosity level, the lower the required bleed rate to obtain the same wall shear stress downstream of the SBLI compared to upstream.

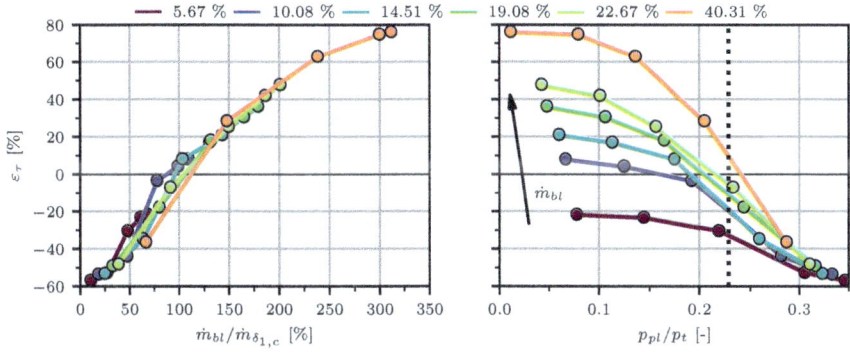

(a) Rise in wall shear stress as a function of the bleed rate

(b) Rise in wall shear stress as a function of the pressure ratio

Fig. 13 Rise in the wall shear stress as a function of the bleed rate $\dot{m}_{bl}$ and the pressure ratio p_{pl}/p_t for varying hole diameters; dotted line in (**b**) highlights isentropic pressure ratio for $M = 1.62$

4.3 Thickness-to-Diameter Ratio

As already observed by Giehler et al. [5], the thickness-to-diameter ratio has no significant effect on the boundary layer. Therefore, only the sonic flow coefficient is regarded in Fig. 14. Like for the other parameters, blowing occurs for small throat ratios, which is unaffected by the thickness-to-diameter ratio. With increasing throat diameter, effects on the sonic flow coefficient become apparent. Upstream of the shock, where the flow is supersonic, large ratios are advantageous if the holes are unchoked. This is in line with the results from Giehler et al. [5]. For low-pressure ratios, this effect inverts, and lower thicknesses are found to be more efficient.

In contrast, downstream of the shock, where the pressure ratio is always lower because of the jump in the external wall pressure, high thickness-to-diameter ratios are observed to negatively affect the sonic flow coefficient. In this area, the tangential flow momentum is very low, resulting in a small size of the separated region inside the holes. Thus, the flow is fully attached inside the holes, and friction increases with further hole length, i.e., plate thickness. Grzelak et al. [13] also found a degradation of the bleeding performance for $T/D \geq 1.5$. However, no separated region was induced by external flow in their study, and the observed hole diameters and porosity levels were smaller.

The difference between subsonic and supersonic conditions complicates the definition of an optimal geometry for the control of SBLIs. However, thickness-to-diameter ratios $T/D > 4$ result in high friction losses and should be avoided. Since the highest mass fluxes are achieved downstream of the shock, a higher sonic flow coefficient in this area is advantageous. Thus, the thickness-to-diameter ratio should not be higher than $T/D = 2$.

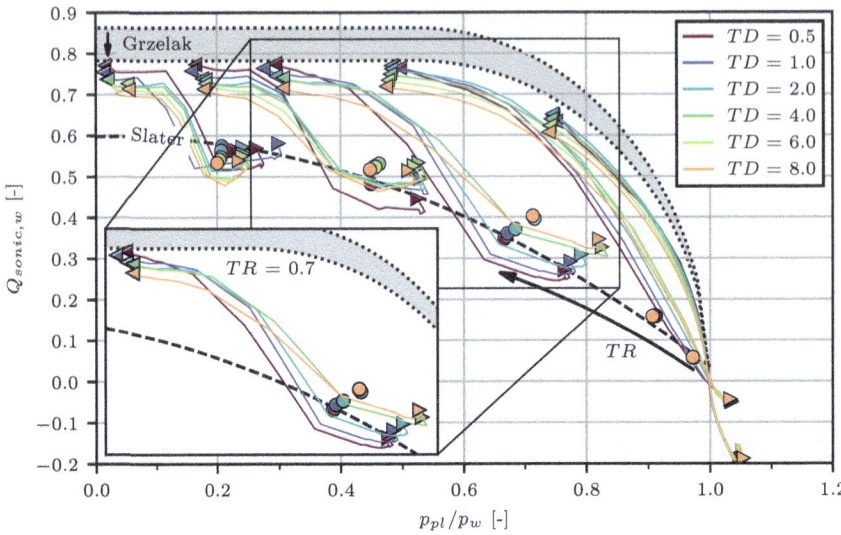

Fig. 14 Variation of the surface sonic flow coefficient for different thickness-to-diameter ratios for the SBLI

4.4 Stagger Angle

The last investigated geometrical parameter is the stagger angle, which is unconsidered in all bleed models. Its effect on the sonic flow coefficient is shown in Fig. 15. The findings are similar to those from Giehler et al. [5]: for subsonic conditions, no difference in the sonic flow coefficient is apparent. For supersonic conditions, the stagger angles $\beta = 30°$ and $45°$ are preferable to use since the pressure losses are lower. Overall, the effects are in a negligible range.

More interesting is the effect on the SBLI control. Therefore, the streamwise wall shear stress is extracted and illustrated in Fig. 16 below the shock foot. Thus, the size of the separation bubble between the holes can be estimated. From top to bottom, the stagger angle is increased, leading to higher spanwise but lower streamwise distances between the holes.

For the stagger angles $\beta = 30°$ and $45°$, the region of reversed flow is very small. The high density of holes in the spanwise direction prevents the occurrence of one large separated area, as the flow reattaches directly downstream of the bleed holes because of the high transpiration flow, leading to an energizing of the boundary layer in this region. On the contrary, a large area of reversed flow is found for large stagger angles, as the spanwise distance between the bleed holes is larger, and the near-wall momentum is low. Thus, the effectiveness of the bleed is expected to be smaller.

The rise in the wall shear stress is illustrated in Fig. 17. Surprisingly, no significant differences are observed if the rise in the wall shear stress is shown as a function of the bleed rate, as seen in Fig. 17a. However, a higher effectiveness is achieved for

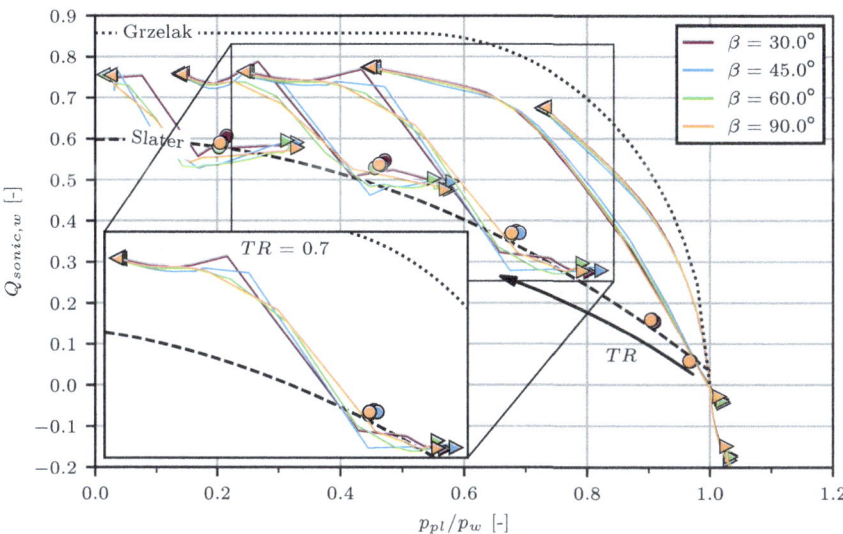

Fig. 15 Variation of the surface sonic flow coefficient for different stagger angles for the SBLI

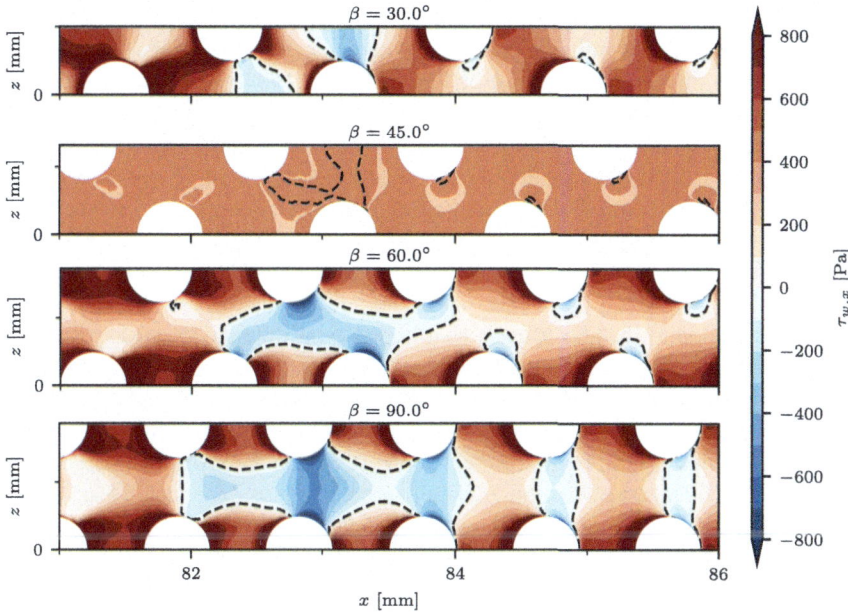

Fig. 16 Streamwise wall shear stress below the shock foot for different stagger angles; dashed lines mark $\tau_{w,x} = 0$ contour line

$\beta = 45°$ if the pressure ratio p_{pl}/p_t is kept constant. Thus, it can be stated that a 45° staggering has the biggest control effect on an SBLI.

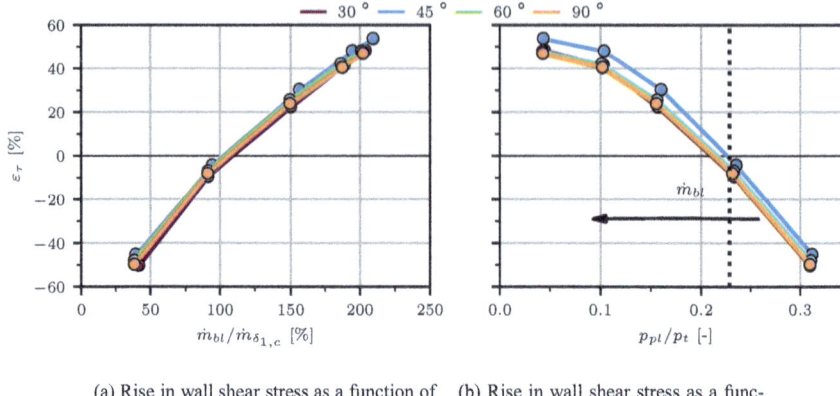

(a) Rise in wall shear stress as a function of the bleed rate

(b) Rise in wall shear stress as a function of the pressure ratio

Fig. 17 Rise in the wall shear stress as a function of the bleed rate $\dot{m}_{bl}$ and the pressure ratio p_{pl}/p_t for varying stagger angles; dotted line in (**b**) highlights isentropic pressure ratio for $M = 1.62$

5 Conclusion

This chapter studies the influence of the hole diameter, the porosity, the thickness-to-diameter ratio, and the stagger angle on the performance of a porous bleed to control a shock-wave/boundary-layer interaction. A comprehensive numerical database is generated, and the bleed efficiency and the control effectiveness are evaluated. The selected flow case is an irregular shock reflection at the wall—a so-called Mach reflection. In the uncontrolled case, a separation bubble is observed below the shock foot.

The bleed control significantly changes the flow field. The first observed effect is linked to the different bleed rates upstream and downstream of the shock. The pressure rise due to the shock leads to a higher wall pressure downstream of the shock, which results in higher suction rates. Additionally, the flow downstream of the shock is subsonic, leading to an increased bleed efficiency. Hence, the tangential flow velocity decreases along the shock, while the transpiration velocity increases. The consequence is a different flow angle downstream of the shock compared to the flow upstream. Thus, a deflection angle along the shock is generated, reducing the shock intensity. The simulations show a significant reduction in the height of the Mach stem, which implies a "more regular" shock reflection.

Overall, smaller hole diameters are found to enhance bleed efficiency and control effectiveness. The smaller the holes, the smaller the size of the interaction length. The porosity level shows opposing trends for bleed efficiency and control effectiveness. High porosity levels are beneficial in the control effectiveness as more mass can be removed, even though the sonic flow coefficient is slightly lower. The thickness-to-diameter ratio is only linked to the bleed efficiency as it influences the flow field inside the holes. Large ratios are advantageous upstream of the shock, where the bleed rates are low, and small ratios are better in the subsonic region. Thus, a medium value may be the optimum. Finally, the stagger angle shows only small influences. An angle of 45° is the optimum since the separated area below the shock foot is the lowest.

Competing Interests The authors have no conflicts of interest to declare that are relevant to the content of this chapter.

References

1. Fukuda, M.K., Hingst, W.R., Reshotko, E.: J. Aircr. **14**, 151 (1977)
2. Babinsky, H., Ogawa, H.: Shock Waves **18**, 89 (2008)
3. Babinsky, H., Harvey, J.K.: Shock Wave-Boundary-Layer Interactions. Cambridge University Press, Cambridge (2011)
4. Giehler, J., Grenson, P., Bur, R.: Flow, Turbul. Combust. **111**, 1139 (2023)
5. Giehler, J., Grenson, P., Bur, R.: J. Propul. Power **40**, 74 (2024)
6. Sutherland, W.: Lond. Edinburgh Dublin Philosoph. Mag. J. Sci. **36**, 507 (1893)
7. Spalart, P.R.: Int. J. Heat Fluid Flow **21**, 252 (2000)
8. Cambier, L., Heib, S., Plot, S.: Mech. & Ind. **14**, 159 (2013)
9. Giehler, J., Leudiere, T., Morgadinho, R.S., Grenson, P., Bur, R.: Aerosp. Sci. Technol. **147**, 109062 (2024)
10. Benoit, C., Péron, S., Landier, S.: Aerosp. Sci. Technol. **45**, 272 (2015)
11. Giehler, J.: Ph.D. Thesis, Institut Polytechnique de Paris, Palaiseau (2024c)
12. Slater, J.W.: J. Propul. Power **28**, 773 (2012)
13. Grzelak, J., Doerffer, P., Lewandowski, T.: Aerosp. Sci. Technol. **110**, 106494 (2021)

Open Access This chapter is licensed under the terms of the Creative Commons Attribution 4.0 International License (http://creativecommons.org/licenses/by/4.0/), which permits use, sharing, adaptation, distribution and reproduction in any medium or format, as long as you give appropriate credit to the original author(s) and the source, provide a link to the Creative Commons license and indicate if changes were made.

The images or other third party material in this chapter are included in the chapter's Creative Commons license, unless indicated otherwise in a credit line to the material. If material is not included in the chapter's Creative Commons license and your intended use is not permitted by statutory regulation or exceeds the permitted use, you will need to obtain permission directly from the copyright holder.

Unsteady Three-Dimensional Oblique Shock Wave Boundary-Layer Interactions

Timothy Missing and Holger Babinsky

Abstract Turbulent oblique Shock Wave Boundary Layer Interactions (SBLIs) were investigated experimentally in two rectangular test section blow down-type supersonic wind tunnels, to examine three-dimensionality induced by the presence of side-walls, as well as the low frequency separation bubble breathing oscillation. Testing was performed at Mach 2.5 and 2, with incident shock deflection angles of 8° and 12° at the Cambridge University (UCAM) and TU Delft (TUD) supersonic wind tunnel facilities respectively. In the UCAM facility, Conical shaped artificial corner separation bodies were used to generate corner waves, similar to those produced by corner separations, and vary their location with respect to the primary interaction. This resulted in a wide range of separation geometries underneath the primary interaction. Correlations between the separation length and pressure rise through interaction along streamwise strips revealed a quasi-2D relationship. The separation length was primarily correlated with the pressure rise from separation to reattachment. A secondary relationship was observed between the separation length and the pressure rise induced upstream of the interaction by corner waves. Corner waves modify the pressure rise in the interaction and this can lead to a significant reduction/elimination of separation in some regions. This strong control authority of pressure waves on the separation length informed the design of shock control bumps. Separation-bubble-shaped shock control bumps were tested in both test facilities with the goal of reducing separation, and dampening the low frequency bubble breathing oscillation. It was shown that these bumps are capable of significantly reducing and even eliminating flow separation. They also significantly dampened/eliminated the low frequency oscillation.

Keywords Supersonic flow · Shock wave—boundary layer interaction · Shock induced separation · Corner separation · Shock control bump

---

T. Missing (✉) · H. Babinsky
Cambridge University, Cambridge, UK
e-mail: tm668@cam.ac.uk

H. Babinsky
e-mail: hb@eng.cam.ac.uk

8.1 Introduction

Oblique Shock Wave Boundary Layer Interactions (SBLIs) occur in many aerospace applications. If the incident shock wave is sufficiently strong, it can cause the boundary layer to separate. Figure 8.1 shows a schematic of this interaction.

Oblique SBLIs can have significant detrimental effects on the downstream boundary layer, resulting in flow distortion, increased turbulent fluctuations, and pressure losses [1]. Furthermore, the separation bubble exhibits a low frequency breathing mode, characterised by periodic growth and shrinkage, which may initiate intake buzz or even engine unstart in supersonic mixed compression jet intakes [2]. This oscillation has a non-dimensional frequency (Strouhal number) of $St_L = \frac{fL}{U_\infty} \approx 0.03$ [6], in terms of the interaction length L (from the separation shock foot to the inviscid shock reflection) and free stream velocity U_∞.

Oblique SBLIs often occur near to 'side-walls' such as in rectangular supersonic jet intakes. The presence of a side wall means that a thick corner boundary layer forms, where the floor and wall boundary layers meet. Due to the combined viscous effect of both walls, this boundary layer has low flow momentum and is more susceptible to separation than the floor boundary layer. Therefore, it separates further upstream than the primary separation on the floor due to the adverse pressure gradient imposed by the reflecting incident shock wave.

The resulting corner separations produces waves which propagate into the primary interaction, resulting in a spanwise-varying separation shape [4]. Xiang and Babinsky [3] used block bodies to artificially enlarge and move the corner separations, which

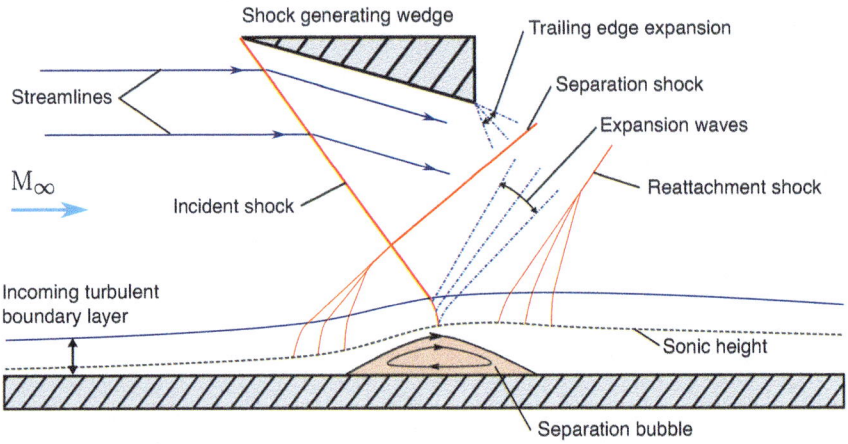

Fig. 8.1 Turbulent oblique SBLI schematic

showed that corner compression and expansion waves can significantly change the shape and size of the primary separation depending on where they arrived over the interaction.

Many techniques have been developed to control SBLIs with the aim to reduce separations, limit oscillations or reduce overall unsteadiness. One method which has had some success is vortex generators. These devices entrain high momentum flow into the boundary layer, thus making it more resistant to separation. Vortex generators have been successful in reducing the size of separations and reducing pressure losses, but do not fully eliminate separation [5]. Another flow control method are shock control bumps SCBs which have been used for controlling normal shock interactions on transonic wings, where they act to smear the shock into a series of compression waves or weaker shocks. This can reduce pressure losses and drag, as well as delay the onset of buffet by effectively anchoring the leading edge shock foot [7]. However, SCBs have not been applied to oblique SBLIs.

8.2 Methodology

Corner waves have been linked to changes in separation length and shape. However, no quantitative correlations have been made. The first part of this investigation utilises conical artificial corner separation bodies to produce corner compression and expansion waves which interact with the primary oblique SBLI. These corner bodies are shifted in the streamwise direction to vary location of the corner waves relative to the interaction, with the aim of producing a varying corner wave influence and identifying the effects of compression and expansion waves on the interaction. A better understanding of the effects of these waves on the main SBLI motivates novel control devices which are explored in part 2.

The second part of the investigation implements a shock control bump with the aim of reducing/eliminating separation and dampening/eliminating the low frequency bubble breathing oscillation. Similarly to corner separations, the control bump produces compression and expansion waves in the viscinity of the interaction, but these waves are generated on the floor.

8.2.1 Test Facilities

Experimental tests were run in two blow-down type supersonic wind tunnel facilities. Initial testing took place in the supersonic wind tunnel no. 1 at the University of Cambridge (UCAM), and subsequently in the ST15 at TU Delft (TUD).

The corner wave investigation was performed solely at UCAM, and the control bump tests were performed in both facilities, with different bump geometries designed for the respective test cases.

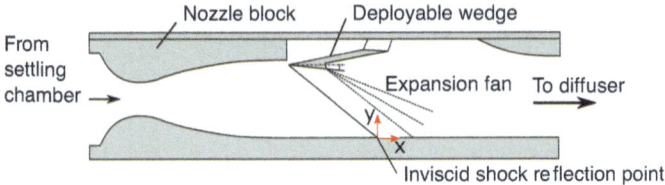

Fig. 8.2 Test section schematic

Table 8.1 Test conditions in the supersonic wind tunnel no. 1 at the University of Cambridge (UCAM), and the ST15 at TU Delft (TUD)

		UCAM	TUD
Upstream Mach number	M_∞	2.5	2
Shock deflection angle	θ	8°	12°
Upstream boundary layer thickness	δ	7.4	5.1
Aspect ratio (tunnel width/δ)	$\frac{w}{\delta}$	15.4	24.6
Reynolds number	Re_δ	1.8×10^5	2.3×10^5

Both facilities used a similar setup which is shown schematically in Fig. 8.2. Incident shock waves are generated by wedges mounted at the top of the test section, with a gap between the top nozzle and the leading edge, so that the leading edge is in the free stream flow. Expansion fans are generated at the trailing edge of the wedge which meet the tunnel floor downstream of the incident shock wave. The wedges are designed to be sufficiently long so that these downstream waves do not affect the oblique SBLIs. An origin is defined at the inviscid shock reflection point. This is the default origin used throughout this chapter.

The flow conditions used in these facilities are shown in Table 8.1. The Mach number is fixed by the nozzle block shape, and since these facilities did not share a common Mach number setting, tests were run at different Mach numbers.

8.2.2 Artificial Corner Separation Bodies

Conical quarter-revolution cone bodies were designed to be placed in the streamwise-corners of the UCAM wind tunnel to artificially replicate the effects of corner separations. Figure 8.3 shows a schematic of these corner cones in the wind tunnel. The semi-apex angle, of 9.2° was chosen to produce a similar inviscid cone-shock pressure rise to that produced by artificial corner block bodies used in a previous investigation by Xiang and Babinsky [3]. The maximum height was 2.91 δ, where δ is the incoming centreline boundary layer thickness.

These cones are placed at a series of 15 streamwise locations, illustrated in Fig. 8.3, with respect to the incident shock wave in order to shift the arrival of corner waves and

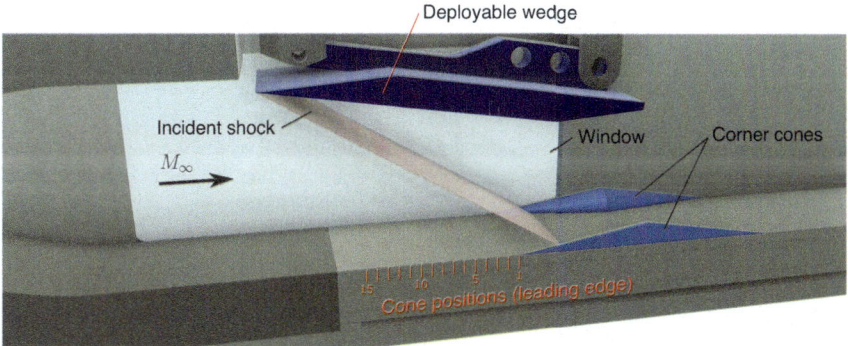

Fig. 8.3 Corner cones mounted in test section, CAD image. Leading edge positions 1–15 shown

thus the produce a varied effect on the interaction. These locations can be described in terms of the virtual crossing point x_c of inviscid cone shocks, emanating from the leading edges, on the centreline, with respect to the origin at the inviscid oblique shock reflection location.

The cone locations vary from $x_c = 99.75$ mm $= 14.05\,\delta$ (position 1) at the most downstream setting, to $x_c = -40.25$ mm $= -5.67\,\delta$ (position 15) at the furthest upstream. Where δ is the upstream boundary layer thickness.

8.2.3 Control Bumps

The control bumps were designed to match the baseline (without flow control) mean separation bubble shape. The spanwise-varying bubble shape was extracted using oil flow visualisation. The bump cross section profile was designed to match the mean dividing streamline. The design process and 3D bump are shown in Fig. 8.4.

For the initial test case at UCAM, PIV data was not available. Instead the separation bubble cross-section was estimated from the mean dividing streamline on a similar interaction reported by Piponniau et al. [8] for a Mach 2.3, 9° deflection incident shock wave. The shape was 3D printed in PLA and stuck to the tunnel surface.

In the second set of experiments at TUD, PIV data was available from a previous investigation in the same facility, with the same flow conditions. The bump was also designed and printed to higher tolerance, compared to the UCAM bump, better matching the mean separation bubble shape. The bump dimensions for each test case are shown in Table 8.2.

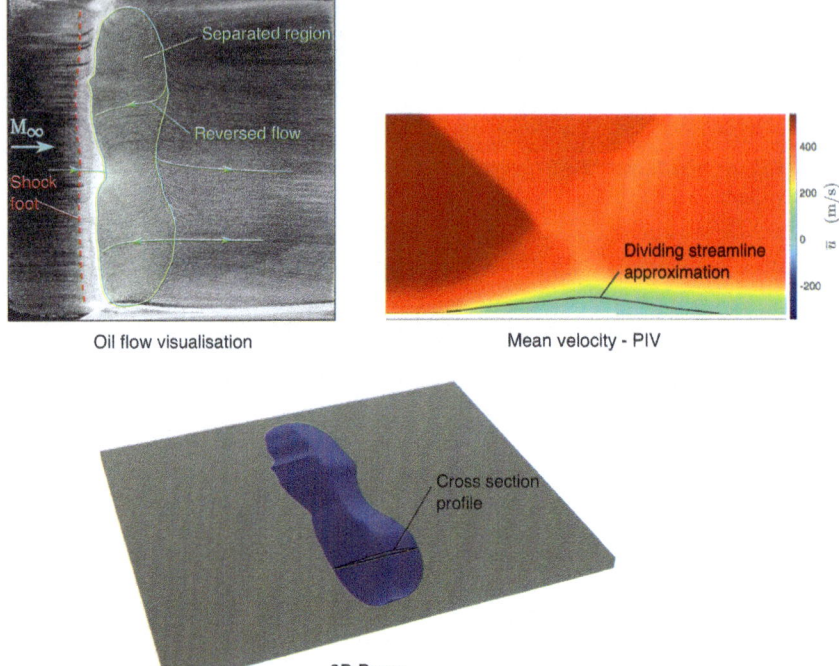

Fig. 8.4 Control bump design. The bump floor footprint follows the separation region extracted from oil flow visualisation (shown for the TUD test case). The cross-section profile matches the mean dividing streamline shape, approximated from the centre-plane mean streamwise velocity, from PIV data in the TUD test case

Table 8.2 Control bump dimensions. δ is the incoming boundary layer thickness

		UCAM	TUD
Streamwise length	L_{max}/δ	2.95	7.96
Wall-normal height	h_{max}/δ	0.35	0.40
Spanwise width	w_{max}/δ	6.44	26.59

8.2.4 Measurement Techniques

In both experimental investigations, two-mirror z-type shadowgraph flow visualisation configurations were used. At UCAM, images were recorded using a Photron CMOS FASTCAM Nova S6 high speed camera recording at 50 kHz, for 2 s, with a full frame shutter speed. At TUD, a Photron FASTCAM-SA1 CMOS camera was used, recording at 40 kHz, for 1 s, with full frame shutter speed. In each case, the recording time was set to maximum based on the available buffer memory. Power spectral density distributions of pixel intensity were estimated using Welch's method, using a window size of 500 and a 50% overlap.

Surface oil flow visualisation was performed on the tunnel floor to examine the separation topology. A mixture of kerosene, titanium dioxide powder and oleic acid is used. The oil streak patterns are videoed during the interaction to avoid issues with smearing during the tunnel shutdown process. In an adverse pressure gradient, the oil tends to stop moving slightly before the actual stagnation point. Therefore, there is expected to be an error of approximately $0.2\,\delta$ in the location of separation lines [9].

Time-averaged pressure measurements were obtained via pressure tappings in the tunnel floor—connected to a NetScanner 9116 via neoprene tubing, with an overall uncertainty of $\pm 0.5\%$. Pressure Sensitive Paint (PSP) was used on the tunnel floor to obtain time-averaged pressure distributions. The paint was illuminated by ultraviolet light and photographed during the test run. The ratio of pixel illumination during the test and with the tunnel off were calibrated in-situ using the point pressure measurements. The resulting PSP pressure measurements have an uncertainty of approximately $\pm 3\%$. Due to the presence of the corner cones, the regions near the side-walls are not captured.

Two-component Laser Doppler Velocimetry (LDV) data was acquired in the UCAM facility to measure velocities in a streamwise-vertical plane. The probe volume was approximately 0.1 mm in diameter, with a spanwise length of 1.4 mm. The air is seeded with parafin oil droplets approximately $0.5\,\mu$m in diameter. Boundary layer profiles were captured by traversing in the wall-normal direction. Points are binned in 0.15 mm segments on which a weighted mean, accounting for velocity bias, is applied to estimate a mean velocity profile. No data is captured within approximately 0.18 mm of the wall, due to reflections and limited particle seeding near to the wall. A wall-wake velocity profile was then fitted to estimate boundary layer properties, and extrapolate the mean profiles down to the wall. Further detail on this method can be found in Xiang and Babinsky [3].

Particle Image Velocimetry (PIV) data was acquired in the TUD facility along the centre-plane of the interaction. The flow was seeded using DEHS oil by a high pressure aerosolisation system. A Quatronix Darwin-Duo diode pumped dual oscillator laser was used to create a vertical light sheet of thickness approximately 2 mm to illuminate the particles. Two Photron FASTCAM-SA1 CMOS cameras were used to acquire high speed images of the particles. The overall time resolution of the system, between successive velocity measurements, was 7.47 kHz. Captured PIV images were processed using LaVision's DaVis 10.4 software. A Butterworth filter, of length 11 frames, was used to normalise the pixel intensity. Correlation was performed in a 4-step process, with an initial-pass window size of 64×64 pixels, and a final pass window size of 24×24. This resulted in a horizontal and vertical spatial resolution of approximately 2 mm. The resulting velocity uncertainty was approximately 1% of the incoming free stream velocity, based on the mean cross-correlation accuracy.

8.3 Results

The results are presented in the following order: firstly the baseline flow fields for the TUD and UCAM test facilities. Next, the artificial corner separation results performed at UCAM, followed by the control bump results for the test cases at both UCAM and TUD.

8.3.1 Baseline Flow Fields

Time-averaged shadowgraph images of the baseline (uncontrolled) test cases are shown in Fig. 8.5. The incoming and outgoing boundary layers are visible as well as the incident, separation, and reattachment shocks.

Oil flow visualisation images of the baseline tests are shown in Fig. 8.6. Both cases are fully separated with a large primary separation region in the centre of the tunnel floor, and corner separations in the streamwise corners of the tunnel.

In each case, the separation shock foot is visible, as an accumulation of oil-dye mixture, upstream of the separation/control bump.

The floor pressure distribution for the UCAM test case is presented in Fig. 8.7. Along the centreline, the pressure rises sharply at the separation shock foot, until the separation point. Between separation and reattachment, the pressure rises more gently, after which the pressure begins to level off. Along an off-centre streamwise strip, near to the corner separation, the pressure begins to rise further upstream than the centreline shock foot location. It then rises more gently throughout.

The off-centre strip is located in a region between the primary and corner separations where the flow remains attached despite the presence of a considerable adverse pressure gradient imposed by the SBLI. Xiang and Babinsky [3] attributed this attached flow to the pressure smearing effect of the corner waves.

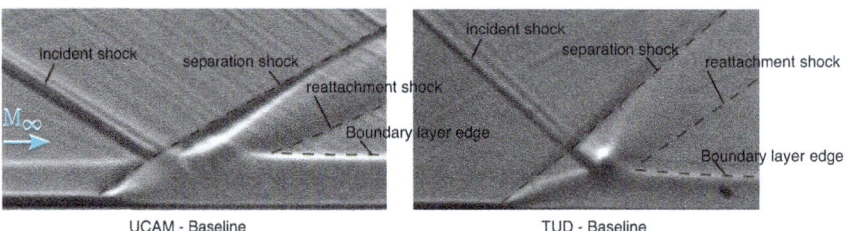

Fig. 8.5 Baseline time-averaged shadowgraph images for the UCAM and TUD test cases, at Mach 2.5 and 2, with incident shock deflection angles 8° and 12° respectively

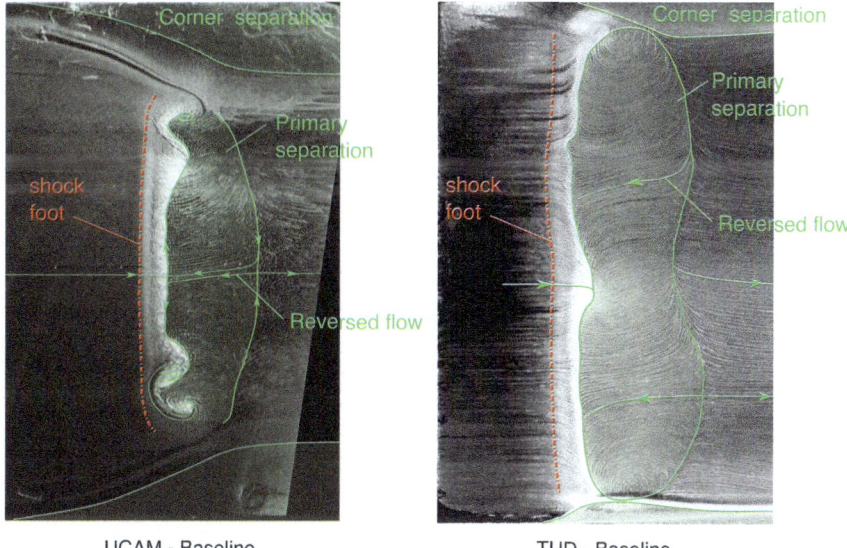

Fig. 8.6 Baseline oil flow visualisation of the UCAM and TUD test cases

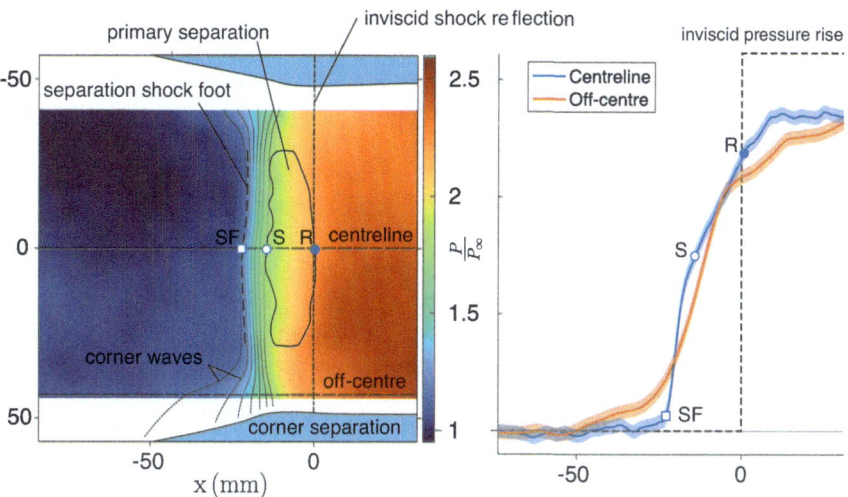

Fig. 8.7 Baseline pressure distribution for the UCAM test case, measured using Pressure Sensitive Paint. Left: Surface pressure distribution. Separation topology extracted from oil flow visualisation. Right: Corresponding pressure traces along the centreline and off-centre. SF = Separation shock Foot, S = Separation, R = Reattachment

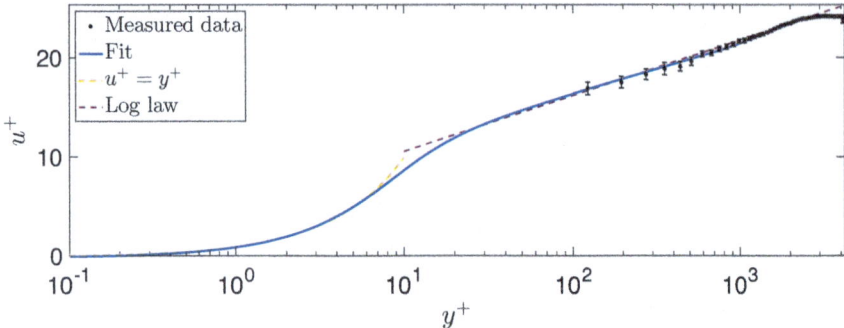

Fig. 8.8 UCAM baseline upstream boundary layer profile, measured $2 \times L_{sep}$ upstream of separation

8.3.2 Artificial Corner Separation Results

The corner cones were first tested without an incident shock wave to examine the resulting corner wave pattern in isolation. Figure 8.9 shows the surface pressure distribution. The inviscid cone shocks, as well as the inviscid oblique shock reflection pressure jump are shown for reference. The pressure is seen to rise smoothly along the streamwise direction to a maximum, near the mid-chord of the cones. This pressure rise can be attributed to compression waves which form along the leading faces of the cones and propagate across the span of the flow field. This pressure rise does not occur as a sharp jump, as would be expected from the inviscid cone shock solution. Instead, the presence of the floor boundary layer acts to spread out this pressure rise. Expansion waves emanate from the crest, which act to turn the flow back towards the streamwise-corners, and reduce the pressure below the upstream level P_∞. Further downstream, beyond the field of view, another set of corner compression waves are expected to form near the cone trailing edges, to turn the flow parallel to the streamwise direction. The maximum centreline pressure rise is approximately 18% of the inviscid oblique shock reflection pressure rise.

The corner cones were then implemented in the presence of the incident oblique shock wave to examine their effects on the oblique SBLI. Figure 8.10 shows time-averaged shadowgraph images for cone positions 1, 8, 12, and 15. The corner cones block the view of a large portion of the images. However, the separation and reattachment shocks are still visible above the cones. In cone position 1, the separation and reattachment shocks are shifted slightly upstream. In cone positions 8 and 12, the separation shock is shifted significantly upstream. In position 12, the reattachment shock is shifted upstream by a similar distance, while in position 8, the reattachment shock is not clearly visible. A 'rear' shock is observed downstream of the reattachment shocks in cone positions 12 and 15. This shock appears to shift with the corner cones, but its origin is unclear.

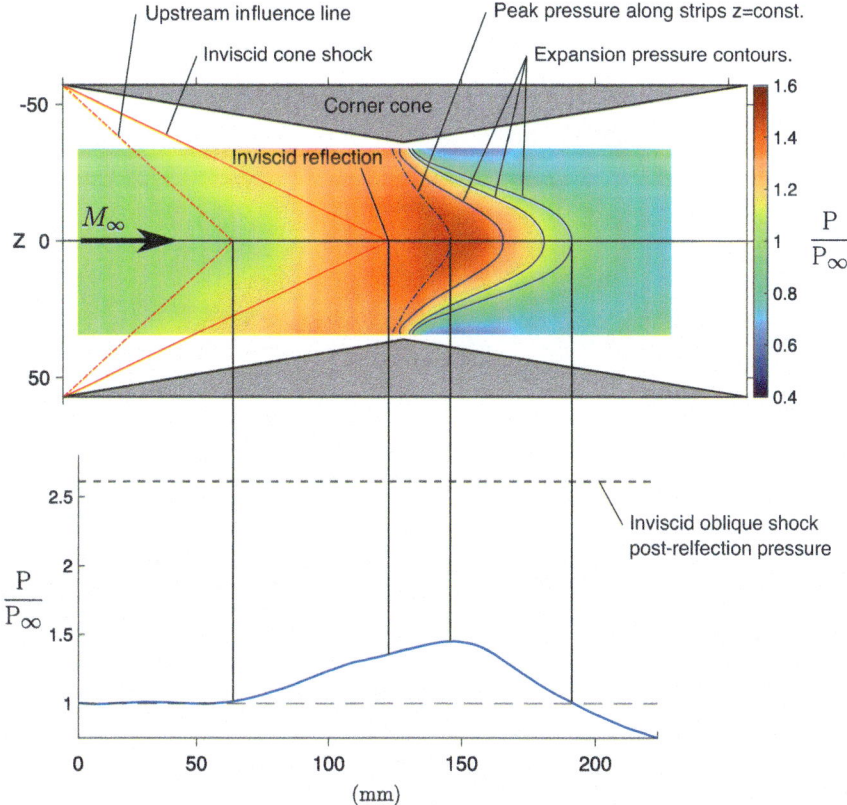

Fig. 8.9 Corner cones only—with no incident shock wave—surface pressure distribution (top), centreline pressure distribution (bottom), for the UCAM test case

Oil flow visualisations of the corner cone positions corresponding to the time-averaged shadowgraph images are shown in Fig. 8.11. In cone position 1, the centreline separation length is slightly increased compared to the baseline. The shock foot and reattachment are shifted upstream relative to the baseline, as observed in the shadowgraph images. In position 8, the separation length is significantly increased, by 2.8 times the baseline centre separation length. In position 12, the centre separation length is similar to the baseline, but the shock foot is shifted upstream, as observed in the shadowgraph. In position 15, the separation length is significantly reduced—to approximately 17% of the baseline length. However, larger separation zones are observed on the rear faces of the corner cones.

The surface pressure distributions for cone locations 1, 8, 12, and 15 are shown in Fig. 8.12, with outlines of the separation topology from the oil flow visualisations (Fig. 8.11). The inviscid cone shock, upstream pressure rise, peak pressure contour, and expansion pressure contours from the cone-only test case (Fig. 8.9) are highlighted.

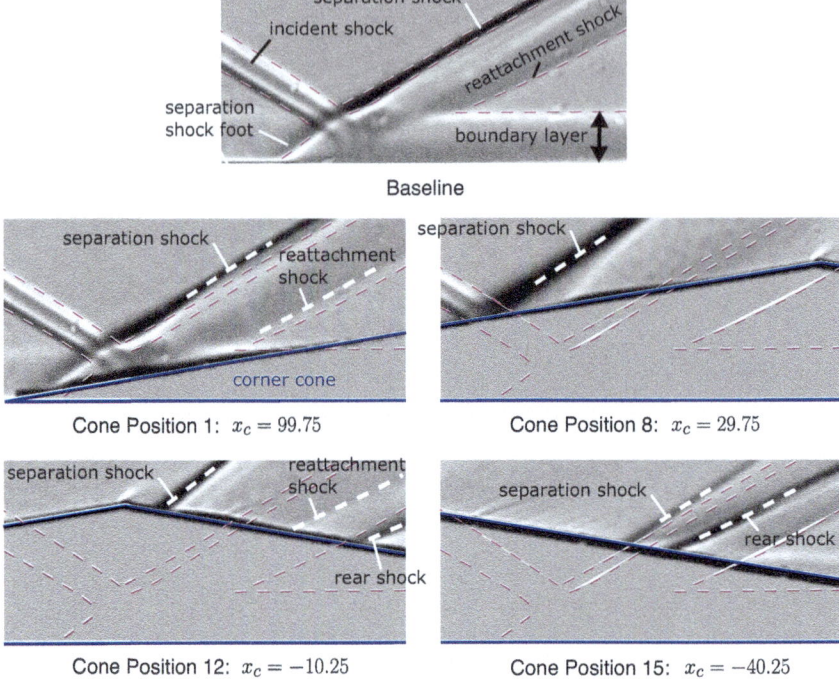

Fig. 8.10 Time-averaged shadowgraph for the UCAM Baseline and cone positions 1 (furthest downstream), 8 (maximum separation length), 12, and 15 (furthest upstream, and minimum separation length). x_c is the location of the inviscid cone shock reflection on the centreline, with respect to the inviscid oblique shock reflection location, measured positive downstream. The incident shock, shock foot, separation/reflected shock, reattachment shock, and apparent downstream boundary layer edge are outlined by dotted lines on the baseline case and overlaid on the corner cone cases for reference

In cone position 1, the leading corner compression waves are expected to arrive downstream of the separation. Nevertheless, we observe a slight increase in the separation length, and a slight shift in the separation location. In cone position 8, with the largest centreline separation length, a large portion of the corner compression waves arrive over the interaction. Whereas in cone position 15, with the shortest separation length, corner expansion waves arrive over the separation. In cone position 12, the separation zone is relatively near to the peak pressure contour for the cone-only test case. The pressure distribution on either side of the peak showed only weak changes. Thus, only a few corner waves arrive in this region on the centreline and the resulting effect on the main separation is small.

The increased separation length observed in cone position 8, and the decreased length observed in cone position 15, can be attributed to the adverse and favourable effects of corner compression and expansion waves arriving over the separation region, respectively—as suggested by Xiang and Babinsky [3]. The change in the

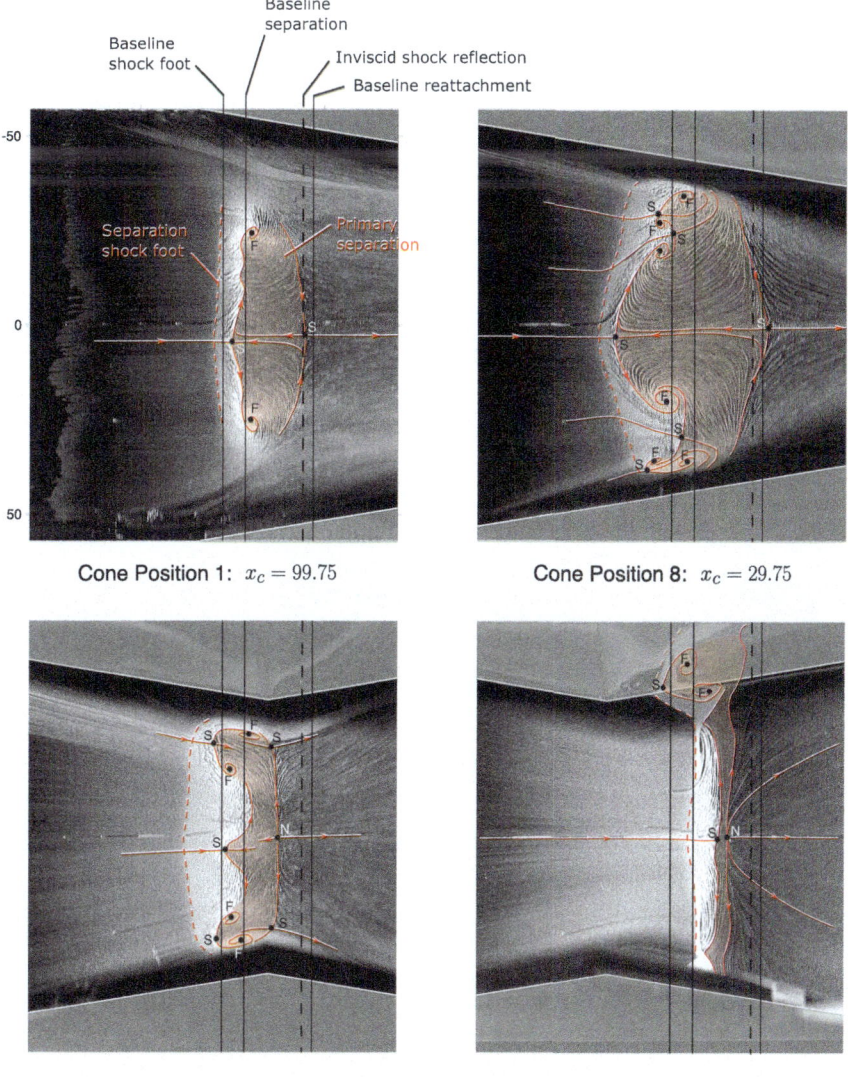

Fig. 8.11 Oil flow visualisations for cone positions 1, 8, 12, and 15—UCAM test case. The primary separation region is highlighted in red. Streamlines are shown passing through critical points: S = Saddle point, F = Focus, N = Node. The baseline shock foot, separation, and reattachment (along the centreline), as well as the inviscid shock reflection line are overlaid for reference

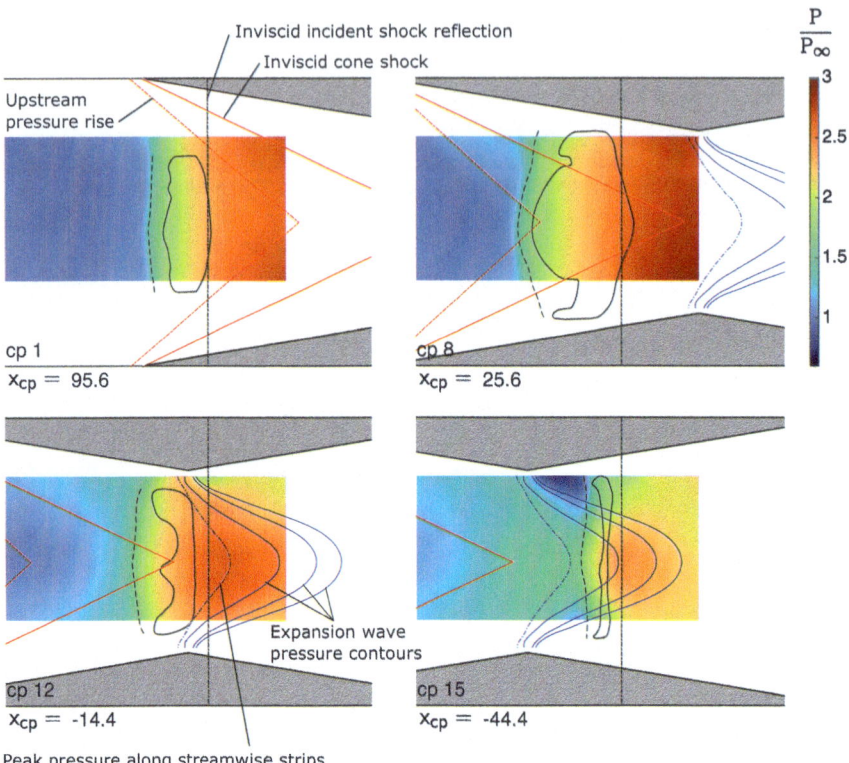

Fig. 8.12 Separation topology and pressure distributions for cone positions 1, 8, 12, and 15—UCAM test case

separation length observed in cone position 1, despite the corner waves being predicted to arrive downstream of the interaction, suggests that the upstream influence of the leading corner compression waves may be modified by the presence of the oblique SBLI. However, further work will be required to explore this.

The effects of corner waves on the spanwise variation of separation shape are examined by correlating the pressure rise and separation length along streamwise strips through the interaction. This is done for a range of spanwise locations—approximately 200 strips per test case. A sample strip pressure distribution is shown in Fig. 8.13 for the centreline of cone position 12, along with the corresponding cone-only pressure distribution. The pressure rise is broken down into various key contributions: firstly, the pressure rises upstream of the interaction by ΔP_u due to corner waves arriving ahead of the separation. Then the pressure rises rapidly by ΔP_S through the separation process—between the shock foot and separation point. Next, the pressure increases more gently between separation and reattachment ΔP_{SR}. Thereafter the pressure continues to change due to corner waves and expansion waves

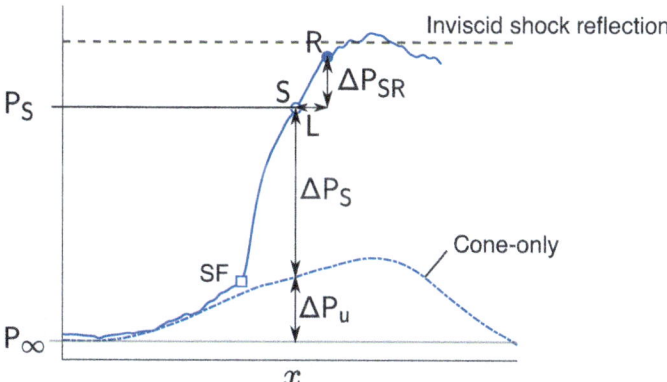

Fig. 8.13 Interaction pressure-rise schematic, from the centreline pressure in cone position 12. SF = separation Shock Foot, S = Separation, R = Reattachment, P_S = Pressure at separation, P_∞ = Upstream pressure. ΔP_u = Upstream pressure rise due to corner waves. ΔP_S = separation pressure rise. ΔP_{SR} = separation to reattachment pressure rise

from the trailing edge of the shock generator until it reaches a more steady level downstream of the interaction.

Figure 8.14 shows the separation pressure rise across all strips and all test cases plotted against the local separation length. It is remarkably constant, despite the wide range of separation lengths. The local separation length is then plotted versus the pressure rise from separation to reattachment ΔP_{SR} in Fig. 8.15 (left). There is a clear positive correlation. This correlation can be improved if the upstream pressure rise is taken into account by defining a modified pressure rise parameter R, as seen in Fig. 8.15 (right). This collapses the data onto a power law fit such that the scattered data lies approximately within the measurement uncertainty from this curve.

This result shows that there is a quasi-2D relationship between the local separation length and pressure rise along streamwise strips through the interaction, even for

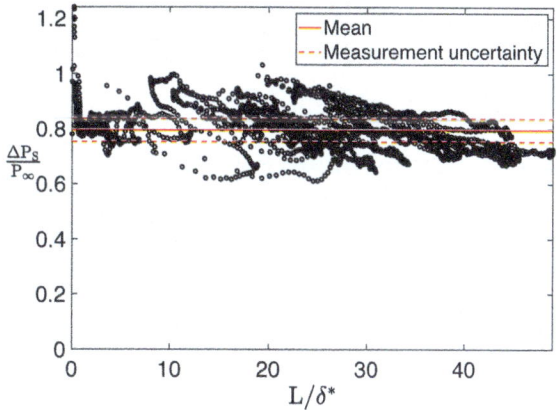

Fig. 8.14 Separation pressure rise versus separation length

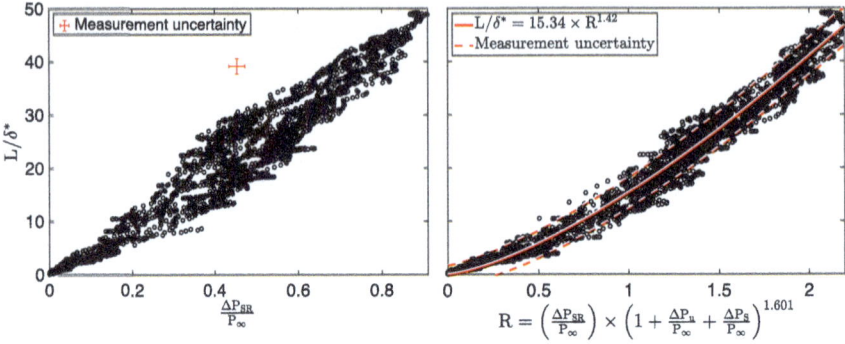

Fig. 8.15 Separation length versus pressure rise from separation to reattachment ΔP_{SR} (left). Separation length versus modified pressure rise (right)

highly 3D separation shapes. The primary effect on the separation length is the pressure rise from separation to reattachment. Compression waves arriving over the interaction increase this pressure rise and thus the separation length. Expansion waves have the opposite effect. Pressure waves arriving upstream of the interaction ΔP_u have a secondary effect on the separation length. An increase in the upstream pressure causes an increase in the separation length. This might be explained by upstream waves modifying the upstream boundary layer, but further work will be needed to examine this.

These results demonstrate that relatively weak corner waves, as observed in this experiment inducing a maximum pressure rise of 18% of the inviscid oblique shock reflection pressure rise, can have a significant effect on the separation length—varying between approximately -80 to $+180\%$ of the baseline centre separation length. Furthermore, it is observed in these experiments and in literature that channels of attached flow form between the primary and corner separations. This demonstrates that corner waves have a strong control authority over the separation length and shape, and can even prevent separation. This motivates a flow control method which utilises secondary pressure waves to control the separation. The approach used in this investigation is to produce pressure waves on the surface by contouring it by means of a control bump placed in the location of the primary separation.

8.3.3 Control Bump Results

Time-averaged shadowgraph images of the baseline (uncontrolled) and control bump cases for both test cases—at the UCAM and TUD test facilities—are shown in Fig. 8.16. The separation and reattachment shocks, as well as the (apparent) downstream boundary layer edge observed in the baseline cases are overlaid to highlight the

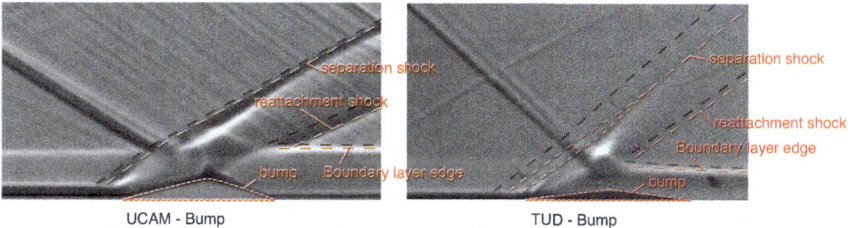

Fig. 8.16 Control bump time-averaged Shadowgraph images with key shock waves and the apparent downstream boundary layer edge highlighted to compare between corresponding baseline and bump cases, with black and red dotted lines used for the baseline and control bump cases respectively

changes due to the presence of the bump. In the controlled cases, the corresponding lines are red-dotted.

The controlled and uncontrolled shadowgraph images look qualitatively similar—indicating that the presence of the bumps does not drastically alter the outer flow field. However, the separation and reattachment shocks move markedly downstream in both control bump cases, and the downstream boundary layer appears slightly thinner, which suggests a possible positive effect of the control bump. However, the downstream boundary layer profile will need to be examined in detail from PIV data to confirm any favourable effects.

Oil flow visualisations of the baseline and control bump cases for both investigations are shown in Fig. 8.17. In the baseline cases, large separation regions are visible. In the UCAM control bump case, a significant portion of the separation over the bump has been eliminated, but smaller separations on the sides and rear persist. In the TUD bump case, the separation is almost entirely eliminated, except for a small separation on one side of the bump. This suprising result demonstrates that control bumps have the potential to eliminate flow separation in oblique SBLIs.

The separation shock foot in each case is visible, as an accumulation of oil-dye mixture, upstream of the separation/control bump. The TUD baseline shock foot is transposed onto the corresponding bump case. It is clear that the shock foot has moved markedly downstream, almost to the bump leading edge, in the controlled case. This agrees with the downstream shift of the separation shock observed in the mean shadowgraph images. The shock foot also appears to bend, following the contour of the bump foot, whereas the uncontrolled shock foot appears relatively straight. This may account for the apparent split shock foot observed in the time-averaged shadowgraph image of the TUD bump case.

The low frequency unsteadiness of the interaction is examined using the high speed shadowgraph images. Pixel boxes were extracted in the vicinity of the separation shock foot for the baseline and control bump test cases. They are skewed to match the local shock angle, so that the shock section is vertical in the unskewed window. These windows are then averaged vertically to a single pixel height. This process is shown in Fig. 8.18. The power spectral density of each pixel is calculated using the

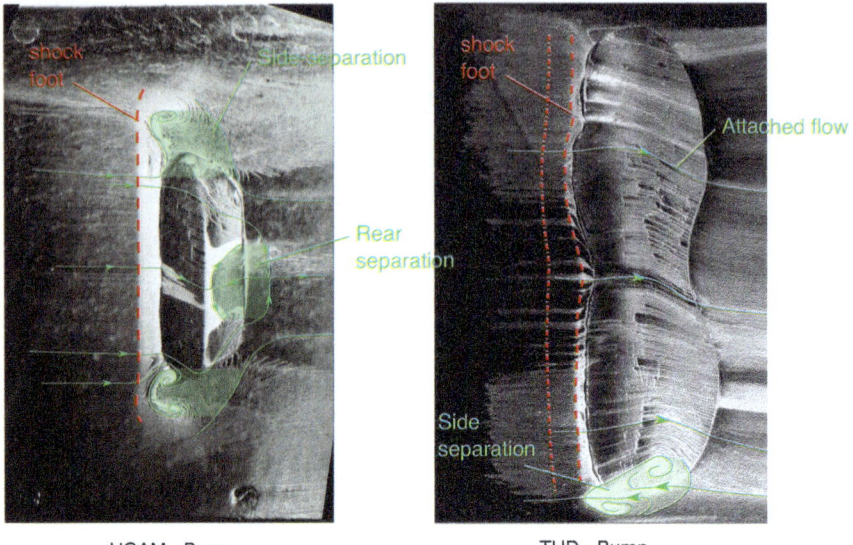

Fig. 8.17 Shock control bump oil flow visualisation of the UCAM and TUD test cases

method described in Sect. 8.2.4. The premultiplied power spectral density $f \times G(f)$ maps of these strips are shown for each case in Fig. 8.18.

In the baseline test cases, low frequency peaks are visible. In the UCAM case, this peak occurs between $St_L = 0.01-0.02$. This is somewhat lower than the expected non-dimensional frequency for the low frequency breathing mode $St_L \approx 0.03$ [11]. In the TUD case, the peak appears to be more spread out $St_L = 0.04-0.09$ and extends to a frequency somewhat higher than expected.

In the UCAM control bump case, the low frequency peak is still visible but appears notably reduced in intensity. The persistence of this peak may be due to the presence of separated regions on the sides of the bump. In the TUD control bump case, no low frequency peak is observed. This indicates that the low frequency oscillation has been suppressed.

Centreline mean and RMS velocity profiles, extracted from PIV data at two locations downstream of the interaction, are shown for the TUD baseline and control bump cases in Fig. 8.19. At the nearest downstream location (0.6δ downstream of reattachment), the control bump boundary layer profile appears to be fuller with a maximum velocity increase of 6.7%. At the furthest downstream location, the profiles appear to collapse once again.

The controlled case exhibits a decreased RMS intensity in the downstream boundary layer, which persists relatively far downstream, beyond $6.4\,\delta$ from reattachment. The maximum decrease in RMS intensity in the furthest downstream profile was approximately 27%. This decreased RMS intensity may be a further consequence

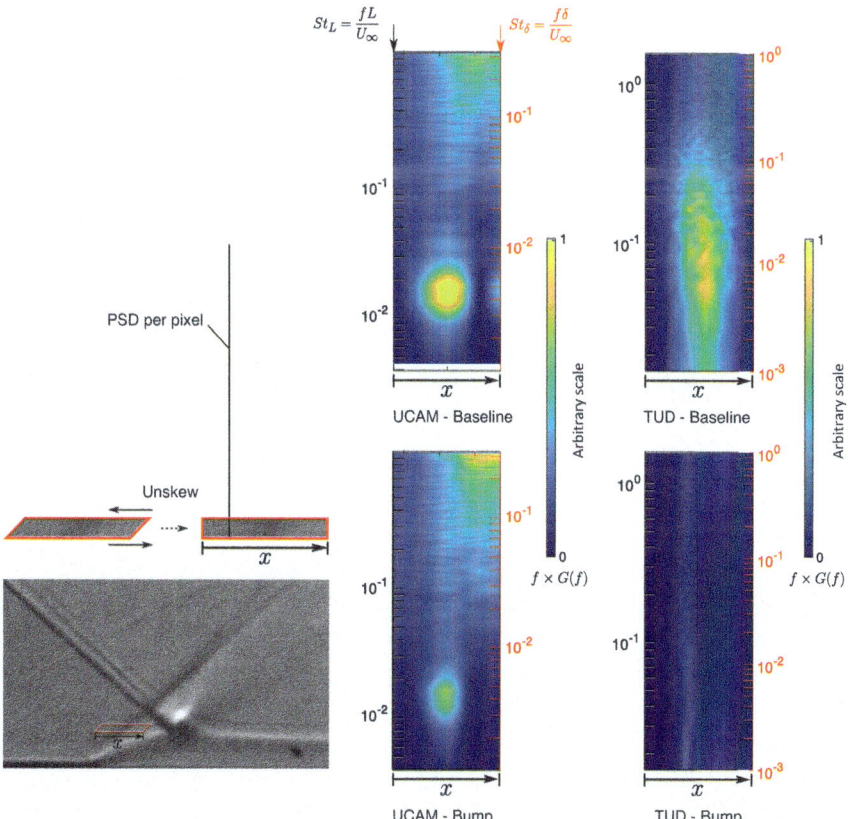

Fig. 8.18 Baseline and control bump high speed shadowgraph separation shock Power Spectral Density (PSD) maps. A skewed pixel box (left) is extracted around the shock foot in each case, unskewed, and averaged to 1-pixel in height. The (frequency premultiplied) PSD is calculated for each pixel and plotted versus streamwise location within the box, and non-dimensional frequencies $St_L = \frac{fL}{U_\infty}$ and $St_\delta = \frac{f\delta}{U_\infty}$

of the bubble breathing mode being dampened/eliminated and it could indicate a favourable effect of the control bump on the downstream boundary layer velocity fluctuations.

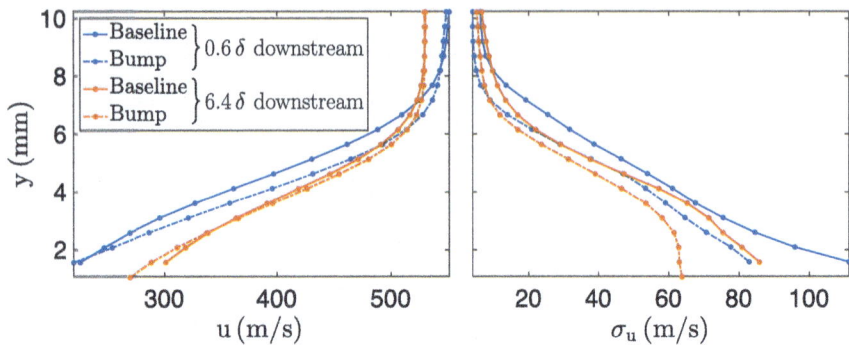

Fig. 8.19 Mean and RMS streamwise velocity boundary layer profiles of the TUD baseline and control bump cases, at two locations downstream of reattachment, from PIV data

8.4 Conclusions

- Experiments were conducted on turbulent oblique SBLIs in two supersonic wind tunnel test facilities (the supersonic wind tunnel no. 1 at the Univeristy of Cambridge—UCAM, and ST15 at TU Delft—TUD) at different flow conditions (Mach 2.5, and 2; and flow deflection angles 8° and 12° respectively).
- At UCAM, conical artificial corner-separation bodies were used to replicate the displacement effect of streamwise corner separations on the interaction. These cones were placed in the streamwise corners of the working section—at the intersections of the tunnel floor and side walls. They produce compression waves along the leading edge, followed by expansion waves emanating from the mid-chord, which turn the flow away from and then towards the streamwise corners respectively.
- When tested without an incident shock wave, corner cones produced a relatively gentle and almost linear pressure rise along the tunnel floor centreline, reaching a maximum of 18% of the inviscid shock reflection pressure jump. This is followed a pressure drop below the freestream pressure, due to the corner expansion waves. Unlike block shaped bodies which have been used previously [3], the conical bodies clearly separate the corner compression and expansion waves, so that their effects on the primary separation can be more clearly identified.
- The corner cones were placed in a range of streamwise locations with respect to the incident shock wave in order to modify the location of the induced corner waves and thus induce a varying influence on the interaction. The varying corner waves resulted in a significant variation in the primary separation topology and length—between approximately −80 to +180% of the baseline centre separation length.
- Streamwise strips were taken through the separation region at a range of spanwise positions to examine the local separation length and pressure rise. It was found that the separation pressure rise ΔP_S was the same, regardless of the local separation length (over a wide range of separation shapes).

- The local separation length was found to depend primarily on the pressure rise from separation to reattachment. Corner compression and expansion waves which arrive over the separation region can therefore alter the local separation length by modifying this pressure rise.
- Corner waves arriving outside the separation had very little effect on the separation length except for a subtle (higher-order) influence on the relationship between the pressure jump and separation length, when waves arrived ahead of the interaction. One possible explanation for this is that the boundary layer profile is altered, but this needs more work to prove.
- The strong control authority of corner waves on the separation length informed the design of shock control bumps which produce pressure waves on the floor surface instead of in the streamwise-corners. Control bumps were designed to match the mean separation bubble shape, with the goal of reducing/eliminating separation, and dampening/eliminating the low frequency separation bubble breathing oscillation. Testing was performed in both test facilities, with the bumps designed for the respective flow conditions.
- In the first investigation, at UCAM, the bump design relied on approximating the mean dividing streamline shape from previous investigations at slightly different flow conditions. This control bump was also produced as a rapid prototype with relatively low tolerance. Thus it was a rough approximation of the mean separation bubble shape. Nevertheless, this control bump was effective in keeping most of the flow attached, except for relatively small regions on the downstream face and on either side of the bump.
- In the second investigation, at TUD, the bump was designed to match the mean separation bubble shape more closely, and with higher tolerances. The mean dividing streamline was approximated from previous PIV data obtained in the same facility with the same flow conditions. This bump almost completely eliminated the shock induced separation.
- High speed shadowgraph was used to examine the unsteadiness of the separation bubble by analysing the fluctuating pixel intensities in a box containing the separation shock. Premultiplied power spectral density maps of these intensities revealed the low frequency bubble-breathing mode in the baseline (uncontrolled) test cases, in both experimental investigations. In the UCAM investigation, the control bump significantly reduced the intensity of this low frequency mode, but did not eliminate it. In the TUD investigation, the control bump apparently eliminated this mode.
- Mean streamwise velocity profiles indicated a slightly more full velocity profile immediately downstream $(0.6\,\delta)$ of reattachment in the controlled case (with a peak velocity increase of 6.7%). However, further downstream $(6.4\,\delta)$ the profiles were almost identical. The RMS streamwise velocity was reduced in the downstream boundary layer, by up to 27%, at $6.4\,\delta$ downstream of reattachment. This indicates a possible favourable effect on the mean boundary layer profile immediately downstream of the interaction, as well as a reduction of streamwise velocity fluctuations relatively far downstream of the interaction.

References

1. Délery, J.M.: Shock wave/turbulent boundary layer interaction and its control. Prog. Aerosp. Sci. **22**(4), 209–280 (1985)
2. Babinsky, H., Harvey, J.K.: Shock Wave-Boundary-Layer Interactions. Cambridge University Press (2011)
3. Xiang, X., Babinsky, H.: Corner effects for oblique shock wave/turbulent boundary layer interactions in rectangular channels. J. Fluid Mech. **862**, 1060–1083 (2019). https://doi.org/10.1017/jfm.2018.983
4. Babinsky, H., Oorebeek, J., Cottingham, T.: Corner effects in reflecting oblique shock-wave/boundary-layer interactions. In: 51st AIAA Aerospace Sciences Meeting including the New Horizons Forum and Aerospace Exposition, p. 859 (2013)
5. Titchener, N., Babinsky, H.: A review of the use of vortex generators for mitigating shock-induced separation. Shock Waves **25**(5), 473–494 (2015)
6. Clemens, N.T., Narayanaswamy, V.: Low-frequency unsteadiness of shock wave/turbulent boundary layer interactions. Ann. Rev. Fluid Mech. **46**, 469–492 (2014)
7. Bruce, P., Colliss, S.P.: Review of research into shock control bumps. Shock Waves **25**(5), 451–471 (2015). https://doi.org/10.1007/s00193-014-0533-4
8. Piponniau, S., et al.: A simple model for low-frequency unsteadiness in shock-induced separation. J. Fluid Mech. **629**, 87–108 (2009)
9. Squire, L.C.: The motion of a thin oil sheet under the steady boundary layer on a body. J. Fluid Mech. **11**(2), 161–179 (1961)
10. Ragni, D., et al.: Particle tracer response across shocks measured by PIV. Exp. Fluids **50**(1), 53–64 (2011). https://doi.org/10.1007/s00348-010-0892-2
11. Dussauge, J.-P., Dupont, P., Debiève, J.-F.: Unsteadiness in shock wave boundary layer interactions with separation. Aerosp. Sci. Technol. **10**(2), 85–91 (2006)

Open Access This chapter is licensed under the terms of the Creative Commons Attribution 4.0 International License (http://creativecommons.org/licenses/by/4.0/), which permits use, sharing, adaptation, distribution and reproduction in any medium or format, as long as you give appropriate credit to the original author(s) and the source, provide a link to the Creative Commons license and indicate if changes were made.

The images or other third party material in this chapter are included in the chapter's Creative Commons license, unless indicated otherwise in a credit line to the material. If material is not included in the chapter's Creative Commons license and your intended use is not permitted by statutory regulation or exceeds the permitted use, you will need to obtain permission directly from the copyright holder.

Oblique-Shock Wave Boundary Layer Interactions Control: Shock Control Bumps

Jane Bulut, Ferry Schrijer, and Bas van Oudheusden

Abstract In this experimental study the effect of three-dimensional shock control bumps (SCB) on oblique shock wave/boundary layer interactions is investigated as a passive control method. It aims to develop an understanding of the effect of such devices on the interaction structure by means of studying the influence of the position of the bump with respect of the shock-impingement location, as well as the effect of the bump geometry (more in particular, the ramp section of the bump). The experiments were conducted in the ST-15 wind-tunnel at the Delft University of Technology for fully developed turbulent boundary layer conditions with Re_θ of $21.8 \cdot 10^3$ and freestream Mach number of 2.0. The control effectiveness is assessed from the size of the separated flow region, as well as the downstream boundary layer velocity profile. For this, PIV is employed as the main diagnostic method to characterise the flow field. In addition to this, high-speed Schlieren and oil flow measurements were performed to asses the effect of the SCB on the overall interaction structure.

Keywords Supersonic flow · Shock induced separation · Flow control · Particle image velocimetry · Measurements

---

J. Bulut · F. Schrijer (✉) · B. van Oudheusden
Delft University of Technology, Delft, The Netherlands
e-mail: F.F.J.Schrijer@tudelft.nl

J. Bulut
e-mail: J.Bulut@tudelft.nl

B. van Oudheusden
e-mail: B.W.vanOudheusden@tudelft.nl

1 Introduction

The occurrence of shock waves is a prominent feature of high-speed flight. There are many potential consequences of the shock wave formation for the vehicle performance. For instance, adverse pressure gradient caused by shock impingement on the boundary layer can result in a partly flow reversal in the boundary layer flow. Such separation bubbles typically cause a drastic increase in drag while leading to high total pressure losses. Moreover, they introduce high levels of unsteadiness in the flow.

Over the years, developing control methods to mitigate the undesired effects of shock wave boundary layer interactions (SWBLI) has been in the focus of many researchers. Most of these methods aim to act on the low momentum region of the boundary layer. Some, such as boundary layer bleed, aim to mitigate the negative effects by removing the low momentum portion of the boundary layer, thus making it fuller in shape and hence more resistant to separation. Others target to energize the boundary layer by increased mixing. This can be achieved by introducing counter rotating vortices; hence, such devices are referred to as vortex generators. Especially, sub-boundary layer vortex generators have been widely investigated over the years due to their efficiency in promoting the momentum of the boundary layer without introducing strong adverse effects on drag. They have been shown to be effective in controlling SWBLI in a variety of applications, including aircraft wings, engine inlets, and hypersonic vehicles.

An alternative method to alleviate the unfavourable effects of the shock wave boundary layer interactions is acting on the formed shocks to reduce their strength. The strength of a shock is commonly defined as the pressure jump across it. The entropy increase across the shock wave is related to the total pressure drop and the shock strength. A decrease in the latter one will result in a decrease in entropy and an increase in total pressure drop. Therefore, shock control techniques in general aim at replacing a strong shock by a weaker one following upstream isentropic compression waves. Due to their geometry, shock control bumps (SCBs) suit well for this purpose when they are placed underneath the interaction. They have been used to reduce the wave drag of transonic wings, especially by modifying the shock structure [1–3]. The bump geometry generates quasi-isentropic compression waves upstream of the normal shock wave resulting in a λ-shock configuration. The flow passing through these isentropic compression waves gradually decreases its velocity; thus, the bump partly eliminates the abrupt effects caused by a single strong normal shock wave.

Ashill et al. [4] introduced the concept of 2D SCBs to reduce the wave drag of the airfoil. Subsequently, various studies have assessed the effectiveness of shock control bumps for normal shock wave boundary layer interactions control. Despite the promising results, SCBs have been found to have a quite narrow effective operational range. Moreover, in off-design conditions, they might even have negative effect on performance. The use of an array of 3D SCBs has been suggested to localize the regions affected by undesired conditions while operating at off-design conditions [5, 6]. Consequently, the effective operational range of the bump is aimed to be extended. Investigations on 3D SCBs indicated strong evidence for streamwise vortex pair formation downstream of the ramp part of the bump [7–9].

In their investigation of the performance of different 3D bump geometries, Ogawa et al. [10] observed the dependence of the shock structure in relation to the impingement location on the bump surface. It was found that when the shock impinges on the upstream part of the bump, re-accelerated flow can create a "supersonic tongue" downstream and thus undesired secondary shock structures and an increase in the wave drag might occur [10]. In contrast, downstream impinging shock can cause a secondary λ-shock structure formation due to the expansion of the flow passing over the crest of the bump. This type of interaction would result in an increase in the total pressure loss [10]. From this, one can conclude that the shock impingement position has a significant importance for mitigating the stagnation pressure losses.

Whereas most studies on the effects of SCBs have involved normal shock wave boundary layer interactions, relatively limited research has focused on the flow structures generated by the combination of oblique SWBLIs and SCBs. Given the shape of the SCB, its effect is expected to downsize the separation by "filling" it with the bump structure. This is also expected to restrict the dynamics of the flow separation region, thereby reducing the overall unsteadiness of the interaction. In addition, the streamwise vortices introduced into the flow downstream of the interaction by the 3D SCB would promote a faster recovery of the boundary layer flow. Moreover, the gradual flow expansion caused by the tail portion of the bump also contributes to a faster flow recovery and providing a fuller velocity profile compared to the uncontrolled case.

In general, the use of SCBs is a promising approach for controlling oblique SWBLIs. Introducing a SCB geometry in the interaction region directly modifies the interaction structure such that the undesired consequences of SWBLI could diminish. This chapter represents a brief summary of the investigation of the effect of the ramp angle on the interaction structure and flow physics as well as the effect of the incident shock impingement location on the bump effectiveness.

2 Experimental Arrangement

2.1 Flow Facility and Experimental Investigation

The experiments were carried out in the ST-15 blow-down supersonic wind tunnel of the Delft University of Technology. The test section of the tunnel has dimensions of 150 mm × 150 mm and has glass windows in the side walls that allow optical access. In the experiments, the tunnel is operated at a Mach number of 2, a total pressure P_0 of 3 bars and a total temperature T_0 of approximately 290 K. In Giepman et al. [11] investigated the effects of location and size of micro-ramp vortex generators in the same facility at similar operating conditions and have also documented undisturbed

Table 1 Experimental conditions and undisturbed boundary layer properties [11]

Parameter	Value	Parameter	Value
M_∞	2.0	δ_i	0.63
U_∞ (m/s)	520	θ_i	0.52
P_0 (N/m^2)	3×10^5	H_i	1.23
T_0 (K)	290	Re	42×10^6
δ_{99} (mm)	5.2	Re_θ	21.8×10^3

boundary layer parameters. A summary of the main flow conditions is given in Table 1.

The bottom wall of the tunnel, where the boundary layer thickness is approximately $\delta_{99} = 5.2$ mm [11], is used to assess the uncontrolled SWBLI flow field. Subsequently, a shock control bump is installed on this wall by using double sided adhesive tape. For all the cases, the incident shock is generated by installing a 12° shock generator.

2.2 Shock Control Bump Specifications

Introducing a SCB geometry in the interaction region directly modifies the interaction structure (see Fig. 1). A typical SCB consists of a ramp, a short crest and a tail part. As mentioned before, in a highly supersonic flow the initial ramp part of the bump acts as a compression ramp that introduces a secondary shock into the interaction. This secondary shock, which originates at the leading edge of the bump, decelerates the flow upstream of the impinging shock. This could alleviate the detrimental effects of a strong impinging-separation shock system and possibly even removing any separated region from the flow.

For the investigation of the effect of the bump geometry, the shock control bump ramp angle is chosen as the main parameter. In case the bump geometry has a high ramp angle it would cause a strong compression corner interaction at the leading edge of the bump and this might induce an undesired separation of the flow. Therefore, it is important to know the maximum shock intensity that the incoming flow boundary layer could withstand without separating. This limits the choice of the ramp angle.

Additionally, the SCB ramp angle for the baseline geometry is chosen to minimize the undesired effects of the shock wave in the flow. It is expected to have a total pressure loss when a shock forms in a supersonic flow. Therefore, it is crucial to optimize the modified shock system for minimizing the total pressure loss. Interpreting the SCB leading edge shock and the separation shock of the SWBLI as similar set of oblique shocks as occurs in a supersonic inlet, ensuring the minimization of the total pressure loss would be achieved by a SCB geometry where the resulting leading edge shock and the following separation shock would have same strength. In other

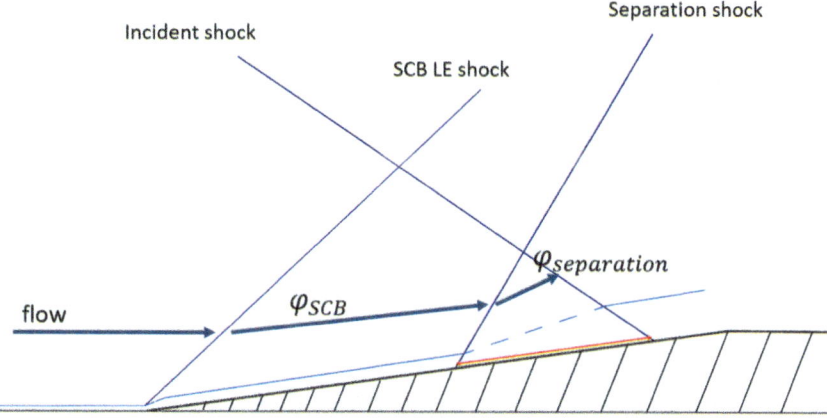

Fig. 1 Schematic representation of the interaction structure with the bump geometry

Table 2 Geometrical parameters of SCBs

Configuration	Height (mm)	θ_{ramp}	θ_{tail}
r4_h5	5	4°	6.25°
4p6_h5	5	4.6°	6.25°
r5p4_h5	5	5.4°	6.25°
r6p25_h5	5	6.25°	6.25°
r7p25_h5	5	7.25°	6.25°

words, by introducing the bump geometry in the flow, the separation shock is aimed to be broken down into two oblique shock waves with equal strength. Therefore, a baseline geometry is designed such that the leading edge shock would be in the same strength of the separation shock wave in the modified interaction structure. The dimensions for the different SCB geometries are shown in Table 2. Overall 5 different bump geometries are tested where the tail portion of the bump and the height of the crest is kept constant and the ramp angle, θ_{ramp}, is set in different values around baseline angle.

Bruce and Colliss indicated in their review [12] that most of the research on SCB agrees that the optimum performance is achieved when the bump height is of the order of the boundary layer thickness. While designing all the geometries presented in this chapter, this guideline has taken into consideration.

The second part of the investigation is focused on the effect of the shock impingement location on the shock control bump effectiveness. he shock impingement location on the bump surface has been introduced as another crucial criteria to determine the efficiency of the bump. Ogawa et al. [10] defined an optimum location for shock impingement as the one that results in the largest λ-shock structure without introducing secondary shocks. When the shock is positioned further upstream of this location

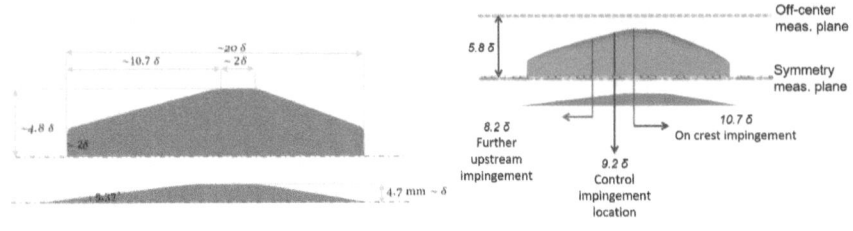

(a) *Geometrical parameters of SCB.* (b) *Measurement planes.*

Fig. 2 Shock Control Bump (SCB)

it is expected to introduce a secondary shock system as a result of the re-expansion of the flow after the initial shock interaction. On the contrary, when the shock impinges downstream of the defined optimum location, the supersonic flow region behind the λ-shock foot accelerates due to the expansion over the bump crest which can lead to secondary shock structures as well as flow separation.

For this part, three inviscid impingement locations on the bump were defined (see Fig. 2b). The height at the bump crest has been selected approximately as δ_{99} and the ramp angle is set approximately to 5.7°. A first control impingement location is defined 1.5 δ upstream of the first point where the bump has a maximum height. This corresponds to 9.2 δ distance from the leading edge of the bump. Additionally, impingement locations 1.5 δ downstream and 1 δ upstream of this point are investigated. These will be referred to as downstream impingement location and upstream impingement location in the following of the discussion. The incident shock impingement locations were decided regarding the previously mentioned studies which have been conducted on normal shock-boundary layer interactions [7, 13]. In this way, the sensitivity of the shock control bump location for flow control can be investigated.

3 Results

3.1 Interaction Structure with and Without SCB

In Fig. 3 images obtained from Schlieren visualization show the global features of the uncontrolled and controlled interactions. The incident shock is generated by a 12° shock generator and reflected shock is formed approximately 6δ upstream of the impinging location of the incident shock on the wall of the uncontrolled interaction. The small distance caused by the insert on the tunnel wall results in very weak two consecutive mach waves upstream of the interaction, where it is confirmed that they have no significant effect on the free stream flow. Since the leading edge of the bump is covering a part of the more downstream side of the insert, second mach wave is losing its strength in the controlled interaction. Downstream of the interaction, it can

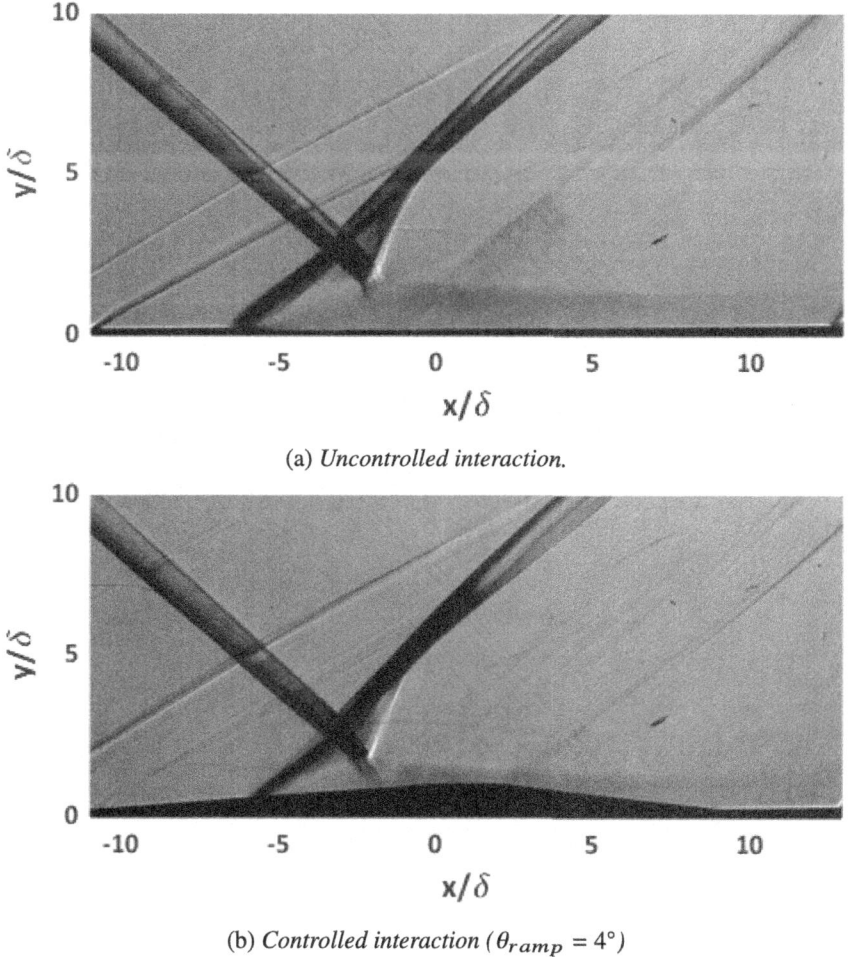

(a) *Uncontrolled interaction.*

(b) *Controlled interaction* ($\theta_{ramp} = 4°$)

Fig. 3 Average flow field obtained from Schlieren measurements [14]

be seen that the boundary layer is growing in thickness, as expected. An expansion fan formation which caused by the expansion of the flow that crosses over the separation bubble in the uncontrolled interaction is observed as a bright region attached to the separation shock. In the Fig. 3a formation of a weak reattachment shock can be seen. It originates from where the inviscid incident shock would have impinged on the wall and bounds the separation bubble.

In Fig. 4 the scheme of the interaction with and without bump is plotted over the wall normal velocity contour. Representative streamline shows the significant flow deflection for the uncontrolled case. It can be seen that strong adverse pressure gradient due to the shock impingement leads to the separation of the flow. The compression caused by the separated shear layer leads to a second shock formation

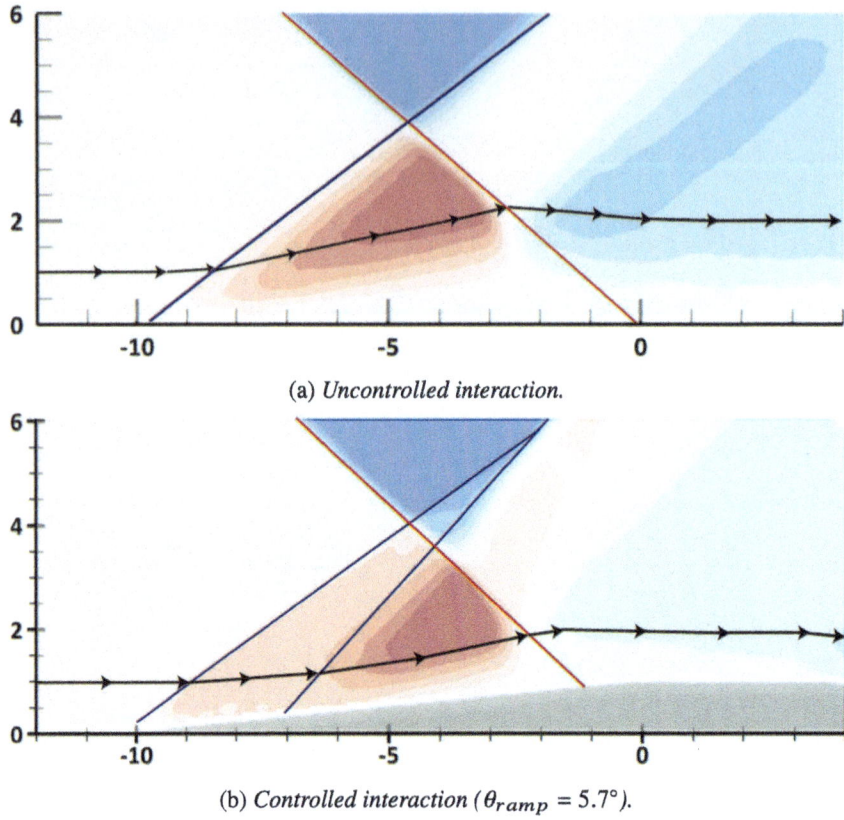

Fig. 4 Interaction structure with and without bump

that is called the separation shock and the flow deflection obtained from streamlines across the separation shock foot is approximately 9.2°. An inviscid flow analysis for the shock reflection at a free stream Mach number of 2.0 and an incident shock formed by a 12° shock generator, predicts a weak solution for a strong interaction with pressure rise across the incident shock P_i/P_0 as 1.89, and the pressure rise over the entire interaction to be approximately $P_i/P_0 = 3.48$. When the measured flow deflection obtained through the velocity field is taken into account in this estimation the value of pressure rise results approximately as $P_i/P_0 = 3.2$. As can be clearly seen in Fig. 10b, flow is exposed to a deflection before reaching to the main interaction when the bump is introduced. For the bump with the ramp angle of 4.6°, estimation of overall pressure rise obtained using the measured flow deflection results in $P_i/P_0 = 1.45$ when passed the reflected shock foot compared to uncontrolled case $P_i/P_0 = 1.64$.

3.2 Effect of the Shock Impingement Location

In Fig. 5 the streamwise velocity component for the uncontrolled case and the control impingement location case are compared, respectively. Measurements are taken in the the symmetry plane of test section and therefore the flow field in the spanwise direction. The region of the immediate vicinity of the bump surface (0.1δ) is blanked out as the reflections did not allow to have a proper measurement. For the same reason, the region following the trailing edge of the bump is also blanked out.

In the uncontrolled case, the familiar mean features of the shock wave boundary layer interaction are observed. The incident shock wave impinges on the boundary layer up to the point where it reaches the sonic line. In the Fig. 5, reversed flow is observed in the mean flow field of the uncontrolled case. In contrast, no separation bubble is observed in the case with bump, when the average flow properties are taken into the consideration.

Overall the interaction structure for the uncontrolled and controlled cases exhibits similar flow characteristics such as the thickening of the boundary layer resulted by an impinging shock wave and formation of the reflected shock. In addition, introducing the bump geometry in the interaction region caused compression of the flow which resulted as compression waves merging as a weak shock wave at the leading edge of the bump. Moreover, the expansion waves introduced by the tail part of the bump helps the boundary flow recover faster. A similar behaviour is also observed in the

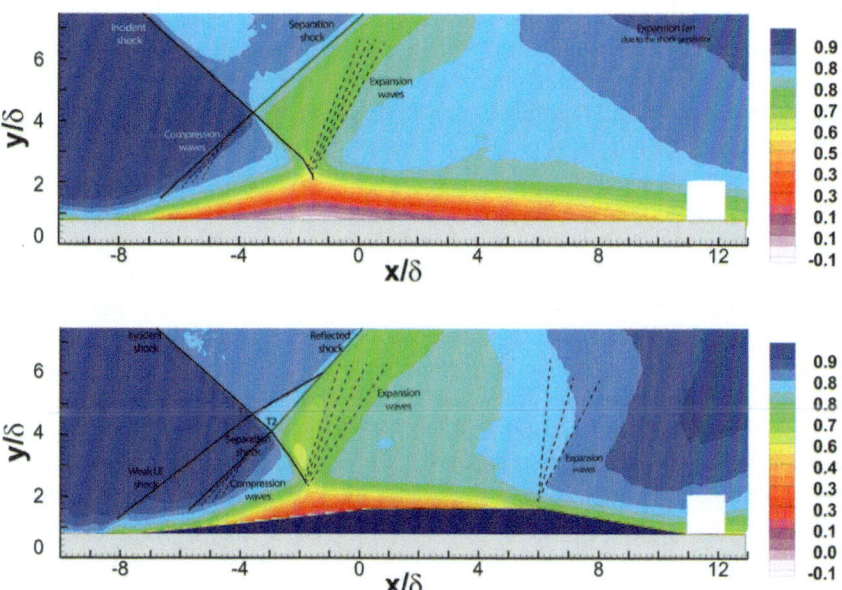

Fig. 5 Average flow field for without control and reference impingement cases (contours represent the streamwise velocity magnitude) [15]

upstream impingement case, the velocity profile over the tail portion of the bump is almost fully recovered. On the other hand, the downstream impingement case demonstrates a steeper velocity profile. Reference and upstream impingement cases reach a fuller boundary layer profile already on the tail portion of the bump. This can be explained by the fact that the interaction flow has been impacted less unfavourably, compared to the downstream-impingement case.

Vertical velocity contour of the averaged flow field is complemented with the streamlines obtained from two dimensional velocity gradients (see Fig. 6). Streamlines don't indicate any reversed flow for the controlled cases, while the streamline layout for the controlled case clearly indicate a flow reversal. In addition, formation of a weak shock originating from the leading edge of the bump is clearly seen especially in the reference impingement location and the downstream impingement location cases. The ramp face of the bump deflects the upstream supersonic flow, hence it is expected to see shock wave formation. When the streamlines are followed, it can be seen that flow passing this shock wave changes its direction upwards. In Fig. 6b and c, respectively the reference and downstream impingement cases, the direction change of the streamlines can be clearly seen. This might be interpreted as, when the shock impinges sufficiently downstream of the leading edge, the effect of this weak shock can be seen in the flow. At the location where leading edge shock meets the incident shock an additional shock-shock interaction occurs in the flow.

Images obtained through high speed Schlieren measurements are used to understand the effect of bump and the shock impingement location on the flow dynamics, in particular the reflected shock unsteadiness. The first step in the analysis of the shock position dynamics is to determine the shock location in each subsequent image. An appropriate interrogation window for the shock detection is selected for each configuration. Both selection windows are placed in the outer flow. Due to the knife-edge orientation it is expected that shocks appear darker in the picture. This corresponds to lower intensity levels in the picture. Schlieren measurements integrate the density gradients over the full span of the test section and some three-dimensional effects, such as side-wall effects, influence the final image. This results in that the minimum intensity pixels cover an interval, corresponding to the appearance of a thick dark line in a Schlieren image. Therefore, the elements that correspond to minimum intensity pixels in each row in the selected window are identified as intervals. To avoid the contamination by noise in the detection process, a threshold is applied for the minimum intensity pixel search and only the largest intensity interval is taken into account for the detection of the shock wave. Once shock points in the selected window are detected, through applying a linear fit the shock line is obtained. Incident shock is expected to behave steady in the outer flow. When the obtained fit is reflected on the wall, inviscid impingement location of the incident shock is obtained. As it is expected, the mean angle and deviation in the impingement point shows little to no difference between the cases. To assess the effectiveness of SCB on reducing the unsteadiness of separation shock, the shock movements are analyzed in terms of both streamwise displacement and change in the shock angle. The results are shown in

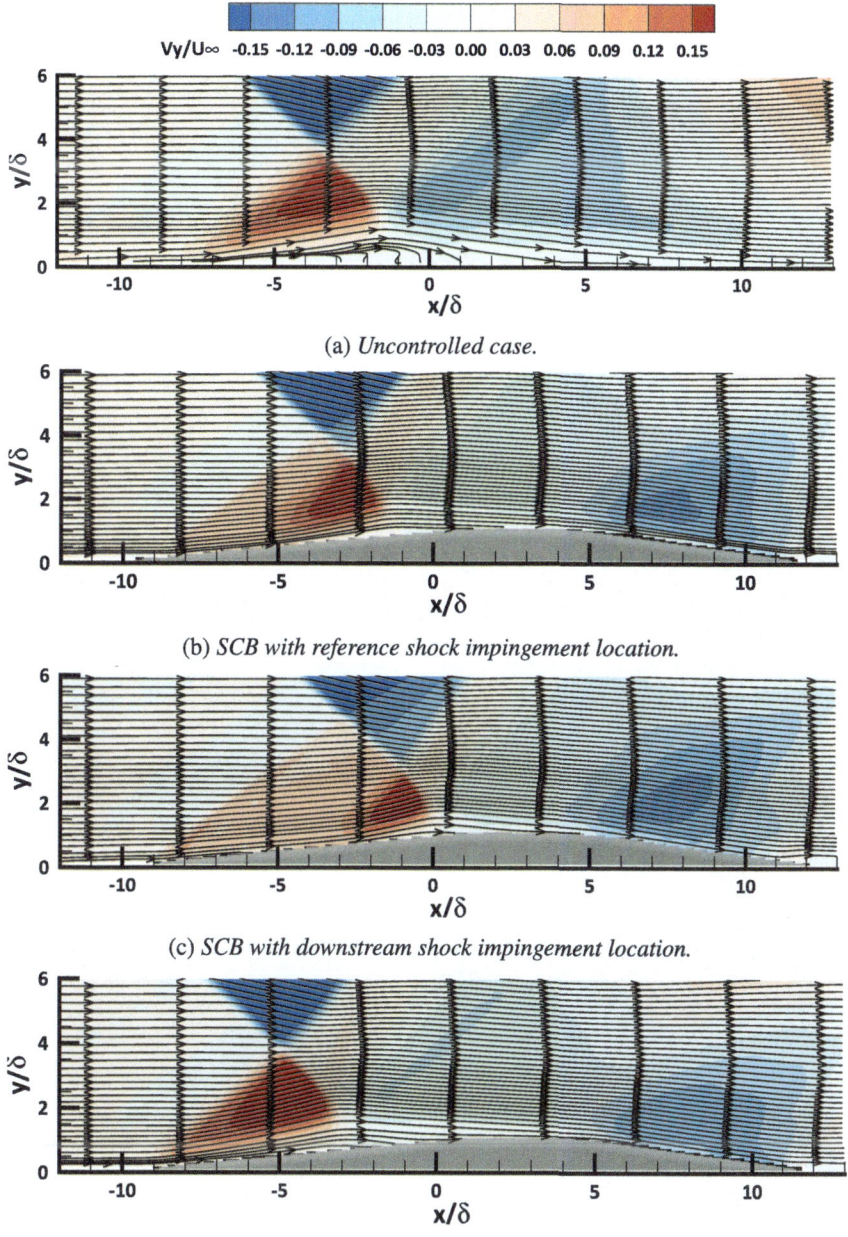

(a) *Uncontrolled case.*

(b) *SCB with reference shock impingement location.*

(c) *SCB with downstream shock impingement location.*

(d) *SCB with upstream shock impingement location.*

Fig. 6 Wall normal velocity contour of mean flow field with streamlines [16]

Table 3 Incident shock statistical properties (mean shock angle, standard deviation in streamwise displacement)

Cases	Incident shock	
	Mean angle	$Std_{imppoint}$
Uncontrolled	−41.0°	0.04δ
Upstream imp.	−40.8°	0.04δ
Control imp.	−40.7°	0.03δ
Downstream imp.	−40.9°	0.03δ

Table 4 Reflected shock statistical properties (mean shock angle, standard deviation in streamwise displacement and angular displacement)

Cases	Separation shock			
	Mean angle	Std_{rot}	Std_{sw}	$Std_{originpoint}$
Uncontrolled	40.7°	0.60°	0.21δ	0.27δ
Upstream imp.	41.1°	0.45°	0.16δ	0.19δ
Control imp.	41.1°	0.40°	0.18δ	0.18δ
Downstream imp.	41.4°	0.47°	0.20δ	0.20δ

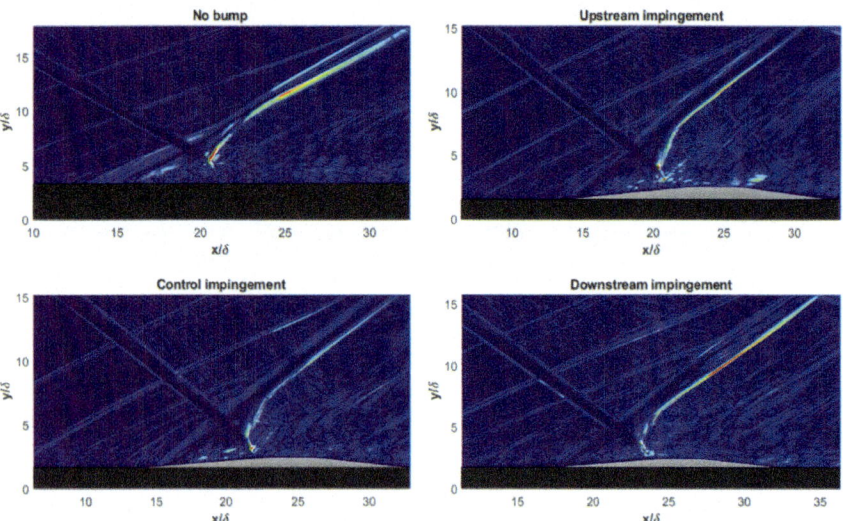

Fig. 7 Standard deviation of the flow field for all cases where the color scale increases from blue to red [15]

the Tables 3 and 4. While the incident shock can be considered as steady, the separation shock exhibits higher oscillatory movements in both streamwise and rotational direction for the uncontrolled case.

In Fig. 7 the overall oscillations in the interaction region for each case is represented by means of the standard deviation of the Schlieren image gray-scale levels. It can be seen that the separation shock oscillations is considerably higher regarding to the cases with the bump installed in the flow. When the inviscid shock impingement location is set as the control point on the bump, the separation shock unsteadiness is reduced significantly particularly on rotational direction. The oscillatory movement of the originating point of the separation shock gives the combined information of the shock movements in both streamwise and rotational direction. While the uncontrolled case exhibits the highest separation shock unsteadiness, in cases with the bump the oscillatory movements of the separation shock is reduced (see Table 4).

3.3 Effect of the Bump Geometry

Surface flow topology with corresponding flow field obtained by Schlieren measurements can be seen in Fig. 8. Both measurements are repeated for all SCB geometries, however only the case with the ramp angle, $\theta_{ramp} = 5.4°$ is represented here. The separation line could be identified by the accumulated oil on a line across the test section span. All images were taken after the wind tunnel runs were completed. Thus, some smearing of the oil due to the shut-off shock can be observed in the images. This has caused the observation of a non quasi one dimensional separation line footprint, especially where the excess oil ad accumulated. Nevertheless, the oil flow visualization for the uncontrolled case results in relatively 2D separation and reattachment lines even though the corner effects and the strong interaction do not allow to have a fully 2D interaction over the span. Moreover, the controlled case shows strong evidence on the highly 3D interaction induced by the 3D shape of the bump geometry. Separation and reattachment lines could be identified in the uncontrolled parts of the test section in the case with bumps, while a conclusive interpretation of the oil

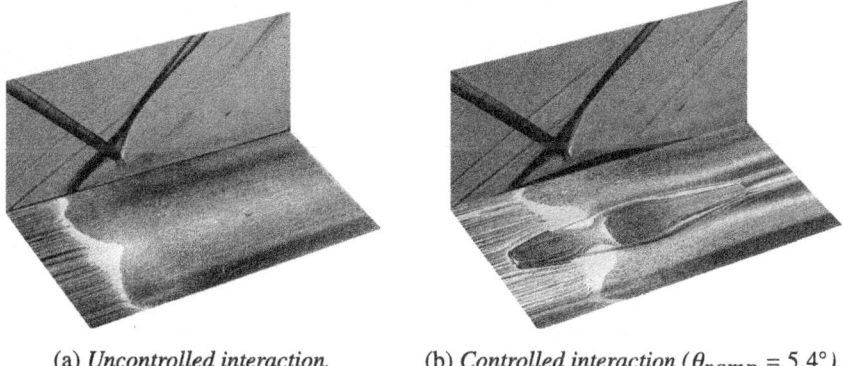

(a) *Uncontrolled interaction.*   (b) *Controlled interaction ($\theta_{ramp} = 5.4°$).*

Fig. 8 Oil flow and Schlieren visualization for uncontrolled and controlled case [14]

pattern over the bump surface is less straightforward. Moreover, the darker regions downstream of the bump in all cases may suggest vortex production in the wake of the SCB. Collis et al. [17] suggested that these vortices originate from the ramp of the bump. This indicates that these vortices might originate from the focal points forming over the referred portion of the bump. Developing a better understanding of the flow topology requires a further analysis with a video recording of the change in oil flow pattern while the wind tunnel is running. Nonetheless, the obtained flow topology indicates strong evidence of the 3D flow structure for the interaction between SCB and SWBLI.

The average flow fields obtained from PIV for vertical velocity components are given in Fig. 10. Streamlines obtained from the velocity gradients are plotted over the mean vertical velocity contour for controlled and uncontrolled cases (see Fig. 10). This confirms that the flow deflection caused by the incident shock is 12°. Nonetheless, the flow deflection obtained from the flow after the separation shock foot is approximately 9°. The interaction structure shows differences depending on the parameters. One could say optimization between the distance between these two interactions plays an important role in the interaction structure hence the effectiveness of the bump. The shock impingement location should be sufficiently away from the corner interaction such that giving enough time to the boundary layer to recover. Therefore, the inviscid shock impingement location is set approximately 1 δ upstream of the start of the bump crest. Additionally, change in the ramp angle modifies the distance between corner and shock incident shock interactions where the maximum height of the bump is set approximately as the upstream boundary layer thickness (5.2 mm).

To understand the effect of this distance on the change of the interaction structure we can look in the two extreme cases, $\theta_{ramp} = 4°$ and $\theta_{ramp} = 7.25°$ (see Figs. 9 and 10). Streamlines show that for the case with $\theta_{ramp} = 4°$, the L_{ramp} allows the flow to change its direction in two steps. Flow passing the bump leading edge is crossing the leading edge shock and directed upwards upstream of the main interaction. While this cannot be seen for the case of $\theta_{ramp} = 7.25°$, Fig. 9 suggest no flow reversal is seen in the average flow unlike the uncontrolled case.

While separation is not present in the average flow field of controlled cases, it can be observed in the instantaneous flow. In order to gather a better insight on the state of the flow separation, the statistical occurrence of local flow reversal is computed for each case. Also called as separation probability [18] is chosen in order to characterize the separation severity because of its practical and less sensitive nature in comparison with identification of the dividing streamline.

In this study, 900 instantaneous images are obtained for each case. From these images the local separation probability is calculated for each point on the selected window on each image. As explained previously, the separation criteria is connected to the occurrence of reversed flow. Figure 11 compares the spatial distribution of the separation probability, P_{sep}, for the bump with different θ_{ramp} as well as for the interaction without bump. The percentages represents time fraction that reversed flow is encountered on that certain location.

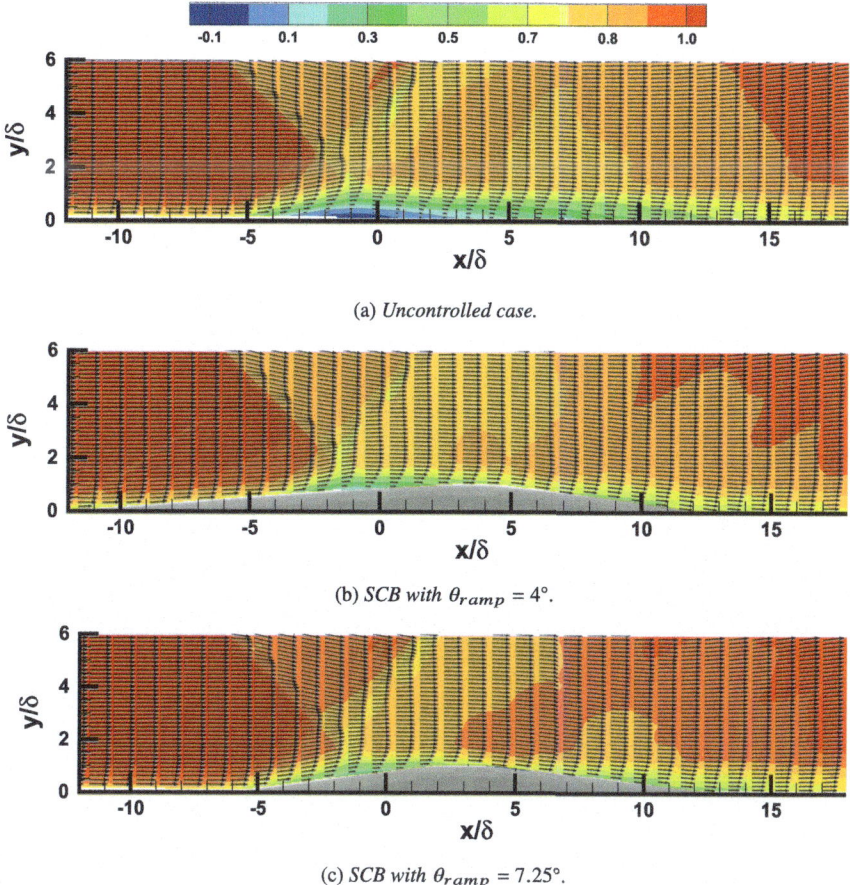

Fig. 9 Velocity vector profiles over the streamwise velocity contour in the interaction region

Without any control method applied, more than 60% of the time flow is reversed in the region adjacent to the wall. Besides this, around 10% of the time reversed flow is observed at the height of approximately 1δ. This maximum height at which reversed flow is encountered in the flow is reduced for all the cases with the bump even though the bump is also contributing to this with its own height. All the cases with a bump exhibit a decrease in the peak value of the P_{sep}. In cases with lower θ_{ramp}, separation is almost removed. For the case with $\theta_{ramp} = 4.6°$, P_{sep} is reduced to maximum of 10% in a quite small region close to the bump surface. Moreover, it could be seen that separation in the cases of interaction with bump is confined in much smaller area.

Figure 12 shows the the higher turbulent intensity levels have been recorded for the uncontrolled case over the shear layer. This could be interpreted as the higher production of the vorticity due to the boundary layer separation. Placement of the SCB has reduced the turbulent intensity for the case with all SCB geometries. Additionally,

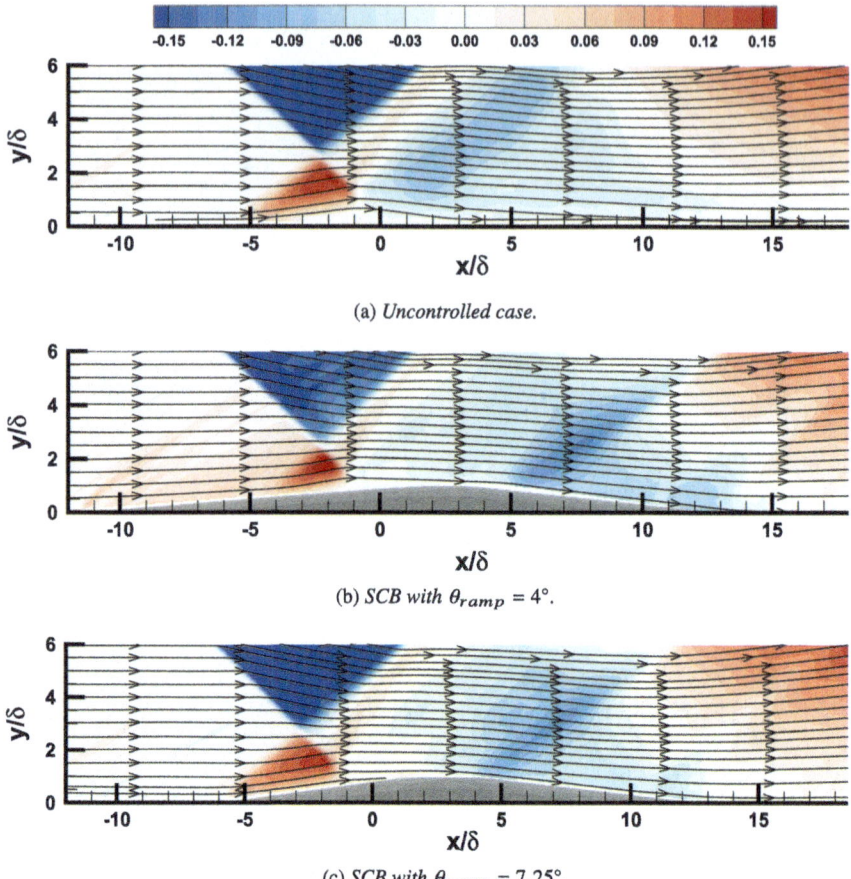

Fig. 10 Streamlines over the vertical velocity contour in the interaction region

the uncontrolled case presents higher turbulence levels over a larger extend in the streamwise direction compared to the cases with the SCB. On contrary, in the cases with the SCB lower turbulence levels are encountered downstream of the interaction. Lower velocity fluctuations could be interpreted as having a more stable velocity profile.

4 Conclusions

In this chapter, a summary of the work done to investigate the interaction between shock control bumps and incident-reflecting type shock wave boundary layer interactions. To do so, an experimental investigation was performed in the facilities of TU

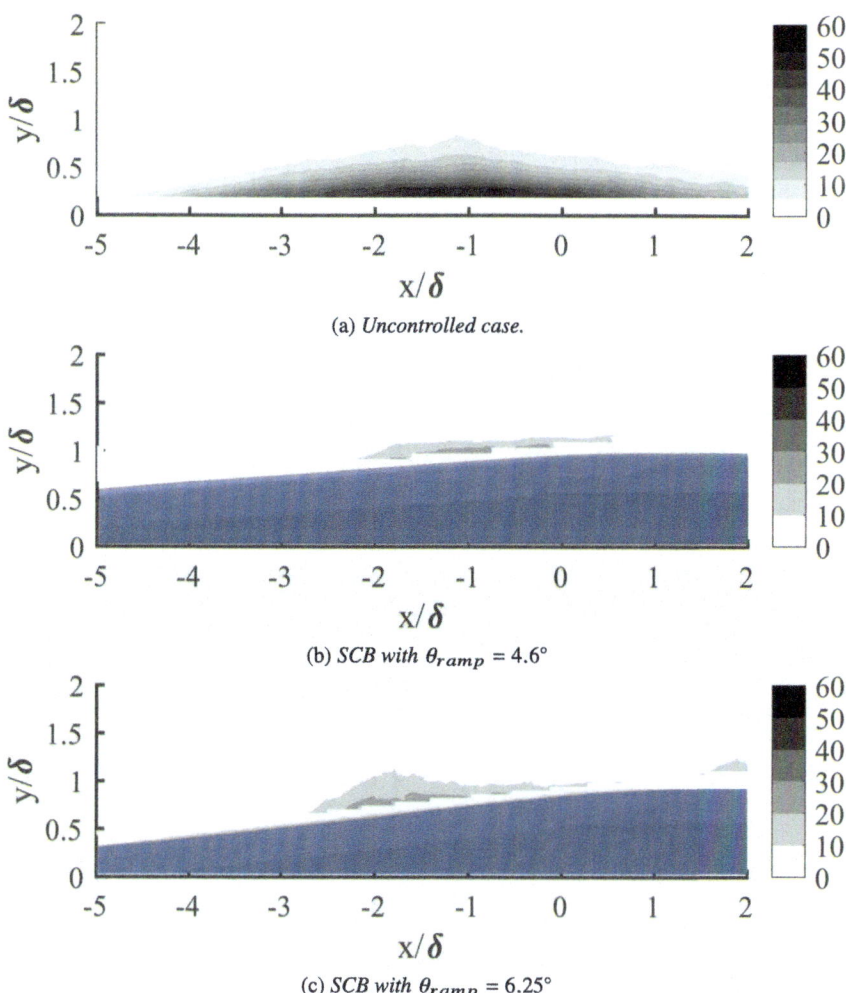

Fig. 11 Separation probability (in %) obtained from instantaneous flow fields

Delft with the freestream flow conditions correspond to a Mach number of 2, a unit Reynolds number Re of 42×10^6, and a momentum thickness Reynolds number Re_θ of 21.8×10^3, while the shock generator's flow deflection angle is set to $12°$.

Initial work carried out to understand the effect of the shock impingement location on the effectiveness of the bump as a control device. High-speed Schlieren measurements are used to investigate the effect of the interaction of SCB and SWBLI on the flow dynamics, in particular the reflected shock unsteadiness. Measurements are performed for three different incident shock impingement locations on the bump surface and for the uncontrolled case. It was found that introducing the bump in the region

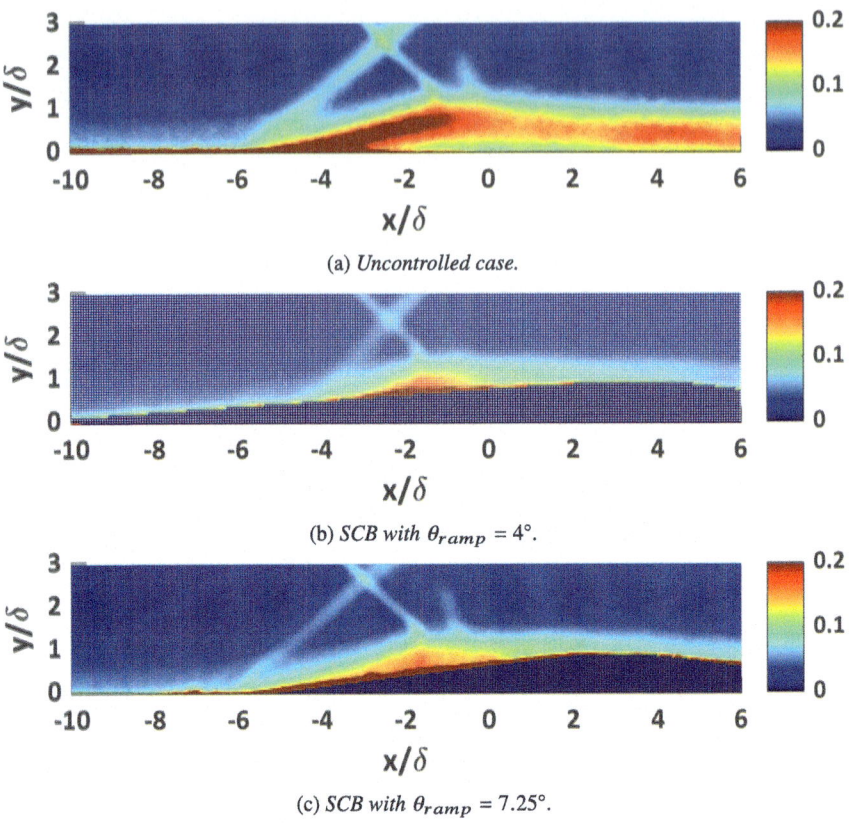

Fig. 12 Turbulent intensity in the interaction region [14]

of the shock wave boundary layer interaction improved the unsteady behaviour of the reflected shock.

Additionally, the effect of bump geometry is studied through Schlieren and PIV measurements. The geometry of bump can be characterized by the design of its ramp and tail part. This investigation is focused on the ramp part and particularly examined the effect of the ramp angle, θ_{ramp}. Overall, 5 different test configurations are derived in addition to the baseline configuration where the uncontrolled interaction is studied. In these test configurations, different geometries of bumps with varying θ_{ramp} and a constant bump height are employed.

Mean flow fields obtained from PIV measurements showed no flow reversal for controlled cases. Moreover, this investigation revealed the distance between the weak compression shock originated from the leading edge of the bump and the main interaction plays a crucial role to optimize the SCB performance. Where the height of the bump is kept constant, cases with the lower ramp angles resulted in lower interaction length values.

Although mean flow did not exhibit any flow reversal for controlled interaction, it could still be observed instantaneously in the flow field. Therefore, the concept of separation probability is used to get further information on the evolution of separation for each case. Lowest maximum P_{sep} is reached where the bump with a ramp angle of $\theta_{ramp} = 4.6°$ is employed as a controlled device. In this case, approximately 80% of mitigation is observed in comparison to the uncontrolled interaction.

References

1. D'aguanno, A., Schrijer, F., van Oudheusden, B.: Investigation of three-dimensional shock control bumps for transonic buffet alleviation. AIAA J. **61**(8) (2023). https://doi.org/10.2514/1.J062633
2. Eastwood, J.P., Jarrett, J.P.: Toward designing with three-dimensional bumps for lift/drag improvement and buffet alleviation. AIAA J. (2012). https://doi.org/10.2514/1.J051740
3. Mayer, R., Lutz, T., Krämer, E., Dandois, J.: Control of transonic buffet by shock control bumps on wing-body configuration. J. Aircr. (2019). https://doi.org/10.2514/1.C034969
4. Ashill, P.R., Fulkner, L.J., Shires, A.: A novel technique for controlling shock strength of laminar flow aerofoil sections. In: DGLR Bericht Proceedings (1992)
5. Holden, H.A., Babinsky, H.: Shock/boundary layer interaction control using 3D devices. In: 41st Aerospace Sciences Meeting and Exhibit (2003). https://doi.org/10.2514/6.2003-447
6. Qin, N., Wong, W.S., Le Moigne, A.: Three-dimensional contour bumps for transonic wing drag reduction. Proc. Inst. Mech. Eng., Part G: J. Aerosp. Eng. **222**(5) (2008). https://doi.org/10.1243/09544100JAERO333
7. Bruce, P.J.K., Babinsky, H.: Experimental study into the flow physics of three-dimensional shock control bumps. J. Aircr. **49**(5) (2012). https://doi.org/10.2514/1.C031341
8. Colliss, S.P., Babinsky, H., Nübler, K., Lutz, T.: Joint experimental and numerical approach to three-dimensional shock control bump research. AIAA J. **52**(2) (2014). https://doi.org/10.2514/1.J052582
9. Ogawa, H., Babinsky, H., Pätzold, M., Lutz, T.: Shock-wave/boundary-layer interaction control using three-dimensional bumps for transonic wings. AIAA J. **46**(6) (2008). https://doi.org/10.2514/1.32049
10. Ogawa, H., Babinsky, H.: Shock/boundary-layer interaction control using three-dimensional bumps in supersonic engine inlets. In: 46th AIAA Aerospace Sciences Meeting and Exhibit (2008). https://doi.org/10.2514/6.2008-599
11. Giepman, R.H.M., Schrijer, F.F.J., van Oudheusden, B.W.: Flow control of an oblique shock wave reflection with micro-ramp vortex generators: effects of location and size. Phys. Fluids **26**(6) (2014). https://doi.org/10.1063/1.4881941
12. Bruce, P.J.K., Colliss, S.P.: Review of research into shock control bumps. In: Shock Waves, vol. 25, issue 5 (2015). https://doi.org/10.1007/s00193-014-0533-4
13. Bruce, P.J.K., Colliss, S.P., Babinsky, H.: Three-dimensional shock control bumps: effects of geometry. In: 52nd Aerospace Sciences Meeting (2014). https://doi.org/10.2514/6.2014-0943
14. Bulut, J., Schrijer, F.F.J., van Oudheusden, B.W.: Investigation of shock control bump geometry variation on oblique shock wave boundary layer interactions. In: 57th 3AF International Conference AERO2023, Bordeaux, France (2023)
15. Bulut, J., Schrijer, F.F.J., van Oudheusden, B.W.: Effect of shock control bumps on oblique shock wave turbulent boundary layer interactions. In: 56th 3AF International Conference AERO2022, Toulouse, France (2022). https://doi.org/10.5281/zenodo.6576217
16. Bulut, J., Schrijer, F., van Oudheusden, B.W.: Impinging shock wave boundary layer interactions control with shock control bumps. In: AIAA Aviation 2024 Forum, Las Vegas, NV, USA (2024)

17. Colliss, S.P., Babinsky, H., Nöbler, K., Lutz, T.: Vortical structures on three-dimensional shock control bumps. J. Aircr. **53**(4) (2016). https://doi.org/10.2514/1.J054669
18. Giepman, R.: Flow control for oblique shock wave reflections. TU Delft University, vol. 1 (2016)

Open Access This chapter is licensed under the terms of the Creative Commons Attribution 4.0 International License (http://creativecommons.org/licenses/by/4.0/), which permits use, sharing, adaptation, distribution and reproduction in any medium or format, as long as you give appropriate credit to the original author(s) and the source, provide a link to the Creative Commons license and indicate if changes were made.

The images or other third party material in this chapter are included in the chapter's Creative Commons license, unless indicated otherwise in a credit line to the material. If material is not included in the chapter's Creative Commons license and your intended use is not permitted by statutory regulation or exceeds the permitted use, you will need to obtain permission directly from the copyright holder.

Airfoil/Wing Configuration

Numerical Study of Unsteady Shock/Boundary Layer Interaction

Andrea Petrocchi, Rene Steijl, and George N. Barakos

Abstract The present work investigates the ability of the Partially-Averaged Navier-Stokes (PANS) method to reproduce transonic buffet, occurring on airfoils and wings at transonic regime under specific flow conditions. The designed test case for this analysis is the OAT15A unswept wing at Mach number $M_\infty = 0.73$ and Reynolds number $Re_c = 3 \times 10^6$. The three-dimensional flow is studied by accounting for the wind tunnel walls in the experiments of Jacquin et al. [1]. The computations on a large-span, confined configuration revealed a strong three-dimensionality of the flow both before and after the buffet onset. The comparison with unsteady Reynolds-averaged Navier Stokes (URANS) results showed the benefits of PANS in resolving flow unsteadiness at different flow resolutions, especially on affordable CFD grids, at limited additional cost.

Keywords Transonic flow · Transonic buffet · Shock induced separation · Numerical simulations · Partially-averaged Navier-Stokes method

10.1 Introduction

At transonic flow conditions, shock wave/boundary layer interaction and the induced, unsteady flow separation lead to a phenomenon known as buffet. It consists of self-sustained shock oscillations around aerofoils and wings. Since the middle of the last century, several experimental and numerical studies have been conducted to shed light on the mechanism driving the shock oscillations [1–5]. Buffet has consequences on the structural aircraft response, as it can lead to structural fatigue. Moreover, it

---

A. Petrocchi · R. Steijl · G. N. Barakos (✉)
CFD Lab, School of Engineering, University of Glasgow, Glasgow, UK
e-mail: george.barakos@glasgow.ac.uk

A. Petrocchi
e-mail: 2613543p@student.gla.ac.uk

R. Steijl
e-mail: rene.steijl@glasgow.ac.uk

deteriorates the aircraft handling qualities and can also cause passenger discomfort. Therefore, an airplane must be free from oscillations at any operating conditions and a buffet boundary is part of the flight envelope boundary. Nonetheless, variations of the flight conditions due to gusts or emergency maneuvers may cause the plane to cross the buffet boundary. Therefore, in recent years, increasing attention has been paid to understanding, predicting, and controlling this phenomenon.

The first explanation of buffet onset was given more than fifty years ago by Pearcey [2] who related the oscillation to bubble bursting due to an increase of Mach number or angle of attack. When the shock was strong enough to cause separation extending from the shock foot to the trailing edge, low frequency, high amplitude fluctuations were measured in the separated region, together with a periodic and self-sustained shock motion. In recent years, that hypothesis was almost completely surpassed (see Iovnovich and Raveh [6]). Tijdeman [3] proposed a characterisation of the shock motion for flap-equipped aerofoils and generalised to plunging or pitching aerofoils. The type A buffet consists of a sinusoidal shock oscillation over the suction side of the aerofoil, with a varying strength and position of the shock throughout the buffet cycle. Type B follows the same pattern of type A, with the shock disappearing during the downstream excursion. Type C sees a strengthening and weakening of the shock as it moves upstream and the propagation into the oncoming flow as a free shock-wave.

One of the most plausible explanations behind buffet is the acoustic feedback mechanism proposed by Lee [4] for the shock motion of type A. The shock oscillation on the upper side of the aerofoil generates pressure waves propagating through the separated flow region extending from the shock to the leading edge. Once the leading edge is reached, another disturbance propagates backwards at the local speed of sound. These waves interact with the shock, and transfer the energy required to sustain the oscillation. The buffet period is the time necessary for a pressure wave to depart from the shock and reach the trailing edge plus the time needed for the disturbance to move backwards and hit the shock. This theory was supported by following numerical [7] and experimental [8, 9] works. In the experimental work of Jacquin et al. [1] on the supercritical aerofoil OAT15A, the authors also observed upstream travelling perturbations on the pressure side that are diffracted at the leading edge, and play a role in the self-sustained motion. The LES computations of Garnier and Deck [10] confirmed this latest findings.

Starting from Crouch et al. [11], buffet has been associated with a global instability mechanism, and studied by means of linear stability analyses and the URANS equations. Buffet is seen as a Hopf bifurcation, for which the least stable eigenvalue of the associated linearised system of equations crosses the imaginary axis of the complex plane and becomes unstable. The eigenmode associated with the unstable eigenvalue is qualitatively different from the mechanism of Lee [4] and similar to that from Jacquin et al. [1]. A pressure disturbance is generated at the shock foot and moves along the shock up to the end of the supersonic region. As the perturbation moves upward, the shock approaches the trailing edge and intensifies. A pressure wave is generated, goes around the trailing edge, propagates forward along the pressure side and, once at the leading edge, is ingested into the sonic zone. The same approach was followed by Sartor et al. [12].

In three-dimensional configurations, the shock dynamics is affected by several factors. Several experimental investigations [13–17] have been carried out on transonic wings, and their findings were confirmed by following numerical simulations. The work of Iovnovich and Raveh [18] reported differences between straight and swept wings, and studied the influence of the sweep angle on the shock dynamics. In general, the 3D buffet is characterised by a broadband spectrum of frequencies rather than a single frequency driving the oscillation. Moreover, the spanwise organisation of the flow in "buffet cells" was shown, characterised by regions of alternated pressure propagating in the spanwise direction, and usually outboard. The approach of Crouch et al. [5, 19] and Timme [20] covered models of infinite and finite swept wings. The onset of buffet on the NASA CRM was studied by means of global stability. The eigenvalues and eigenmodes were analysed to enforce the theory that 3D buffet can be connected to a global instability mechanism. As the angle of attack increases, the eigenvalues cross the imaginary axis and give rise to the buffet.

10.1.1 Experimental Findings

The first experimental studies were conducted on a thick, symmetric aerofoil by McDevitt et al. [21] and Levy [22]. Through unsteady pressure measurements they detected shock oscillations on both sides of the aerofoil, with alternating movement of the shocks on the two sides. The study revealed a periodic motion at a precise frequency occurring together with a significant shock-induced separation region extended from the shock foot to the trailing edge.

Studies on conventional and supercritical aerofoils were carried out afterwards. Roos [23] performed experiments using the NACA0012 and Whitcomb aerofoils, at various Mach numbers and angles of attack. Pressure measurements over the suction side of the aerofoils allowed the detection of SIOs, with the amplitude of the lift oscillations correlating with the trailing edge pressure divergence. Buffet over the NACA0012 was extensively investigated in the work of McDevitt and Okuno [24]. They used the Ames High Reynolds Number Facility at several flow conditions, through unsteady pressure measurements, collected in the broadest available database for buffet onset. Lee and Ohman [25] and Lee [26] performed experiments on the BGK1 aerofoil. The comparison with the WHEA II aerofoil, which is thicker than the BGK1, put into evidence a more significant impact of buffet on thick aerofoils. Following the experimental campaigns, Lee [4] proposed the aforementioned acoustic feedback mechanism.

Among more modern works, the most notorious experiments on buffet was carried out by Jacquin et al. [1] on the supercritical OAT15A aerofoil in the ONERA wind tunnel SCh3. Because of the employed experimental apparatus, some good data were provided for comparison with CFD results. The experiment revealed a 2D behaviour on the wing, and different frequencies were detected in the shock, and at the separation regions. A well defined peak in the spectrum was associated with 2D shock oscillations. By computing the correlation between pressure signals on both sides of the aerofoil, the presence of upstream travelling waves on both side was observed. The spectral analysis showed the independence of the fundamental buffet frequency on the spanwise and streamwise coordinates, and the buffet evolution was deemed to be of purely 2D nature. Nevertheless, the oil flow visualisation detected the presence of mushroom-shaped vortical regions of the separated boundary layer. Their presence resulted in a non uniform shock front.

The following experimental campaigns of Hartmann et al. [27, 28] and Feldhusen-Hoffmann et al. [8, 29] studied the flow around the DRA2303 aerofoil. The upstream shock motion was associated with the upstream-propagating pressure disturbance generated at the trailing edge. The mechanism held until the sound pressure level was too low, the extent of separation too large, and the shock no longer influenced by the disturbances. They also introduced artificial disturbances to reproduce the disturbances at the trailing edge emanated during the shock buffet period. The emission of sound waves from a horn placed behind the airfoil resulted in a shift of the buffet fundamental frequency according to that of the loudspeaker, underlining the fundamental role of pressure feedback mechanism in 2D buffet flows. Modal analysis via DMD distinguished two main modes. A low-frequency buffet mode saw the coupling of the shock wave oscillation, and the changes in size of the separated flow region, while the high-frequency vortex mode saw the downstream propagation of vortices in the separated region. Once they passed over the trailing edge, the interaction with it resulted in sound generation. The strengths of the downstream propagating vortices varied according to the shock strength, i.e. it was modulated with a frequency equal to the buffet frequency. This reinforced the idea of Lee [4] for which buffet is sustained by sound waves of different SPLs.

The OAT15A aerofoil was later investigated by D'Aguanno et al. [30, 31]. The adoption of a reduced wing aspect ratio (2.8 against 3.4 in Jacquin et al. [1]) resulted in the absence of 3D mushroom-shaped structures. The POD modes distinguished two types of modes, similar to those of [29]. In [31], the spanwise organisation of UTWs was studied. Their formation coincides with the 3D turbulent structures reaching the trailing edge. Their strength and velocity, varied with the shock phase, confirming that these waves travel at the local sound speed [30]. Moreover, the strength of the UTWs is higher when the shock is at its most downstream position. In this view, the structures causing the propagation of UTWs may be formed in the separated TE area, rather that at the shock foot.

The same test case was later analysed by Accorinti et al. [32] by means of BOS. It was found that the shock inversion is a necessary but not sufficient condition for the buffet onset. Nevertheless, the delay between the shock motion inversion and the buffet onset, was not explained in their work. In a following work [33], the role of the wind tunnel walls was investigated. At pre-buffet conditions, the shock was weakened by the sidewall corner separation, and shifted upstream. At fully developed buffet conditions, the side-wall influence decreased. The shock was weaker only in a small spanwise portion and the amplitude of the shock oscillation became uniform for large parts of the span. Corresponding to the corner separation, a low-frequency content was detected in both pre- and post-buffet. An even lower frequency was observed in the separated flow region around the centreplane at buffet conditions. This was attributed by the authors to the interaction between the separated boundary layer and the shock, i.e. the one forming mushroom-like structures in Jacquin et al. [1], Sugioka et al. [34]. Nevertheless, such structures were not present in that campaign, mainly because of the reduced wing aspect ratio with respect to the other cases (2 compared to 3.4 [1] and 4.5 [34]).

Experimental investigations of buffet were extended on a 30° swept configuration (BUFET'N Co) by Molton et al. [13]. A large separated flow region was observed for the cases at buffet conditions, and the flow topology revealed to be strongly three-dimensional, with streamlines deflecting outboard. At the highest angle of attack, the separated flow region spanned from the shock foot to the trailing edge. Dandois et al. [14] determined a dual frequency content in a study of the AVERT wing mode. They are associated with the large amplitude motion at low frequency, and high frequency Kelvin-Helmoltz instabilities. Unlike 2D buffet cases, showing a distinct peak was in the spectrum, the swept wing spectrum exhibited an extended region of frequency content, centered around the main buffet frequency. The simple configuration displayed a convection velocity directed outboard and towards the trailing edge, associated with buffet cells.

Investigations on the NASA CRM were carried out by Koike et al. [15, 16] and Sugioka et al. [35]. In these works, the aerodynamic coefficients were measured at several angles of attack to detect the buffet onset. PSP [35] provided a clear visualisation of the evolving pressure field over the wing. A new low-frequency content at higher angles of attack stemming from the interaction with the trailing edge separation was detected. This finding was confirmed by the work of Sugioka et al. [36]. When the angle exceeded the buffet onset, buffet cells with an outboard propagation velocity were observed.

The work of Lawson et al. [17] studied the flow around the RBC12 wing of ARA using dynamic PSP. Their results were used by Masini et al. [37] who performed modal decomposition through POD and DMD. In their work, a new, low-frequency contribution, characterised by in inboard-propagating pressure disturbances, was observed at low angles of attack. At higher incidence, the outboard-propagating cells were also detected.

Although the majority of the experiments showed the presence of buffet cells, the flow over the ONERA-M4 wing showed a different pattern [38]. The reduced wing aspect ratio led to a different buffet development with respect to the previous works. At high incidence, the analysed wing displayed a dual shock topology and boundary layer separation starting from the wingtip. Still, some oscillations were detected. At low angles, if the separation reached the leading edge, the shock oscillated with large amplitude and frequency. At higher angles, the shock oscillated with slightly higher frequencies and the fluctuations propagated outboard. One can speculate that this case, for geometric reasons, the exhibited features are intermediate between 2D and "common" 3D buffet.

In this view, Sugioka et al. [34] tried to shed light on the differences between swept and unswept wings by carrying out experiments on unswept and low-sweep wings extruded from a CRM profile. The employed wing aspect ratio ($AR = 4.5$) allowed for the development of three dimensional structures even on the unswept configuration. In the swept case, the oil flow showed a strong flow three-dimensionality, culminating in spanwise-directed flow. Two concurrent mechanisms were observed. At low frequency, the reconstructed mode showed large-amplitude shock oscillations in the longitudinal direction and pressure disturbances propagating inboard. At higher frequencies, the location of the oscillation was shifted in the second half-span with outboard-travelling disturbances. The comparison of the local Strouhal number for different sweep angles from different works, showed a clear correlation between the two quantities, although a physical explanation was not given.

10.1.2 Transonic Buffet CFD Results

The prediction of buffet by means of CFD is still challenging. The difficulties stems from the presence of shock waves and boundary layer separation, culminating in a self-induced instability. This has introduced questions about the performance of different turbulence models in the URANS context, and led to the use of hybrid modelling approaches.

A number of authors investigated the potential of this approach to capture buffet around two- and three-dimensional configurations [6, 39–46]. Still, there is no consensus on the ability of URANS to predict buffet Turbulence modelling, numerical schemes, spatio-temporal discretisation, and wind tunnel geometry are the main identified infuelnces.

Successful URANS computations were used to corroborate existing experimental results. Xiao et al. [7] investigated the flow around the BGK1 aerofoil. By using cross-correlation on the aerofoil suction side boundary layer, the presence of downstream pressure wave propagation was observed. The same tool allowed for the detection of upstream travelling waves outside of the boundary layer, as hinted by Lee [4]. Following the experiments of Jacquin et al. [1] and the works of Crouch et al. [5] and Garnier and Deck [10], pressure disturbances propagating towards the leading edge on the pressure side were observed. To exclude the impact of such disturbances, Memmolo and Pirozzoli [47] introduced a sponge region before the trailing edge on the pressure side of a V2C aerofoil. In their study, this effect was found to have limited impact on the overall buffet prediction. Iovnovich and Raveh [6] focused on the behaviour at the buffet onset and offset, for different aerofoils. The first was reconnected to the unsteady interaction between the shock and the separation bubble. At the onset, the shock was located after the point of maximum curvature on the upper surface, where flow separation is promoted. Once the shock foot and the trailing edge separations merge, buffet takes place. After the offset, the flow is stalled and the shock assumes a forward position. A similar study was conducted by Giannelis et al. [45] on the OAT15A aerofoil. At particular flow conditions [48], the shock oscillation was a combination of types A and C. Unlike the canonical type A, the aerodynamic coefficients saw a strongly irregular behaviour, caused by the ingestion of the shock in the free-stream region.

Hybrid RANS-LES approaches showed an improvement in the prediction of buffet, representing more accurately the flow physics of transonic buffet. The main drawback of these methods is the considerably higher CPU costs. The test case of Jacquin et al. [1] was looked with particular interest for this purpose, because of the large variety of results obtained in the experimental campaign carried out at ONERA. Thiery and Coustols [42] claimed that the addition of the lateral, upper, and lower walls did not affect the ability of turbulence modelling to predict buffet. Following their study, the majority of the works adopted a 2D, free stream approach, using periodic boundary conditions in the spanwise dimension. Deck [49] was the first to claim the need for ZDES to overcome the issues encountered from standard DES in the framework of thin layer separation. The standard DES was not able to predict the self-sustained shock motion, even well beyond the experimental onset. The predictions did not lead to a satisfactory agreement with the experiment, because of the over-prediction of the pressure fluctuations, and the mis-prediction of the mean shock position. The DDES simulation of Grossi et al. [44] showed a slight improvement with respect to ZDES. Still, a too high level of unsteadiness was predicted at the aerofoil trailing edge. The high-resolution simulation emphasised the formation of three-dimensional structures at the trailing edge, and an irregular separation region. The DDES also suffered from a deficit in the mean velocity and a small gap in the pressure coefficient plateaux. Huang et al. [50] employed IDDES, using a timestep of more than an order of magnitude higher than the previous cases. This reflected in a poor prediction of the shock position. The large employed timestep prevented the resolution of smaller structures, and the reduced eddy viscosity promoted a too upstream shock position. An improvement was obtained with the use of OES by Szubert et al. [51]. Spectral analysis underlined the presence of three distinct contributions: (i) a low-frequency contribution representing the shock motion, (ii) the Von Karmann shedding at the trailing edge, and (iii) the Kelvin-Helmholts instability in the shear layer. Bonnifet et al. [52] tried to simulate buffet on the OAT15A aerofoil using a PANS-RSM approach. PANS, unlike RANS, was able to model the self-sustained buffet but failed to compare well with experiments. Their work suggested the use of a variable f_k to improve the prediction of buffet.

The issue of CPU cost became even worse when LES were performed (see, e.g., the work of Garnier and Deck [10]). Fukushima et al. [53] employed WMLES to reduce the costs for resolving the boundary layer. They showed that their method worked well for the OAT15A aerofoil, and the accuracy was higher than previous computations using hybrid RANS-LES methods. To reduce the computational costs associated with WMLES, Kojima and Hashimoto [54, 55] employed embedded LES, where turbulence was artificially generated at the RANS-LES interface. The results were comparable with the WMLES of Fukushima et al. [53]. Embedded LES delivered a reduction in the required CPU. However, the location of the RANS-LES interface played an important role and had a significant impact on the overall prediction.

Some researchers performed global stability analysis using RANS simulations as base-flow. Crouch et al. [5, 11] linearised the RANS equations around the equilibrium position, and solved the associated eigenproblem. The pressure propagation mechanism of the unstable buffet mode was qualitatively different from that of Lee [4], and more similar to Jacquin et al. [1]. Poplingher et al. [56] obtained similar results by using DMD reconstruction from TMSs. Sartor et al. [57] adopted the same approach for the flow around the OAT15A aerofoil, and also solved the adjoint equation to study the receptivity of the buffeting flow to the location of flow control. The distinction between global modes and adjoint modes allowed to determine where the unstable mode was more energetic, and where it was more receptive to flow control, respectively. They [57] stated that the buffet flow is mostly sensitive in the leading edge region and downstream of the shock. This means that the influence of incoming turbulence is not negligible and that the disturbances propagating downstream of the shock can affect the global flow structure because of the waves scattering from the trailing edge and propagating in the subsonic free-stream.

3D buffet cases have only been recently investigated, because of the high CPU costs. Few successful applications of RANS were found in literature [18, 57, 58]. The early computations of Brunet and Deck [59] on the CAT3D wing-body model put forward difficulties of the SA model to capture buffet, and the authors opted for ZDES (similarly to [49]). The results of ZDES correlated well with the experiments of Caruana et al. [60], and backward-propagating waves were detected in the longitudinal direction. In addition, a spanwise component was found, hinting for the first time at differences between the 2D and 3D cases. Sartor and Timme [57] adopted the SST and SA models to predict buffet on the RBC12 wing. Buffet started from a region close to the wing tip, with the amplitude of the aerodynamic coefficients increasing with the angle of attack. Although it showed some periodicity, the dynamics could not be identified as perfectly periodic. The work confirmed the modification in the frequency content observed in previous experimental works [61].

Iovnovich and Raveh [18] studied the effect on buffet of the sweep angle, and of the aspect ratio on a RA16SC1-based wing using the SA model. They first observed alternated positive-negative pressure fluctuations propagating outboard from the λ-shock structure at the wing root of swept wings. These structures were named *buffet cells*. These, could only manifest on high aspect ratio wings. In the 2.5D computations of Plante et al. [62], the frequency content associated with buffet cells was found to be independent of the sweep angle. Therefore, the Strouhal-sweep dependency suggested in [34] is only valid for finite wings. Nevertheless, with increasing sweep angles, the convection velocity of the buffet cells increased. At $\Lambda = 20°$, the buffet flow was seen as a superposition of a 2D mode and a high-frequency mode associated with buffet cells. This latter only manifested with $\Lambda \neq 0$.

The DDES of Sartor and Timme [63] partly confirmed the need for SRSs for a better characterisation of buffet, showing the non-periodic behaviour of the aerodynamic coefficients, and the broadband peak in the spectral visualisation. The higher resolution of DDES allowed the detection of small scales pressure propagations, that are crucial in the understanding of 3D buffet. POD and DMD et al. [64–66] confirmed the development of two concurrent phenomena found in the work of Masini et al. [37].

A similar computational campaign was carried out at JAXA on the CRM wing-body model [67, 68] and modal analysis was performed [69, 70]. The adopted ZDES proved to be extremely sensitive to the height of the transition from RANS to LES. The outcomes of the modal analysis were the same as the aforementioned works on the RBC12 configuration. The comparison with the experiments of Koike et al. [16] showed large discrepancies in the mean shock position. Indeed, a too upstream shock position and a higher pressure RMS peak were predicted. A modification of ZDES, called AZDES, was proposed by Ehrle et al. [71]. The method was tested on the CRM configuration and showed intermediate results between URANS and DDES, being able to delay the transition to SRS in the separated flow region. This avoided having the excessive separation predicted by DDES, which led to a misprediction of frequency and convection velocity associated with the buffet cells.

Stability analyses conducted on finite wings [19, 20, 72, 73] enforced the theory that 3D buffet is also the result of a global instability mechanism. As the angle of attack increases [72], one eigenvalue crosses the imaginary axis and gives rise to the flow instability. A number of other modes with similar frequencies become unstable shortly after the onset, suggesting that the canonical bump in the pressure spectra is connected to such modes becoming unstable. Crouch et al. [19] decomposed the perturbation vector into an in-plane part and a spanwise component, for a flow around an untapered, infinite wing. For the unswept case, two classes of modes were found: **(i)** oscillating modes, responsible for the instability, and **(ii)** stationary modes, adding the spanwise component. For swept cases, the first mode remained, while the second became outboard-travelling. Moreover, **(iii)** a high-frequency, short-wavelength mode related to turbulence in the shear layer was found. Sansica et al. [73] investigated the same confined configuration of et al. [34]. The presence of the wind tunnel sidewalls triggered the convection of three-dimensional structures towards the centreplane, even for the unswept configuration. On the swept configuration, the much larger separated flow region on the outboard wall caused the perturbation propagation direction to reverse and counteract the main cross-flow introduced by the sweep angle. This might explain the presence of inboard-travelling disturbances at the root of swept wings as a consequence of flow separation, confirming the findings of [37, 66, 70].

After studying the available literature, questions about the exact buffet mechanism remain. Moreover, the prediction through CFD of SIOs is still influenced by the adopted simulation strategy. In this paper, we carried out simulations around the OAT15A wing using a PANS method. Attention was paid to the ability of PANS to predict the 3D flow topology observed in Jacquin et al. [1], to shed light on the buffet dynamics. In Sect. 10.2, the compressible PANS formulation is presented with a focus on the different strategies adopted to estimate the PANS model parameters; in Sect. 10.3 the results for the pre-buffet and post-onset flows are presented before drawing some conclusions in Sect. 10.4.

10.2 PANS Formulation

The partially-averaged Navier-Stokes (PANS) formulation was first introduced by Girimaji et al. [74] as a bridging model between RANS and DNS. This method is based on a RANS paradigm, where the blending is obtained by means of the user-prescribed unresolved-to-total ratios of turbulent kinetic energy f_k and dissipation f_ϵ, bounded between 0 and 1, acting on the turbulence closure equations. They read $f_k = k_u/k$ and $f_\epsilon = \epsilon_u/\epsilon$, where the u subscripts stands for unresolved, and the quantities at the denominator are the total ones. The PANS method was initially derived for k-ϵ closures and then extended to the Wilcox k-ω model [75] by Lakshmipathy et al. [76] and to the Menter SST model [77] by Luo et al. [78]. In k-ω based formulations the parameter f_ϵ is replaced by the unresolved-to-total turbulence frequency f_ω through the following relation: $f_\omega = \omega_u/\omega = f_\epsilon/f_k$. These formulations inherit from the parent RANS models an eddy viscosity, based on a Boussinesq approximation, that is reduced with respect to the RANS case because of the effects of the f_k parameter: since only a fraction of the turbulent kinetic energy is modelled, the corresponding value of the eddy viscosity is reduced. This gives the possibility for the turbulent structures to be resolved. Alternative formulations like those in Refs. [79, 80], based on a k-ϵ-ζ-f model, and the PANS-RSM approach of Bonnifet et al. [52].

In this work the SST-PANS formulation is adopted:

$$\frac{\partial (\rho k)}{\partial t} + \frac{\partial \left(\rho U_j k\right)}{\partial x_j} = P_k - \beta^* \rho k \omega + \frac{\partial}{\partial x_j}\left[\left(\mu + \mu_t \sigma_k \frac{f_\omega}{f_k}\right)\frac{\partial k}{\partial x_j}\right], \quad (10.1)$$

$$\frac{\partial (\rho \omega)}{\partial t} + \frac{\partial \left(\rho U_j \omega\right)}{\partial x_j} = \frac{\gamma}{\nu_t} P_k - \beta' \rho \omega^2 + \frac{\partial}{\partial x_j}\left[\left(\mu + \mu_t \sigma_\omega \frac{f_\omega}{f_k}\right)\frac{\partial \omega}{\partial x_j}\right] + \\ + 2\frac{f_\omega}{f_k}(1 - F_1)\frac{\rho \sigma_{\omega 2}}{\omega}\frac{\partial k}{\partial x_j}\frac{\partial \omega}{\partial x_j}, \quad (10.2)$$

where ρ is the density, U_j is the flow velocity, μ is the dynamic molecular viscosity, and μ_t is the turbulent viscosity. Here, the turbulent kinetic energy k and frequency ω are the modelled, or unresolved, fractions. In the ω-equation, $\beta' = \left(\gamma \beta^* - \frac{\gamma \beta^*}{f_\omega} + \frac{\beta}{f_\omega}\right)$; F_1 is the SST blending function while $\gamma, \beta, \beta*, \sigma_k, \sigma_\omega$ are the model coefficients, calculated as in reference [77]. The turbulent viscosity is calculated as

$$\mu_t = \min\left(\frac{\rho k}{\omega}; \frac{\rho a_1 k}{SF_2}\right), \quad (10.3)$$

where S is the main strain rate tensor, F_2 is the second SST blending function and a_1 is equal to 0.31.

The f_ϵ parameter is usually set to one, under the assumption that all dissipative scales are not resolved in the computation. This approach is suitable for high

Reynolds numbers for which there is a net separation between energy-containing and dissipative scales [74]. The estimates of f_k adopted in this work are:

$$f_k = C_{\text{PANS}} \left(\frac{\Delta}{L_t}\right)^{2/3}, \qquad (10.4)$$

$$f_k = \frac{1 + \tanh(2\pi(\Lambda - 0.5))}{2}, \quad \Lambda = \frac{1}{1 + \left(\frac{L_t}{\Delta}\right)^{4/3}}, \qquad (10.5)$$

where $L_t = \sqrt{k}/(C_\mu \omega)$ is the local turbulent length scale, and Δ is the local grid size. The C_{PANS} coefficient is reduced with respect to the value of $1/\sqrt{C_\mu}$ prescribed for static estimates. Since the turbulent length scale is not based on total quantities, like in the case of estimates based on preliminary RANS calculations, it is reduced and, in turn, f_k is overly increased.

10.3 Application of PANS to Buffet

10.3.1 Test Case, Grids, and Numerical Setup

In the present study, the flow around the supercritical aerofoil OAT15A is investigated. This was studied experimentally by Jacquin et al. [1] in the S3Ch wind tunnel at ONERA. Measurements were collected at free-stream Mach numbers in the range of 0.7–0.75, a chord-based Reynolds number of $Re_c = 3 \times 10^6$, and angle of attack in the range of 1.36–3.9°C. Flow unsteadiness was detected at an angle of attack of 3.1°C and self-sustained SIO at an angle of 3.5°C.

In this work, the study is mainly focused on conditions of Mach number $M = 0.73$ and angles of attack of $\alpha = 2.5$, 3.5 and 3.9°, representative of a statistically steady flow and two fully-established buffet flows, respectively. Two configurations were analyzed. The first one, denoted 3D in Table 10.2, is a reproduction of half of the wind tunnel used in the experimental campaign (see Fig. 10.1). The grids adopted for this configuration are indicated as C3, M3, and F3 in Table 10.1.

Table 10.1 Main features of the CFD grids used for computations.

#	N_a	N_z	$\Delta z_w/c$	$\Delta z_{max}/c$	N_y	$\Delta n_a/c$	N_{wake}	$\Delta x_{TE}/c$	$N_{tot}[\times 10^6]$
M2	470	–	–	–	115	5×10^{-6}	220	5×10^{-6}	0.11
C3	405	76	2×10^{-6}	0.033	102	5×10^{-6}	86	5×10^{-6}	5.03
M3	500	100	2×10^{-6}	0.025	114	5×10^{-6}	120	5×10^{-6}	10
F3	510	164	2×10^{-6}	0.015	128	5×10^{-6}	150	5×10^{-6}	16.7

Table 10.2 Computations performed at different angles of attack, on different grids, and using different timesteps. All computations are at $M_\infty = 0.73$ and $Re_c = 3 \times 10^6$. U: URANS; P: PANS; FT: fully turbulent.

Run #	α	2D/3D	Mesh	f_k	$f_{k,\text{inf}}$	Δt	x_{tr}/c	P_out/P_in	Buffet
R2M25	2.5	2D	M2	–	–	Steady	FT	1.0	No
R3C25	2.5	3D	C3	–	–	Steady	FT	1.0	No
R3M25	2.5	3D	M3	–	–	Steady	FT	1.0	No
R3F25	2.5	3D	F3	–	–	Steady	FT	1.0	No
P2M25	2.5	2D	M2	0.7	–	0.1	FT	–	No
P3C25a	2.5	3D	C3	Eq. (10.4)	0.6	0.1	FT	1.0	No
P3C25b	2.5	3D	C3	Eq. (10.5)	0.6	0.1	FT	1.0	No
P3C25c	2.5	3D	C3	Eq. (10.5)	0.6	0.025	FT	1.0	No
P3C25d	2.5	3D	C3	Eq. (10.5)	0.6	0.025	FT	0.99	No
U3M25	2.5	3D	M3	1.0		0.025	FT	1.0	No
P3M25a	2.5	3D	M3	Eq. (10.5)	0.6	0.025	FT	1.0	No
P3M25b	2.5	3D	M3	Eq. (10.5)	0.6	0.01	FT	1.0	No
P3M25c	2.5	3D	M3	Eq. (10.5)	0.6	0.01	FT	0.99	No
U3F25	2.5	3D	F3	1.0		0.025	FT	1.0	No
P3F25a	2.5	3D	F3	Eq. (10.5)	0.6	0.1	FT	1.0	No
P3M25b	2.5	3D	F3	Eq. (10.5)	0.6	0.01	FT	1.0	No
P3M25c	2.5	3D	F3	Eq. (10.5)	0.6	0.005	FT	1.0	No
U3C35a	3.5	3D	C3	–	–	0.01	FT	–	No
U3C35b	3.5	3D	C3	–	–	0.01	0.07	0.99	No
P2M35a	3.5	2D	M2	0.7	–	0.01	FT	–	Yes
P2M35b	3.5	2D	M2	0.7	–	0.01	0.07	–	Yes
P3C35a	3.5	3D	C3	Eq. (10.4)	0.4	0.01	FT	1.0	Yes
P3C35b	3.5	3D	C3	Eq. (10.5)	0.4	0.01	FT	1.0	Yes
P3C35c	3.5	3D	C3	Eq. (10.5)	0.4	0.01	FT	0.99	Yes
P3C35d	3.5	3D	C3	Eq. (10.5)	0.4	0.01	0.07	0.99	Yes
P3C35d	3.5	3D	C3	Eq. (10.5)	0.4	0.01	0.25	0.99	Yes
P3C35e	3.5	3D	C3	Eq. (10.5)	0.6	0.01	0.07	0.99	Yes
P3M35a	3.5	3D	M3	Eq. (10.5)	0.6	0.01	0.07	0.99	Yes
P3M35b	3.5	3D	M3	Eq. (10.5)	0.6	0.005	0.07	0.99	Yes
P3F35	3.5	3D	F3	Eq. (10.5)	0.6	0.005	0.07	0.99	Yes
U2M39	3.9	2D	M2	–	–	0.1	FT	–	No
P2M39	3.9	2D	M2	0.7	-	0.1	FT	–	Yes
P3C39a	3.9	3D	C3	Eq. (10.5)	0.4	0.01	0.07	0.99	Yes
P3C39b	3.9	3D	C3	Eq. (10.5)	0.6	0.01	0.07	0.99	Yes

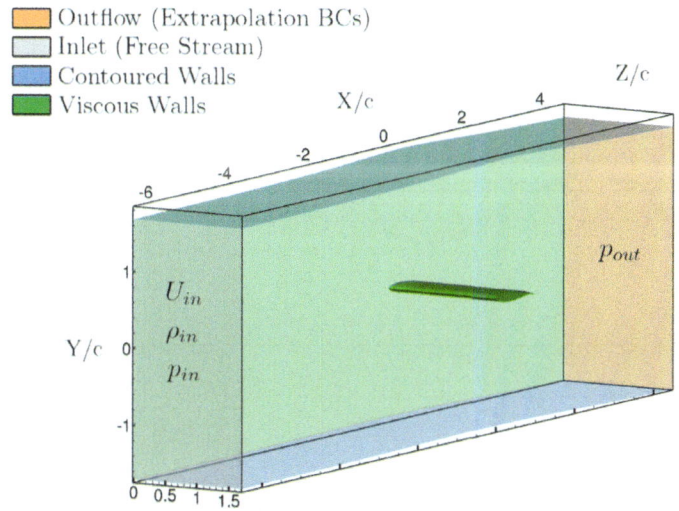

Fig. 10.1 Computational domain with coloured by boundary conditions. The symmetry plane at $z/c = 1.7$ is not coloured

The adopted boundary conditions are indicated in Fig. 10.1. The upper and lower walls were modelled as slip-walls, and the shape was extracted from a preliminary 2D RANS simulation to replicate the results of the adaptive technique used in the experimental campaign.

The second configuration is called 2D in Table 10.2 and M2 in Table 10.1, and it was used for 2D preliminary computations. The aerofoil is no longer confined and free-stream values are applied at the far field, distant 80c from the aerofoil.

The grid spacing was chosen to satisfy the condition of $\Delta y^+ \leq 1$ on the solid walls, and provide adequate resolution of the trailing edge vortex detachment. For the unsteady simulations, the convergence of the implicit scheme was based on the reduction of the flow field residual by 3 orders of magnitude with respect to the previous step. A maximum of 150 inner iterations were computed for each timestep.

10.3.2 Initial 2D Study

An initial study on the 2D configuration was carried out using $f_k = 0.7$. This helped the prediction of the shock oscillation, compensating for the inability of the SST model to predict buffet in this case and condition [81, 82]. The effect of the inflow turbulence value, sustainability term, and laminar-to-turbulence transition were investigated. Figure 10.2 reveals how the addition of the transition at the 7% of the chord allowed for a better prediction of the RMS peak and trailing edge RMS. The introduction of the sustainability term of Spalart and Rumsey [83], on the other hand, led

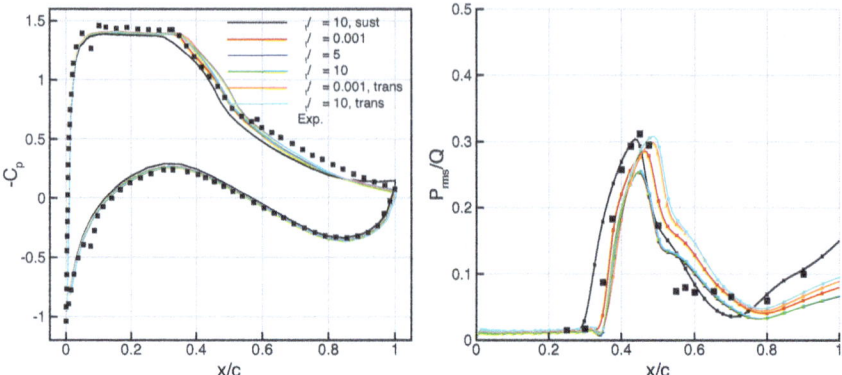

Fig. 10.2 Left: time-averaged pressure coefficient at $\alpha = 3.5°$; right: root means square of the pressure. Experiments from Jacquin et al. [1]

to a slightly upstream shock position and a high level of unsteadiness at the trailing edge. Because of the limited extent of the wind tunnel configuration, this correction was not applied in the following 3D study. The level of free-stream turbulence did not play a major role in the prediction of the mean shock position, and slightly influenced the RMS peak value. In conjunction with transition fixed at $x_{tr}/c = 0.07$, a turbulent-to-molecular eddy viscosity ratio of 10 led to the best agreement with the experiments and was used for the following 3D campaign.

10.3.3 Pre-Buffet Flow

For the statistically steady flow at $\alpha = 2.5°$, the aim is to recover the RANS solution, within a certain tolerance, with the use of PANS.

Figure 10.3, right plot, is taken from the converged URANS solution. It underlines one of the challenges of this test case by displaying the friction lines and surface pressure contours. The shock position can be detected by the negative values of the

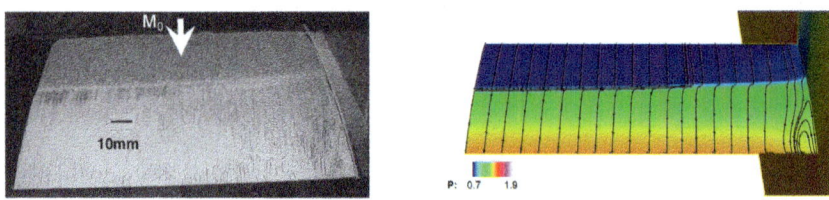

Fig. 10.3 Left: oil flow visualization around the OAT15A (photograph from Jacquin et al. [1]) at $\alpha = 2.5°$; right: friction lines and surface pressure contour from CFD

streamwise component of the stress tensor at the surface. As the flow exhibits a small separation at the shock foot, the pressure contour helps locate the shock. The shock bends when approaching the sidewalls because of viscous effects. Moreover, a region of separated flow is confined within 10% of the span, in agreement with the experimental results in the left image.

Because of the viscous sidewall, the turbulence level in the boundary layer grows as it approaches the aerofoil. This coincides with a growth in the turbulence length scales and a following reduction of f_k at the wall. In principle, the behaviour of the method is correct since f_k is lowered in regions of high turbulence content, but in this case, it would require a smaller timestep to resolve all flow structures developing at the wall-aerofoil junction. Therefore, we exploited the multi-block grid to impose a RANS treatment in the first layer of blocks closer to the wall. This avoids unnecessary costs to discretise the sidewall region, which plays a secondary role in the buffet dynamics. This is further justified by the experiments, stating that the separated flow region extent was almost constant over a buffet period.

Figure 10.4, left plot, shows the comparison of the pressure coefficient C_p around the aerofoil with the experiments for different grids using URANS. Figure 10.4, right plot, shows the same quantities for PANS simulations using the estimate in Eq. (10.5) with $f_{k,\text{inf}} = 0.6$ and different grids. The back pressure was here reduced by 1% with respect to the inlet value. This choice is motivated by the work of Thiery and Coustols [42], which shows experimental and numerical results for the case under analysis. In the current work, the shock position is slightly influenced by the grid adopted. As the adopted estimate accounts for the local grid size, f_k is lowered for finer grids. This results in a slight difference in the shock position, as a reduced eddy viscosity promotes boundary layer separation. A lower back pressure results in a pressure gap on both sides of the aerofoil. Nevertheless, a pressure difference between the inlet

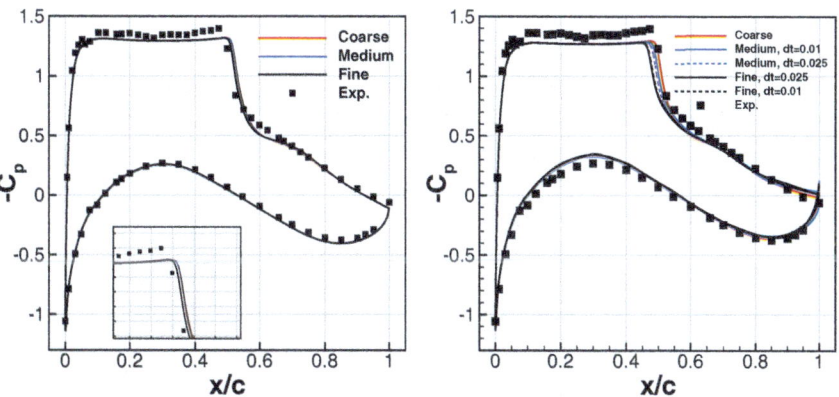

Fig. 10.4 Pressure coefficient around the OAT15A at $\alpha = 2.5°$ using URANS (left) and PANS (right). Experiments from Jacquin et al. [1]

and outlet sections is required to counteract the viscous effect of the wind tunnel walls. By imposing such a difference, shock oscillations are also promoted.

10.3.4 Fully Buffeting Flow

The buffet case is now investigated. Several simulations were performed at $\alpha = 3.5$ and $3.9°C$, using different grids and time steps. The transition location was fixed to $x/c = 0.07$, corresponding to the location of the tripping device used in the experiments. An outlet-to-inlet pressure ratio of 0.99 was used. Different values for both parameters were tested in a preliminary study. Moreover, the comparison between the estimates of f_k introduced in the previous section is shown. Where not specified, the coarser grid in Table 10.1 was used as it proved able to correctly predict the oscillatory behaviour of the flow. A comparison with finer grids is then shown.

f_k investigation

Different f_k distributions were compared. A previous study [82] revealed that the optimal value of C_{PANS} for the actual configuration, and grid size is 0.5 when the estimate of Eq. (10.4) is used. That value allowed for the desired reduction of eddy viscosity in the region around the trailing edge and the formation of two distinct separated flow regions, one at the wing-wall junction and the other at the trailing edge. Higher values of the constant resulted in the absence of boundary layer separation at the centreplane and in an over-prediction of the corner separation size. Figure 10.5 shows the comparison between the f_k distributions provided by the two estimates. The behaviours obtained were similar, at least for the case under analysis. The modelled-to-total ratio of turbulent kinetic energy reacts to regions of high turbulent content and acts by lowering the level of eddy viscosity. On the other hand, it takes values close to 1 in the boundary layer and the far field region, i.e. switching the formulation to RANS. The estimate of Eq. (10.5) also guarantees a slightly later transition from the RANS region in the boundary layer with respect to that of Eq. (10.4). As the desired value of f_k in the boundary layer is 1, this is an improvement. Moreover, the estimate of Eq. (10.4) can exceed 1 in the far field, while that of Eq. (10.5) is bounded between 0 and 1 by definition. The latter is seen as a minor issue as f_k can be clipped. Furthermore, the estimate of Eq. (10.5) did not need the calibration of additional parameters, like Eq. (10.4). For these reasons, the estimate of Eq. (10.5) was adopted in the following sections.

The effect of clipping the f_k parameter was also studied for different angles of attack. Without clipping f_k, the risk of too-low eddy viscosity in some regions of the flow field is high, especially for coarse grids. Too low values of the clip, 0.4 in this case, resulted in a too upstream shock position with respect to the experiments. The

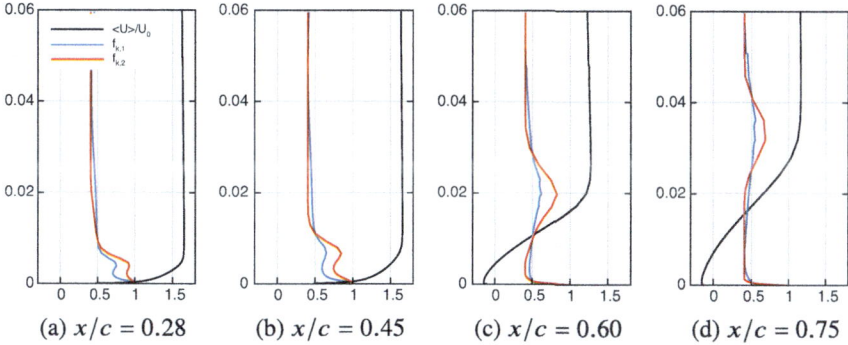

Fig. 10.5 Instantaneous profiles of horizontal velocity, turbulent-to-molecular viscosity, and f_k. The shock position is $x/c \simeq 0.5$. In the plot legend, $f_{k,1}$ and $f_{k,2}$ refer to the estimates of Eqs. (10.4) and (10.5), respectively

low eddy viscosity promoted boundary layer separation and raised the level of flow unsteadiness at the TE. Conversely, adopting the higher value of 0.6, the prediction was improved for $\alpha = 3.5$ and 3.9.

Study on Different Resolutions

Here the role of the mesh is investigated. The Q-Criterion in Fig. 10.6 shows that the approach enabled the unsteadiness associated with buffet, and was able to describe structures with increasing resolution as the grid size increases. The finer grid was employed with a reduction of the time step by 50%. Employing the coarse mesh, the resolution was not enough to capture the motion of the smaller scales. Indeed, the adopted spanwise discretisation did not allow for resolving smaller structures associated with the turbulence in the separated flow region and the wake. Moreover, the adoption of a coarse grid did not allow for the development of any 3D structures in the separated boundary layer. Using the finer grid, it was possible to predict the same flow topology seen in the experiments. From this, we conclude that the spanwise discretisation of the coarse grid was not enough to capture the correct flow physics. Although the boundary layer was separated from the shock foot to the leading edge, the structures generated under the influence of the sidewall could not propagate towards the center of the tunnel. The case is similar to the computation of Thiery and Coustols [42], where the flow oscillation was predicted and just a hint of flow three-dimensionality was shown through friction lines. In that case, possibly because of the use of a RANS-like grid, the mushroom-shaped cells could not be established along the span.

In all cases, the overall agreement with the experiments for C_p and P_{RMS}, shown in Fig. 10.7, was satisfactory. The mean shock position slightly moved upstream when a finer grid was used. This effect was mainly due to the non-uniform shock

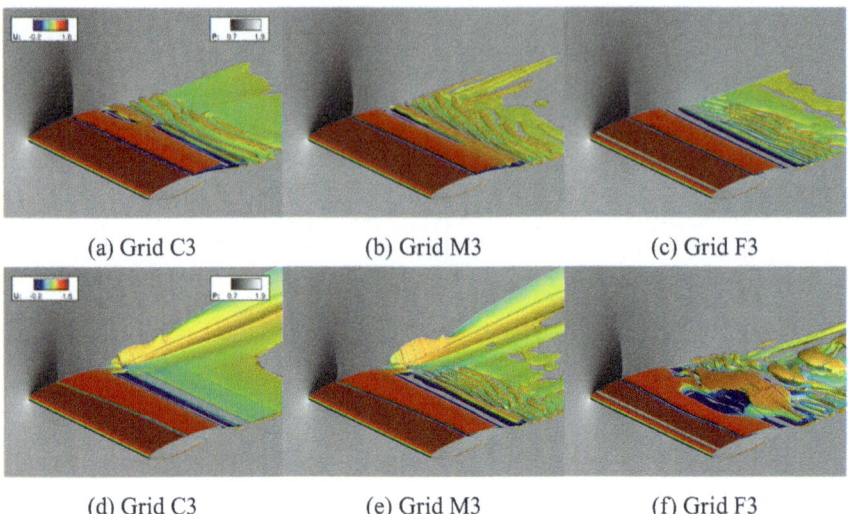

Fig. 10.6 Iso surfaces of Q-Criterion at $Q = 0.1$ for the confined configuration at different grid sizes. Top: most downstream shock position; bottom: most upstream shock position

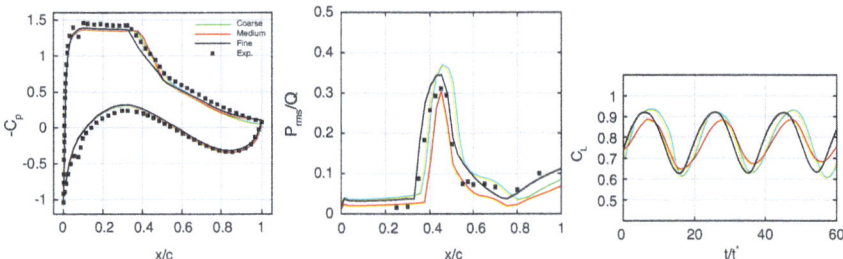

Fig. 10.7 Top: pressure coefficient (left) and RMS (centre) for different grid resolutions; right: lift coefficient history

front and the presence of the stall cells on the suction side of the wing. Moreover, the finer grid adopted at the trailing edge, together with the smaller timestep, allowed for the resolution of smaller scales and an increase in the level of fluctuations. This was also beneficial in terms of agreement of the mean pressure coefficient. A slight gap in terms of pressure coefficient is present on both sides. This is possibly due to the lack of knowledge of the experimental setup and differences in the adaptive wall shape between the experiments and the CFD simulations.

The analysis in the following chapter was performed with data from the simulation P3F35 of Table 10.2.

10.3.4.1 Mean Quantities

Here, the mean longitudinal component of the velocity and its RMS are shown in Figs. 10.8 and 10.9, respectively. The statistics were computed over several buffet periods. The probes were located at the same position as in the LDV measurements, i.e. around $z/c = 1.32$, where $z = 0$ is the sidewall and $z/c = 1.7$ is the symmetry plane. The pressure differences shown in the previous section coincide with the velocity differences outside the boundary layer, evident at $x/c = 0.45$. The slightly upstream shock position reflected in a difference in the RMS peak at $x/c = 0.45$, while the agreement is remarkably good on the aft part of the aerofoil. At $x/c = 0.28$, since the fluctuations in the boundary layer were not resolved, the RMS is practically zero, unlike in the experiments.

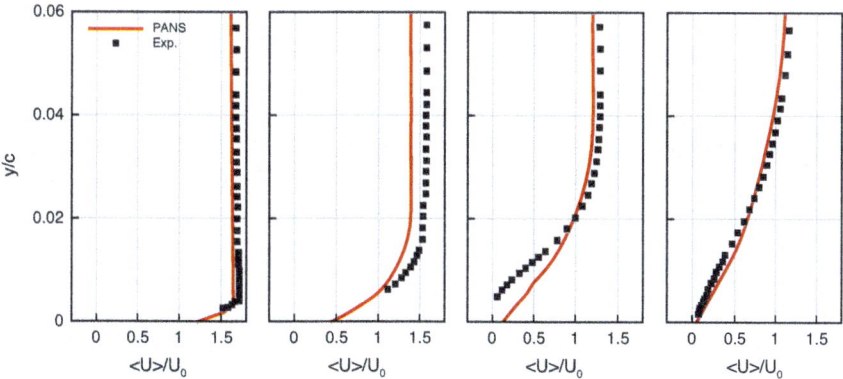

Fig. 10.8 Mean value profiles of the longitudinal velocity component. From left to right: $x/c = 0.28, 0.45, 0.6, 0.75$

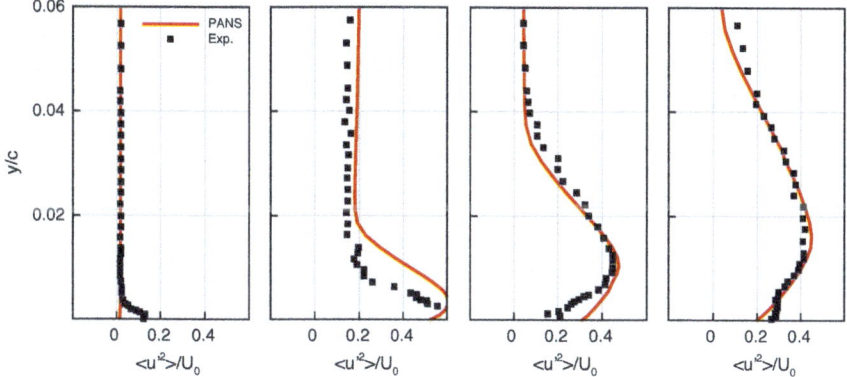

Fig. 10.9 RMS profiles of the longitudinal velocity component. From left to right: $x/c = 0.28, 0.45, 0.6, 0.75$

10.3.4.2 Buffet Dynamics and Flow Topology

Here, further details on the 3D buffet dynamics are given. PANS brought benefits in the prediction of transonic buffet. The URANS simulation converged to a steady state, while PANS was able to predict buffet. In the former case, the corner separation induced a spanwise flow deviation at the trailing edge that prevented the flow from separating. Even if the flow separated at the shock foot, the separation region did not merge with the one at the trailing edge, and buffet was inhibited. The main reasons can be found in the excessive flow separation at the corner, together with a too high level of eddy viscosity given from the URANS in the post-shock region. Therefore, the presence of the tunnel walls did not help the development of the self-sustained shock oscillation. Conversely, the PANS simulation predicted a buffet flow that develops following a precise pattern, illustrated in Fig. 10.10. Approaching the most downstream position (a), the flow is attached and the shock strength increases until the flow separates underneath (b); at the same time, trailing edge separation occurs due to the reduced eddy viscosity. The two regions merge around the centreline (c) inducing a flow acceleration between the corner and central separated regions. The effect of this is the creation of a vortical structure at the interface between the attached and separated boundary layer. When the flow is fully separated after the shock (d), this is affected by the disturbances coming from the trailing edge, and the shock begins to move upstream. In parallel, the aforementioned vortical structures propagate in the separated region and extend over the entire wingspan giving rise to a separated region characterized by large stall cells, in agreement with what was observed during the experiments (see Fig. 10.10d and e). This reflects in a non-uniform shock front. Approaching the most upstream position (f), the shock strength decreases and the flow re-attaches completely (g) so that a new period begins with the shock moving downstream.

The strong flow three-dimensionality was confirmed by the numerical schlieren visualizations in Fig. 10.11. The visualizations were obtained by averaging the density gradient magnitude in the spanwise direction and compared with the experimental images (Fig. 10.11, bottom plots). At the most downstream position, the shock front is straight for the greater part of the span and slightly deflects near the sidewall. At this condition, the boundary layer is attached. On the other hand, at the most upstream position, there is a large flow separation, and the shock front is not uniform. This is confirmed by the wide region of high density gradient (the predominant component is in the longitudinal direction) in the right plots in Fig. 10.11.

10.3.4.3 Spectral Analysis

Figure 10.12, right plot, shows the power spectral density of the pressure signal different locations. The left plot shows the pressure signal at $x/c = 0.45$ and $z/c = 1.6$, compared with the experiments. The fundamental and secondary frequencies are clearly detected, although at slightly lower frequency than the experimental one ($50 \div 60$ Hz against the experimental value of 69 Hz). This is particularly evident

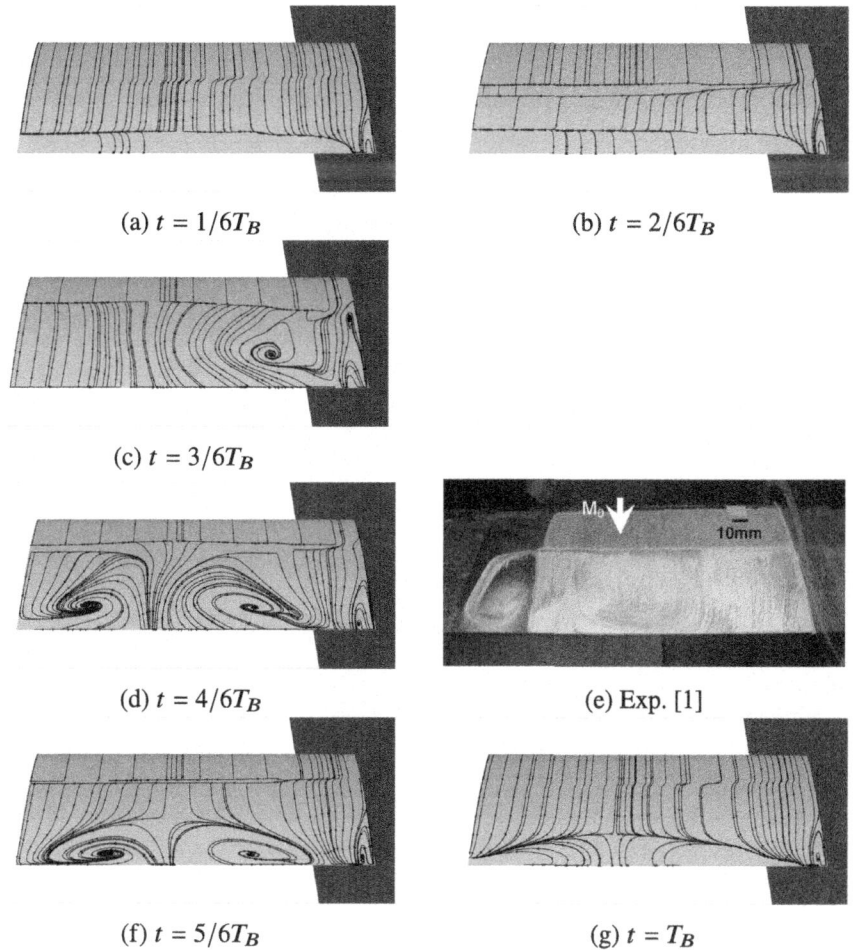

Fig. 10.10 Surface friction lines at different phases of the buffet period T_B. The experimental oil flow visualization was introduced for comparison purposes

looking at the difference in the periods of the pressure signals in the left plot. A slight difference in the mean shock position, shown in Fig. 10.7, left and centre plots, resulted in a better agreement with the pressure signal at $x/c = 0.43$. It has to be noted that the raw experimental data are not available for processing alongside the CFD, adding to the observed differences.

The other probes were located on the model at $x/c = 0.1$ and 0.9, while the fourth probe was located at $x/c = 1.2$ and $y/c = 0.03$. Three mean frequencies are detected: a fundamental buffet frequency around 50–60 Hz with the secondary harmonics, a bump around 1000–2000 Hz, and another in the wake around 5000–6000 Hz. This subdivision is coherent with the analysis of Szubert et al. [51] that associated

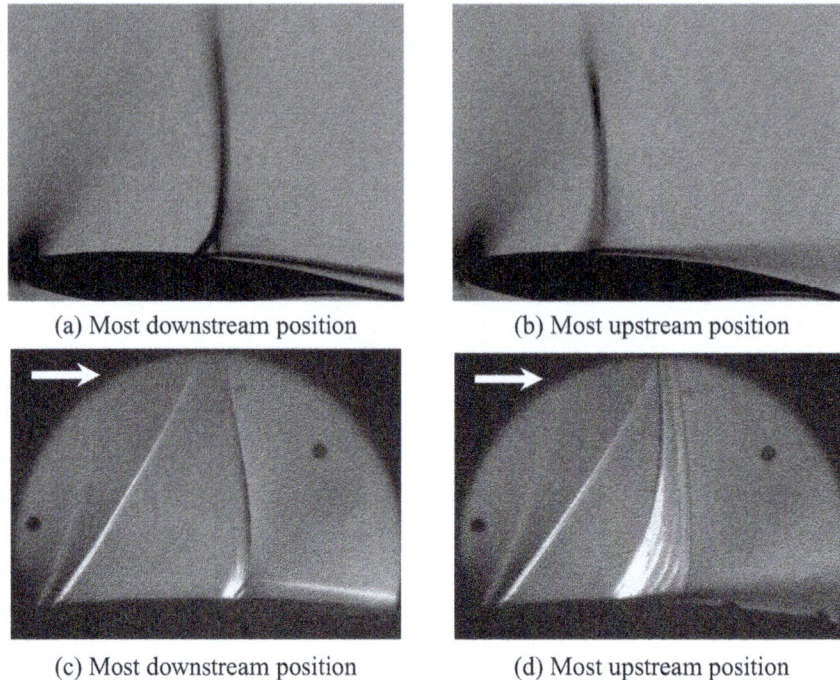

(a) Most downstream position (b) Most upstream position

(c) Most downstream position (d) Most upstream position

Fig. 10.11 Numerical (top) and experimental (bottom) schlieren visualizations. The visualizations were obtained by averaging the density gradient magnitude in the spanwise direction. Experiments taken from Jacquin et al. [1]

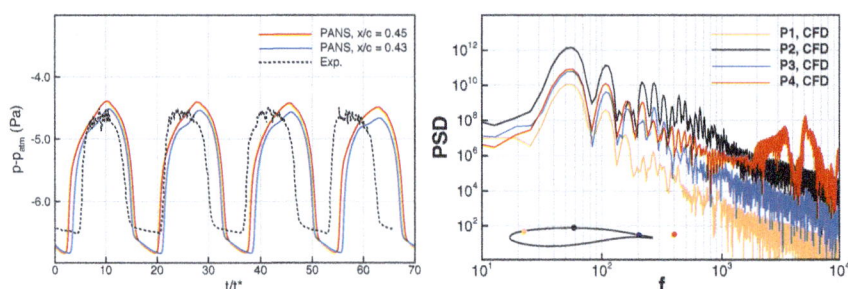

Fig. 10.12 Left: pressure signal at $x/c = 0, 34, 0.45$ and $z/c = 1.6$. Right: power spectral density of the pressure at different points in the domain. The position with respect to the aerofoil is indicated in the bottom left sketch

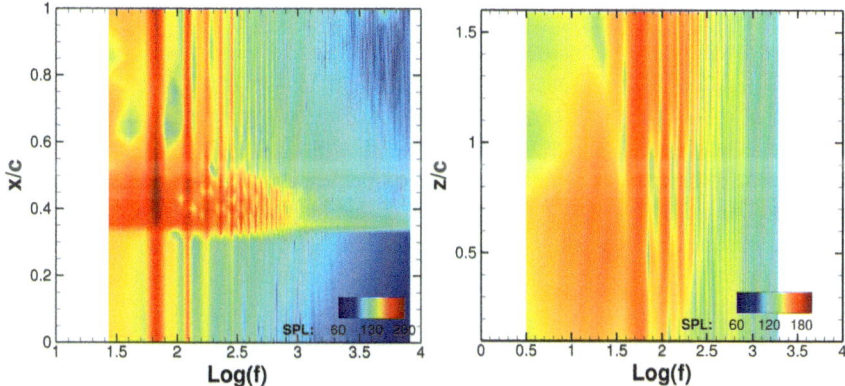

Fig. 10.13 Left: sound pressure levels along the chord on the upper surface at $z/c = 1.6$; right: Sound pressure levels along the span on the upper surface at $x/c = 0.6$

these three values to the buffet motion, Von Karman shedding and Kelvin-Helmholtz instabilities, respectively. While the fundamental and secondary harmonics were detected by every probe, just the last two ones were able to capture the presence of other types of instability.

The dependence of the PSD on the streamwise and spanwise coordinate was studied using sound pressure levels, defined as $\text{SPL} = 20\text{Log}\left(\text{PSD}/2 \times 10^{-5}\right)$. Figure 10.13 shows the distribution of SPL along the streamwise (left plot) and spanwise (right plot) directions. Figure 10.13, left plot, clearly shows the buffet frequency and the secondary harmonics all over the chord. The energy content associated with the pressure signals within the region of shock motion is higher than at other locations because of the higher level of fluctuations. The right plot shows that the energy distribution between frequencies is invariant in the spanwise direction, in spite of the 3D flow topology developed over the buffet period. The same behaviour was shown in the work of Jacquin et al. [1].

10.4 Conclusions

In this work, PANS simulations were conducted for the flow around the OAT15A wing model. The wind tunnel walls were accounted for and symmetry condition at the mid-span of the wing was used. The wind tunnel presence influenced the buffet dynamics by introducing strong three-dimensionality to the flow. At pre-buffet conditions, the shock was influenced by the presence of the wind tunnel and displayed a front bending when approaching the wall, as observed by previous experimental works [1, 33]. PANS recovered the URANS behavioiur, provided a reasonable distribution of the f_k parameter.

After the onset, the interaction with the separated region at the wall-wing junction was found to be crucial in the generation of large 3D vortical structures on the wing. This differs from a typical 2D behaviour with an alternated separation and reattachment of the boundary layer. URANS simulations were not able to predict the buffet, even at an angle of attack beyond its onset, while the use of PANS unlocked the shock oscillations. The flow physics compared well with the experiments, as confirmed by Figs. 10.7, 10.8, 10.9, 10.10, 10.11 and the unsteady loads exhibited periodic oscillations. A preliminary study was required to find the correct setup to correctly predict the mean shock position. The f_k parameter was clipped to a value of 0.6 to avoid an excessive reduction of the eddy viscosity. The PANS approach proved able to work at different grid resolutions. It was successful in predicting the self-sustained oscillation even with a coarse, RANS-like grid, and allowed for scale-resolving when a finer grid was used. In the latter, the description of the buffet dynamics was more accurate since the detection of 3D, vortical cells on the upper surface of the wing was enabled. In comparison with URANS, a low, additional ($\simeq 3\%$) CPU cost was introduced by PANS.

References

1. Jacquin, L., Molton, P., Deck, S., Maury, B., Soulevant, D.: Experimental study of shock oscillation over a transonic supercritical profile. AIAA (2012)
2. Pearcey, H.H.: A method for the prediction of the onset of buffeting and other separation effects from wind-tunnel tests on rigid models. Technical report. AGARD Report 223 (1958)
3. Tijdeman, H.: Investigation of the transonic flow around oscillating airfoils. Technical report, National Aerospace Laborator, The Netherlands. RNLR TR 77090 U (1977)
4. Lee, B., Tang, F.C.: Oscillatory shock motion caused by transonic shock boundary-layer interaction. AIAA J. **28**, 942–944 (1990). https://doi.org/10.2514/3.25144
5. Crouch, J.D., Gargaruk, A., Magidov, D., Travin, A.: Origin of transonic buffet on aerofoils. J. Fluid Mech. **628**, 357–369 (2009). https://doi.org/10.1017/S0022112009006673
6. Iovnovich, M., Raveh, D.E.: Reynolds-averaged Navier-Stokes study of the shock-buffet instability mechanism. AIAA J. **50**, 880–890 (2012). https://doi.org/10.2514/1.J051329
7. Xiao, Q., Tsai, H.M., Liu, F.: Numerical study of transonic buffet on a supercritical airfoil. AIAA J. **44**, 620–628 (2006). https://doi.org/10.2514/1.16658
8. Feldhusen-Hoffmann, A., Statnikov, V., Klaas, M., Schroder, W.: Investigation of shock-acoustic-wave interaction in transonic flow. Exp. Fluids **59**, 1–20 (2018). https://doi.org/10.1007/s00348-017-2466-z
9. D'Aguanno, A., Schrijer, F., van Oudheusden, B.W.: Finite-wing and sweep effects on transonic buffet behavior. AIAA J. **60**, 6715–6725 (2022). https://doi.org/10.2514/1.J061974
10. Garnier, E., Deck, S.: Large-Eddy Simulation of Transonic Buffet Over a Supercritical Airfoil. Direct and Large-Eddy Simulation VII, ERCOFTAC Series, vol. 13, pp. 549–554 (2010)
11. Crouch, J.D., Gargaruk, A., Magidov, D.: Predicting the onset of flow unsteadiness based on global instability. J. Comput. Phys. **224**, 924–940 (2007). https://doi.org/10.1016/j.jcp.2006.10.035
12. Sartor, F., Mettot, C., Sipp, D.: Stability, receptivity, and sensitivity analyses of buffeting transonic flow over a profile. AIAA J. **53**, 1980–1993 (2015). https://doi.org/10.2514/1.J053588
13. Molton, P., Dandois, J., Lepage, A., Brunet, V., Bur, R.: Control of buffet phenomenon on a transonic swept wing. AIAA J. **51**, 761–772 (2013). https://doi.org/10.2514/1.J051000

14. Dandois, J.: Experimental study of transonic buffet phenomenon on a 3D swept wing. Phys. Fluids **28**, 1–23 (2016). https://doi.org/10.1063/1.4937426
15. Koike, S., Nakakita, K., Nakajima, T., Koga, S., Sato, M., Kanda, H., Kusunose, K., Murayama, M., Ito, Y., Yamamoto, K.: Experimental investigation of vortex generator effect on two- and three-dimensional NASA common research models. In: 53rd AIAA Aerospace Sciences Meeting, Kissimmee, FL, USA (2016)
16. Koike, S., Ueno, M., Nakakita, K., Hashimoto, A.: Unsteady pressure measurement of transonic buffet on NASA common research model. In: 34th AIAA Applied Aerodynamics Conference, Washington, DC, USA (2016)
17. Lawson, S.G., Greenwell, D., Quinn, M.: Characterisation of buffet on a civil aircraft wing. In: 54th AIAA Aerospace Sciences Meeting, San Diego, CA, USA (2016)
18. Iovnovich, M., Raveh, D.E.: Numerical study of shock buffet on three-dimensional wings. AIAA J. **53**, 449–463 (2015). https://doi.org/10.2514/1.J053201
19. Crouch, J.D., Gargaruk, A., Strelets, M.: Global instability in the onset of transonic-wing buffet. J. Fluid Mech. **881**, 3–22 (2019). https://doi.org/10.1017/jfm.2019.748
20. Timme, S.: Global shock buffet instability on NASA common research model. In: AIAA SciTech 2019 Forum, San Diego, CA, USA (2019)
21. McDevitt, J.B., Levy L.L., Jr., Deiwert, G.S.: Transonic flow about a thick circular-arc airfoil. AIAA J. **14**, 606–613 (1975). https://doi.org/10.2514/3.61402
22. Levy, L.L., Jr.: Experimental and computational steady and unsteady transonic flows about a thick airfoil. AIAA J. **16**, 564–572 (1978). https://doi.org/10.2514/3.60935
23. Roos, F.W.: Some features of the unsteady pressure field in transonic airfoil buffeting formulation. AIAA J. **17**, 781–788 (1980). https://doi.org/10.2514/6.1979-351
24. McDevitt, J.B., Okuno, A.F.: Static and dynamic pressure measurements on a NACA 0012 airfoil in the Ames high Reynolds number facility. Technical report, National Aeronautics and Space Administration. NASA-TP-2485 (1985)
25. Lee, B., Ohman, L.H.: Unsteady pressure and forces during transonic buffeting on a supercritical airfoil. J. Aircr. **21**, 439–441 (1986). https://doi.org/10.2514/3.45876
26. Lee, B.: Investigation of flow separation on a supercritical airfoil. J. Aircr. **26**, 1032–1037 (1989). https://doi.org/10.2514/3.25144
27. Hartmann, A., Klaas, M., Schroeder, W.: Time-resolved stereo PIV measurements of shock-boundary layer interaction on a supercritical airfoil. Exp. Fluids **52**, 591–604 (2011). https://doi.org/10.1007/s00348-011-1074-6
28. Hartmann, A., Feldhusen, A., Schroeder, W.: On the interaction of shock waves and sound waves in transonic buffet. Phys. Fluids **25**, 026101 (2013). https://doi.org/10.1063/1.4791603
29. Feldhusen-Hoffmann, A., Lagemann, C., Loosen, S., Meysonnat, P., Klaas, M., Schröder, W.: Analysis of transonic buffet using dynamic mode decomposition. Exp. Fluids **62**, 66 (2021). https://doi.org/10.1007/s00348-020-03111-5
30. D'Aguanno, A., Schrijer, F., van Oudheusden, B.W.: Experimental investigation of the transonic buffet cycle on a supercritical airfoil. Exp. Fluids **621**, 214 (2021). https://doi.org/10.1007/s00348-021-03319-z
31. D'Aguanno, A., Schrijer, F.F.J., van Oudheusden, B.W.: Spanwise organization of upstream traveling waves in transonic buffet. Phys. Fluids **33**, 106105 (2021). https://doi.org/10.1063/5.0062729
32. Accorinti, A., Baur, T., Scharnowski, S., Kahler, C.J.: Experimental investigation of transonic shock buffet on an OAT15A profile. AIAA J. (2022). https://doi.org/10.2514/1.J061135
33. Accorinti, A., Korthauer, T., Scharnowski, S., Kahler, C.J.: Characterization of transonic shock oscillations over the span of an OAT15A profile. Exp. Fluids **64**, 61 (2023). https://doi.org/10.1007/s00348-023-03604-z
34. Sugioka, Y., Kouchi, T., Koike, S.: Experimental comparison of shock buffet on unswept and 10-deg swept wings. Exp. Fluids **632**, 132 (2022). https://doi.org/10.1007/s00348-022-03482-x
35. Sugioka, Y., Koike, S., Nakakita, K., Numata, D., Nonomura, T., Asai, K.: Experimental analysis of transonic buffet on a 3D swept wing using fast response pressure-sensitive paint. Exp. Fluids **59**, 1–20 (2018). https://doi.org/10.1007/s00348-018-2565-5

36. Sugioka, Y., Nakakita, K., Koike, S., Nakajima, T., Nonomura, T., Asai, K.: Characteristic unsteady pressure field on a civil aircraft wing related to the onset of transonic buffet. Exp. Fluids **62**, 20 (2021). https://doi.org/10.1007/s00348-020-03118-y
37. Masini, L., Timme, S., Peace, A.J.: Analysis of a civil aircraft wing transonic shock buffet experiment. J. Fluid Mech. **884**, 1–42 (2019). https://doi.org/10.1017/jfm.2019.906
38. Uchida, K., Sugioka, Y., Kasai, M., Saito, Y., Nonomura, T., Asai, K., Nakakita, K., Nishizaki, Y., Shibata, Y., Sonoda, S.: Analysis of transonic buffet on ONERA-M4 model with unsteady pressure-sensitive paint. Exp. Fluids **62**, 133–151 (2021). https://doi.org/10.1007/s00348-021-03228-1
39. Barakos, G., Drikakis, D.: Investigation of onlinear Eddy-Viscosity turbulence models in shock/boundary-layer interaction. AIAA J. **38**, 461–469 (2000). https://doi.org/10.2514/2.983
40. Goncalves, E., Houdeville, R.: Turbulence model and numerical scheme assessment for buffet computations. Int. J. Numer. Meth. Fluids **46**, 1127–1152 (2004). https://doi.org/10.1002/d.777
41. Thiery, M., Coustols, E.: URANS computations of shock-induced oscillations over 2d rigid airfoils: influence of test section geometry. Flow Turbul. Combust. **74**, 331–354 (2005). https://doi.org/10.1007/s10494-005-0557-z
42. Thiery, M., Coustols, E.: Numerical prediction of shock induced oscillations over a 2D airfoil: influence of turbulence modelling and test section walls. Int. J. Heat Fluid Flow **27**, 661–670 (2006). https://doi.org/10.1016/j.ijheatfluidflow.2006.02.013
43. Illi, S., Lutz, T., Kramer, E.: On the capability of unsteady RANS to predict transonic buffet. In: Proceeding of the Third Symposium Simulation of Wing and Nacelle Stall, Braunschweig, Germany (2012)
44. Grossi, F., Braza, M., Hoarau, Y.: Prediction of transonic buffet by delayed detached-eddy simulation. AIAA J. **52**, 2300–2312 (2014). https://doi.org/10.2514/1.J052873
45. Giannelis, N.F., Levinski, O., Vio, G.A.: Influence of Mach number and angle of attack on the two-dimensional transonic buffet phenomenon. Aerosp. Sci. Technol. **78**, 89–101 (2018). https://doi.org/10.1016/j.ast.2018.03.045
46. Zimmermann, D.-M., Mayer, R., Luiz, T., Kramer, E.: Impact of model parameters of SALSA turbulence model on transonic buffet prediction. AIAA J. **56**, 874–877 (2018). https://doi.org/10.2514/1.J056193
47. Memmolo, A., Bernardini, M., Pirozzoli, S.: Scrutiny of buffet mechanisms in transonic flow. Int. J. Numer. Methods Heat Fluid Flow **28**, 1031–1046 (2018). https://doi.org/10.1108/HFF-08-2016-0300
48. Giannelis, N.F., Levinski, O., Vio, G.A.: Origins of atypical shock buffet motions on a supercritical aerofoil. Aerosp. Sci. Technol. **107**, 106304 (2020). https://doi.org/10.1016/j.ast.2020.106304
49. Deck, S.: Numerical simulation of transonic buffet over a supercritical airfoil. AIAA J. **43**, 1556–1566 (2005). https://doi.org/10.2514/1.9885
50. Huang, J.B., Xiao, Z.X., Liu, J., Fu, S.: Simulation of shock wave buffet and its suppression on an OAT15A supercritical airfoil by IDDES. Sci. China: Phys. Mech. Astron. **55**, 260–271 (2012). https://doi.org/10.1007/s11433-011-4601-9
51. Szubert, D., Grossi, F., Garcia, A.J., Hoarau, Y., Hunt, J., Braza, M.: Shock-vortex shear-layer interaction in the transonic flow around a supercritical airfoil at high Reynols number in buffet conditions. J. Fluids Struct. **55**, 276–302 (2020). https://doi.org/10.1016/j.jfluidstructs.2015.03.005
52. Bonnifet, V., Gerolymos, G.A., Vallet, I.: Transonic buffet prediction using partially averaged Navier-Stokes. In: 23rd AIAA Computational Fluid Dynamics Conference, Denver, Colorado, USA (2017)
53. Fukushima, Y., Kawai, S.: Wall-modeled Large-Eddy simulation of transonic airfoil buffet at high Reynolds number. AIAA J. **56**, 2372–2388 (2018)
54. Kojima, Y., Hashimoto, A.: Embedded large eddy simulation of transonic flow over an OAT15A airfoil. In: AIAA SciTech Forum 2022, San Diego, CA, USA and Online (2022)

55. Kojima, Y., Hashimoto, A.: An application of embedded large eddy simulation for transonic buffet prediction. In: AIAA SciTech Forum 2023, National Harbor, MD, USA and Online (2023)
56. Poplingher, L., Raveh, D., Dowell, E.H.: Modal analysis of transonic shock buffet on 2D airfoil. AIAA J. **57**, 2851–2867 (2019). https://doi.org/10.2514/1.J057893
57. Sartor, F., Timme, S.: Reynolds-averaged Navier-Stokes simulations of shock buffet on half wing-body configuration. In: 53rd AIAA Aerospace Sciences Meeting, Kissimmee, FL, USA (2015)
58. Apetrei, R.M.: Numerical prediction and characterization of shock-buffet in transport aircraft. Ph.D. thesis, University of Sheffield (2019)
59. Brunet, V., Deck, S.: Zonal-detached eddy simulation of transonic buffet on a civil aircraft type configuration. In: 38th Fluid Dynamics Conference and Exhibit, Seattle, WA, USA (2008)
60. Caruana, D., Mignosi, A., Robitaillie, C., Correge, M.: Separated flow and buffeting control. Flow Turbul. Combust. **71**, 221–245 (2003). https://doi.org/10.2514/1.40932
61. Paladini, E., Dandois, J., Sipp, D., Robinet, J.-C.: Analysis and comparison of transonic buffet phenomenon over several three-dimensional wings. AIAA J. **57**, 379–396 (2019). https://doi.org/10.2514/1.J056473
62. Plante, F., Dandois, J., Laurendeau, E.: Similarities between cellular patterns occurring in transonic buffet and subsonic stall. AIAA J. **58**, 71–84 (2020). https://doi.org/10.2514/1.J058555
63. Sartor, F., Timme, S.: Delayed detached-Eddy Simulation of shock buffet on half wing-body configuration. AIAA J. **55**, 1230–1240 (2017). https://doi.org/10.2514/1.J055186
64. Masini, L., Peace, A., Timme, S.: Scale-resolving simulation of shock buffet onset physics on a civil aircraft wing. In: Royal Aeronautical Society 2018 Applied Aerodynamics Conference, Bristol, UK (2018)
65. Masini, L., Peace, A., Timme, S.: Reynolds number effects on wing shock buffet unsteadiness. In: AIAA Aviation 2019 Forum, Dallas, TX, USA (2019)
66. Masini, L., Timme, S., Pace, A.J.: Scale-resolving simulations of a civil aircraft wing transonic shock-buffet experiment. AIAA J. **58**, 4322–4338 (2020). https://doi.org/10.2514/1.J059219
67. Ishida, T., Ishiko, K., Hashimoto, A., Aoyama, T., Takekawa, K.: Transonic buffet simulation over supercritical airfoil by unsteady-FaSTAR code. In: 54th AIAA Aerospace Sciences Meeting, San Diego, CA, USA (2016)
68. Ishida, T., Hashimoto, A., Ohimichi, Y., Aoyama, T., Yamamoto, T., Takekawa, K.: Transonic buffet simulation over NASA-CRM by unsteady-FaSTAR Codelf wing-body configuration. In: 55th AIAA Aerospace Sciences Meeting, Grapevine, TX, USA (2017)
69. Ohmichi, Y., Ishida, T., Hashimoto, A.: Numerical investigation of transonic buffet on a three-dimensional wing using incremental mode decomposition. In: AIAA SciTech Forum 2017, Grapevine, TX, USA (2017)
70. Ohmichi, Y., Ishida, T., Hashimoto, A.: Modal decomposition analysis of three-dimensional transonic buffet phenomenon on a swept wing. AIAA J. **56**, 3938–3950 (2018). https://doi.org/10.2514/1.J056855
71. Ehrle, M., Waldmann, A., Lutz, T., Kramer, E.: Simulation of transonic buffet with an automated zonal DES approach. CEAS Aeronaut. J. **11**, 1025–1036 (2020). https://doi.org/10.1007/s13272-020-00466-7
72. Timme, S.: Global instability of wing shock-buffet onset. J. Fluid Mech. **885**, 37 (2020). https://doi.org/10.1017/jfm.2019.1001
73. Sansica, A., Hashimoto, A., Koike, S., Kouchi, T.: Side-wall effects on the global stability of swept and unswept supercritical wings at buffet conditions. In: AIAA SciTech 2022 Forum, San Diego, CA, USA & Online (2022)
74. Girimaji, S.S., Abdol-Hamid, K.S.: Partially-averaged Navier Stokes model for turbulence: implementation and validation. In: AIAA Aerospace Sciences Meeting and Exhibit, Reno, NE, USA (2005)
75. Wilcox, D.C.: Formulation of the k-omega turbulence model revisited. AIAA J. **46**, 2823–2838 (2008). https://doi.org/10.2514/1.36541

76. Lakshimpathy, S., Girimaji, S.S., K.S.: Partially-averaged Navier Stokes method for turbulent flows: k-omega model implementation. In: AIAA Aerospace Sciences Meeting and Exhibit, Reno, NE, USA (2006)
77. Menter, F.R.: Two-equation eddy-viscosity turbulence models for engineering applications. AIAA J. **32**, 1598–1605 (1994). https://doi.org/10.2514/3.121495
78. Luo, D., Yan, C., Wang, X.: Computational study of supersonic turbulent-separated flows using partially averaged Navier-stokes method. Acta Astronaut. **107**, 234–246 (2015). https://doi.org/10.1016/j.actaastro.2014.11.029
79. Krajnovic, S., Larusson, R., Basara, B.: Superiority of PANS compared to LES in predicting a rudimentary landing gear flow with affordable meshes. Int. J. Heat Fluid Flow **37**, 109–122 (2012). https://doi.org/10.1016/j.ijheatuidow.2012.04.013
80. Basara, B.: Fluid flow and conjugate heat transfer in a matrix of surface- mounted cubes: A PANS study. Int. J. Heat Fluid Flow **51**, 166–174 (2015). https://doi.org/10.1016/j.ijheatuidow.2014.10.012
81. Petrocchi, A., Barakos, G.N.: Buffet boundary estimation using a harmonic balance method. Aerosp. Sci. Technol. **132**, 108086 (2023). https://doi.org/10.1016/j.ast.2022.108086
82. Petrocchi, A., Barakos, G.N.: Transonic buffet simulation using a partially-averaged Navier-Stokes approach. In: ECCOMAS 2022, Oslo, Norway (2022)
83. Spalart, P.R., Rumsey, C.L.: Effective inflow conditions for turbulence models in aerodynamic calculations. AIAA J. **45**, 2544–2553 (2007). https://doi.org/10.1025/1.29737

Open Access This chapter is licensed under the terms of the Creative Commons Attribution 4.0 International License (http://creativecommons.org/licenses/by/4.0/), which permits use, sharing, adaptation, distribution and reproduction in any medium or format, as long as you give appropriate credit to the original author(s) and the source, provide a link to the Creative Commons license and indicate if changes were made.

The images or other third party material in this chapter are included in the chapter's Creative Commons license, unless indicated otherwise in a credit line to the material. If material is not included in the chapter's Creative Commons license and your intended use is not permitted by statutory regulation or exceeds the permitted use, you will need to obtain permission directly from the copyright holder.

Numerical Study and Physical Analysis of the Transonic Interaction and Its Modification Through Morphing Around Supercritical Wings at High Reynolds Number

Cesar Jimenez Navarro, Jacques Abou Khalil, Rajaa El Akoury, Abderahmane Marouf, Jean-Baptiste Tô, Yannick Hoarau, Jean-François Rouchon, and Marianna Braza

Abstract This study investigates the Shock-Wave Boundary Layer Interaction (SBLI) around supercritical aerofoils and wings in the Reynolds number range of $(2, 4.5) \times 10^6$. Physical analysis of the transonic buffet and its interaction with the shear-layer and near wake unsteadiness has been carried out in detail, showing the strong inter-dependence of the SBLI with the downstream unsteadiness. The numerical simulations have been performed with the NSMB—Navier Stokes Multi-Block code, using Delayed Detached Eddy Simulation and Organised Eddy Simulation. Improvement in the aerodynamic forces prediction has been discussed by means of a stochastic forcing approach, based on POD low-energy modes, applied in the DDES

C. Jimenez Navarro · J. Abou Khalil · R. El Akoury · J.-B. Tô · M. Braza (✉)
Institut de Mécanique des Fluides de Toulouse (IMFT), Toulouse, France
e-mail: marianna.braza@imft.fr

C. Jimenez Navarro
e-mail: cesar.jimeneznavarro@imft.fr

J. Abou Khalil
e-mail: jacques.aboukhalil@toulouse-inp.fr

R. El Akoury
e-mail: rajaa.elakoury@imft.fr

J.-B. Tô
e-mail: jean-baptiste.to@imft.fr

A. Marouf · Y. Hoarau
ICUBE Laboratoire des sciences de l'Ingénieur, de l'Informatique et de l'Imagerie, CNRS-University of Strasbourg-ENGEES-INSA, Strasbourg, France
e-mail: amarouf@unistra.fr

Y. Hoarau
e-mail: hoarau@unistra.fr

J.-F. Rouchon
LAPLACE Laboratoire Plasma et Conversion d'Energie (INPT), Toulouse, France
e-mail: rouchon@laplace.univ-tlse.fr

simulations around the V2C wing of Dassault Aviation, considering a constant spanwise section. A significant improvement in the aerodynamic performances has been shown and analysed thanks to specific electroactive morphing concepts based on trailing-edge vibration and on travelling waves along the suction side of an A320 morphing prototype. A decrease in the drag reduction in the order of 7% and in lift increase in the order of 3% have been obtained.

Keywords Transonic flow · Shock wave—boundary layer interaction · Shock induced separation · Numerical simulations · Flow control

1 Introduction

Understanding the principal mechanisms related to buffet instabilities and their interaction with the shear layers and the near wake unsteadiness in the transonic regime around a supercritical airfoil, to successfully modify them through electroactive morphing is the main objective of this paper. Under certain conditions, aircraft wings in the transonic Mach number range of (0.7–0.8) may experience an amplification of a low frequency-high amplitude instability of the shock wave that moves periodically along a specific area on the lifting structure. This leads to a drastic drag increase and may create triggering conditions for amplification of harmful dip-flutter modes. Pioneering studies [1–3] made evidence and analysed the transonic buffet phenomenon. This periodic "excursion" of the shock wave occurs along the suction and pressure sides in case of symmetric airfoil sections and along the suction side only in case of supercritical wing sections as the one of the A320 that concerns the present study. A considerable number of numerical studies were devoted in the prediction and physical analysis of the transonic buffet. Numerical simulations in Deck [4], using the Zonal Detached Eddy Simulations (ZDES), were performed on similar configurations. Furthermore, a DDES approach using the SALSA—"Strain adaptive linear Spalart-Allmaras model" for the near wall region was applied in Grossi et al. [5] on the OAT15A wing in the context of the European research program ATAAC, "Advanced Turbulent simulations for Aerodynamic Application Challenges". This study, as well as the study by Deck [4] have shown the reduction of the well known "grey area" effects between the URANS and LES regions that occur in the DDES and create a "Massive Stress Depletion", (MST), thanks to these modelling approaches. The LES simulations by Dandois et al. [6] have shown through phase-averaged fields a linking of the transonic buffet frequency to a separation bubble "breathing" phenomenon associated with a vortex shedding mechanism. Furthermore, Paladini et al. [7] simulated three-dimensional buffet cells along a swept wing configuration. An illustration of the transonic buffet among first experimental visualisations by Schlieren approach is shown in Fig. 1 (left) and through numerical simulation as for example in Fig. 1, right. In both parts of this figure, the Shock-Boundary Layer Interaction (SBLI) is shown forming a "lambda" structure in the shock foot, as well as the development of the separated shear layers containing small-scale shear-layer

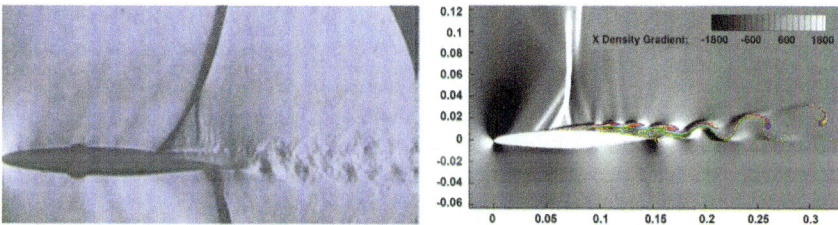

Fig. 1 Normal shock wave and instabilities in transonic flow. Left: Schlieren photograph of the eddying wake following a shock-induced separation (courtesy of National Physical Laboratory, UK [9]). Right: streakline visualization from simulations by Tô et al. [10] illustrating the SBLI, the shear layer (Kelvin-Helmholtz) and von Kármán vortices

vortices, the Kelvin-Helmholtz (KH) ones that lead farther downstream to the alternating von Kármán vortices. In-between these vortex structures, smaller scale chaotic eddies are formed due to the turbulence. There is also shown that secondary waves are formed on the "crests" of these vortices along the shear layers as well as upstream of the shock, because of the thickening of the so-called "effective obstacle", which is not the nominal profile's shape but the one countered by the shearing regions. For the same reasons, an oblique shock just upstream of the lambda foot is created, because of the flow reversal that goes even under the shock's foot towards this upstream area and creates there a local thickening of the boundary layer, thus producing this secondary shock. This pattern was shown in a number of experimental studies in the collected contributions of the European research project TFAST, "Transition Location effect on shock boundary layer interaction", Doerffer et al. [8]. Therefore, a significant *feedback effect* from this downstream unsteadiness towards the area upstream of the shock can be seen.

When critical conditions in terms of Mach number and angle of attack are met, the formed shock wave over the wing becomes unstable and moves. This transonic buffet motion and its underlying governing physics are a long-standing challenge in aeronautical research. These self-sustained shock oscillations cause a premature separation of the boundary layer followed by a massive separation whose dimensions evolve in a cyclic manner due upstream and downstream motion of the shock waves. Whereas the majority of the studies devoted to the transonic interaction deal with the high Reynolds number range, the physical mechanisms of the buffet onset can be studied more easily in the lower Reynolds number range, allowing detection of the physical mechanisms onset, possible through Navier Stokes and Direct Numerical Simulations that are fully established in the high Reynolds range. Bouhadji and Braza [11, 12] and Bourdet et al. [13] have studied the progressive stages of unsteadiness due to compressibility effects over the NACA0012 airfoil in the Mach number range 0.3–1.0 by 2D and 3D simulations in the Reynolds number range of 10,000, as well as the onset of the secondary instability along the span and the evaluation of the buffet frequency by DNS et by stability analysis, Bourdet et al. [13]. The buffet instability and interaction with the von Kármán vortex shedding in the Mach number range

0.75–0.85 were analysed in detail. The von Kármán instability (mode I) is present for the whole range of Mach numbers up to 0.85. The buffet instability (mode II), appears in the Mach number range (0.7–0.8) after the appearance of low frequency regular oscillations of the upstream supersonic regions that strongly interact with the von Kármán instability. Therefore, the buffet *was found to be sustained by mode I*. Numerical simulations in the high-Re range [14] studied the influence of the near-wake unsteadiness on the buffet through a trailing edge extension by a splitter plate of the OAT15A airfoil, a test-case investigated in the EU-UFAST project, Doerffer et al. [15]. Beyond a critical length of the splitter plate, *the ability of a total suppression of the shock wave's* motion was shown because of the von Kármán mode "sweep" much farther downstream. Therefore, the interaction between the shock wave and this mode became very weak as shown in Fig. 2, Szubert et al. [16]. These results highlight the importance of studying the shock-vortex interactions at high Reynolds number and their impact on the upstream instabilities through strong *feedback effects*. These facts enlighten the reasons of appearance and of suppression of the buffet mode and are largely exploited in the next section, aiming at *attenuating this harmful mode through electroactive morphing*.

2 Transonic Buffet Prediction Through Interfacial Shear Layer Investigation

The interfacial shear layer dynamics play an essential role in the development of different classes of coherent vortices, in association with the amplification of the shear layer instabilities. As can be seen in Fig. 2 (right), the thin interfaces that separate the irrotational and sheared regions are known as turbulent/non-turbulent (TNT) interfaces. A wide range of turbulent structures exists between these interfaces. Interaction modifies the various vortical structures in—between and form an inner interface, named TT: Turbulent-Turbulent interface, Fig. 3. The majority of widely used turbulence modelling closures employ concept of a downscale cascade,

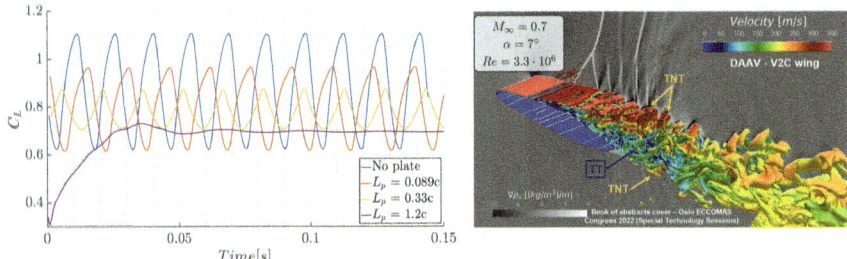

Fig. 2 Left: Time evolution of lift coefficient with different lengths of trailing edge plate. Adapted from Jimenez-Garcia [14]. Right: DDES-OES simulation, V2C wing, Jimenez-Navarro [17]

that produce an excessive level of turbulence diffusion, resulting in a thickening of the shear layers past the SBLI. This results in over-prediction of the drag coefficient and under-prediction of lift, in high-Re simulations. A way to attenuate these effects comes from physical considerations in interfacial shear layers [18] and through morphing of wings [19–21] by means of *eddy-blocking effect*. A first study of turbulence modelling with stochastic forcing was made by introducing smaller-scale vortices than the main shear-layer coherent structures, through appropriate source terms in the turbulence transport equations in 2D [22]. In that study, a small kinetic energy perturbation was generated by means of low-energy POD modes as detailed in this section. A first 3D study was done by Simiriotis [23] where three-dimensional POD modes derived from URANS simulation have been used for reconstruction of the kinetic energy perturbation and implemented in the OES—Organised Eddy Simulation Turbulence modelling. During the TEAMAERO project, in the context of the Ph.D. thesis of César Jimenez-Navarro [17], the kinetic energy perturbation has been reconstructed by 3D POD modes from a simulation using the DDES-OES turbulence modelling approach, Bourguet et al. [24], Marouf [25]. In the experimental context, the stochastic forcing has been materialised by morphing through trailing-edge vibrations [19–21] for low subsonic regimes, around the A320 transonic regime wing prototype (tRS) of the H2020 N°723402 SMS—"Smart Morphing and Sensing for aeronautical configurations" research project, Braza et al. [26], www.smartwing. org/SMS/EU. By means of injection of smaller-scale vortices in the shearing regions, a shear-sheltering effect is produced due to vortex breakdown of the existing natural instability vortices turning in the opposite sense as the new introduced ones and enhancement of those turning in the same sense. These last are reinforced and their diffusive spreading is reduced [27]. This effect, generated either numerically or experimentally through morphing, leads to thinning of the shear layers by modification of the TT interface and enhances *the upscale cascade, i.e. energy transfer from smaller-scales to larger scales*, thus refraining the excessive turbulence diffusion. The stochastic forcing is performed numerically by adding production source terms in the turbulent transport equations that model the effects of energy transfer by small scale fluctuations to the large eddies in cases of non-equilibrium turbulence. To concentrate this effect in the sheared regions, the injected turbulent kinetic energy is obtained from a Proper Orthogonal Decomposition (POD), where high-order low-energy modes associated to chaotic turbulence are used to reconstruct a small kinetic energy amount, K_{POD} scalar field: $K_{POD} = \frac{1}{2}(\overline{u'^2}_{recon} + \overline{v'^2}_{recon} + \overline{w'^2}_{recon})$. The stochastic behaviour of the non-coherent eddies is taken into account by introducing a random-number variation of the forcing over time with the scalar value $\tilde{r}$ that changes from 0 to 1 randomly. The resulting source term is: $S_{POD} = \frac{C_\mu (k_{amb}^2 + \tilde{r} k_{POD}^2)}{\nu_{t\infty}}$. The approach was developed first in Szubert et al. [22]. In the present study, DDES simulations have been done with the Chien's $k - \varepsilon$ model in the URANS part. The POD modes have been generated by these 3D simulations and the K_{POD} has been re-injected in the shearing regions through the above equations. It is worth noticing that the benefits of the POD reconstruction are their action *within the shearing regions only*, Fig. 5 (right). The modified transport equations are given by the expressions

Fig. 3 Turbulent interfaces delimiting the sheared regions

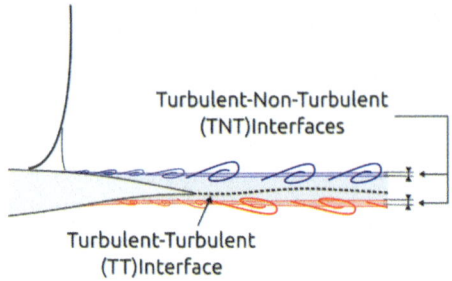

(1). All the simulations have been done with the use of the NSMB (Navier-Stokes Multi-Block) solver, Hoarau et al. [28].

$$\frac{\partial \overline{\rho} k}{\partial t} + \frac{\partial (\overline{\rho} \widetilde{u}_j k)}{\partial x_j} = -\overline{\rho u_i' u_j'} \frac{\partial \bar{u}_i}{\partial x_j} + \frac{\partial}{\partial x_j}\left(\left(\mu + \frac{\mu_t}{\sigma_k}\right)\frac{\partial k}{\partial x_j}\right) - \overline{\rho}\varepsilon \underbrace{- \frac{2\mu k}{d^2}}_{Chien's\ damping} \underbrace{+ S_{POD}}_{Forcing\ term}$$

$$\frac{\partial \overline{\rho}\varepsilon}{\partial t} + \frac{\partial (\overline{\rho}\widetilde{u}_j\varepsilon)}{\partial x_j} = C_{\varepsilon 1}\frac{\varepsilon}{k}\left(-\overline{\rho u_i' u_j'}\right)\frac{\partial \bar{u}_i}{\partial x_j} + \frac{\partial}{\partial x_j}\left(\left(\mu + \frac{\mu_t}{\sigma_\varepsilon}\right)\frac{\partial \varepsilon}{\partial x_j}\right)$$
$$- C_{\varepsilon 2}f\rho\frac{k^2}{\varepsilon} \underbrace{- \frac{2\mu\varepsilon e^{-0.5\bar{\rho}u_\tau d/\mu}}{d^2}}_{Chien's\ damping} \underbrace{+ \frac{C_{\varepsilon 2}S_{POD}^2}{k_{amb}}}_{Forcing\ term}. \quad (1)$$

2.1 The V2C Wing

The 3D simulations around the V2C wing in transonic conditions have been carried out with a chord's length of 0.25m at angle of attack of $\alpha = 7°$, a free-stream Mach number of 0.7 with the value of total pressure set to 100000 Pa and total temperature of of 290 K which leads to values of 72092.79 Pa for the inlet static pressure and 264.12 K for the static temperature, leading to a chord-based Reynolds number of $3.24 \cdot 10^6$. The inlet free-stream turbulence intensity is 10^{-6} m^2 s^{-2}. The DDES $k - \varepsilon$ modelling has been employed. The flow regime shows amplification of the buffet instability with corresponding shock oscillations in the order of 91 Hz. The shock's excursion takes place in the range of x/c = (0.175–0.4) that corresponds to an amplitude of 22.5% of the chord.

The different phases of the shock's motion are illustrated in Fig. 4. The most downstream location of the shock takes place at $t/T_b = 0.00$. As the shock advances ($t/T_b = 0.18$), the lambda shock grows in size while causing an increase of the

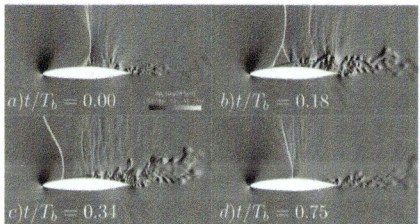 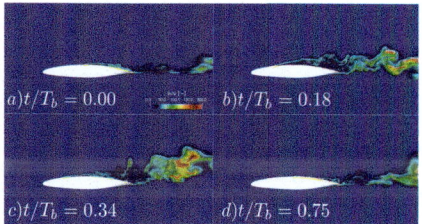

Fig. 4 Left: Instantaneous density gradients in the streamwise direction contours in 4 relevant phases of the shock's motion. The slice is taken at the wing's centerline, $y = b/2$. Right: Instantaneous eddy viscosity ratio (ν_t/ν) contours in 4 relevant phases of the shock's motion

recirculation region. This induces a reduction of the effective angle of attack due to the wake deflection and the increase of the effective geometry of the wing which tends towards a bluff body geometry. Therefore, the shock becomes progressively weaker until it reaches its most upstream location at $t/T_b = 0.34$. At this stage, the shock induced separation is significantly reduced and the shock can move downstream, $t/T_b = 0.75$.

The Proper Orthogonal Decomposition (POD) [29] is used. Within this context, the POD is used to analyze the complex turbulent flow past the V2C wing to gain insights to the underlying the principal dynamics, to find principal natural frequencies of the main coherent structures and to use the low-energy POD modes for the stochastic forcing. The chosen sampling rate for the snapshots has been set to 20 kHz (see the lift signal in Fig. 6).

2.2 Simulations with Stochastic Forcing

Figure 6 shows the lift coefficient variations with different percentages of the k_{POD} stochastic forcing. The cases with higher rates of this parameter re-inject a significant turbulence background and orient the flow system towards a chaotic behavior similar to the one obtained by DNS in the TFAST Book, without sustaining the buffet mode. The reduced rates of the stochastic forcing by 0.001 and 0.0001 are the most successful.

The present study shows that the production of the POD modes from a DDES simulation induce a rather important turbulence injection rate than the study that used 3D POD from a URANS $k - \omega$ SST modelling, injected in the DES approach [23].

One possible solution to address this issue is to limit or trim the k_{POD} field before a certain chord-wise coordinate. This approach can help avoid the contamination of the attached boundary layer when the shock is not at its most upstream location (Figs. 7 and 8). The proposed field is presented in Fig. 9b where the k_{POD} has been set to zero for the cells whose axial location is lower than 55% of the chord ($x/C < 0.55$) which is the approximate most downstream location that the shock reaches

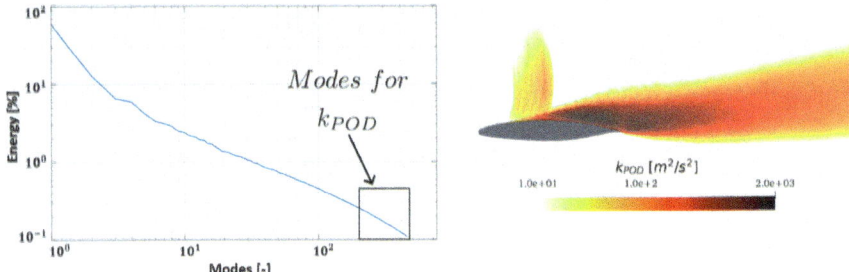

Fig. 5 Left: POD modes selected for K_{POD} reconstruction. Right: Reconstructed k_{POD} field with modes from 200 to 450

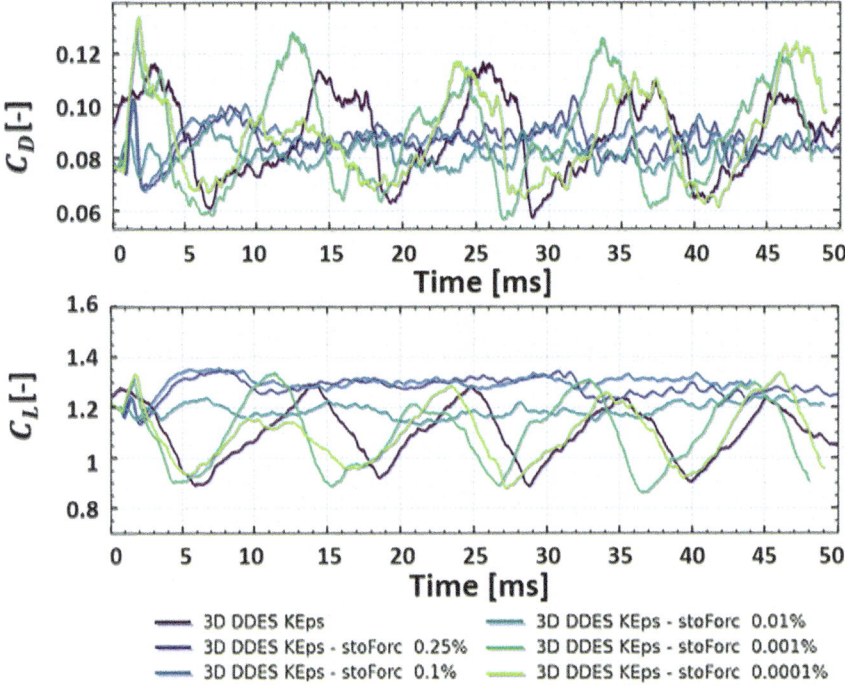

Fig. 6 Time evolution of the lift coefficient for different values of energy injected through stochastic forcing

during the simulations. The new field was substituted in the same simulation and the continuation showed the prediction of a self-sustained shock oscillation as displayed in Fig. 10. Therefore, this could confirm that the source of the problem associated with the prediction of the shock's oscillation is indeed the miss-prediction of the boundary layer characteristics by an incorrect energy injection upstream the shock.

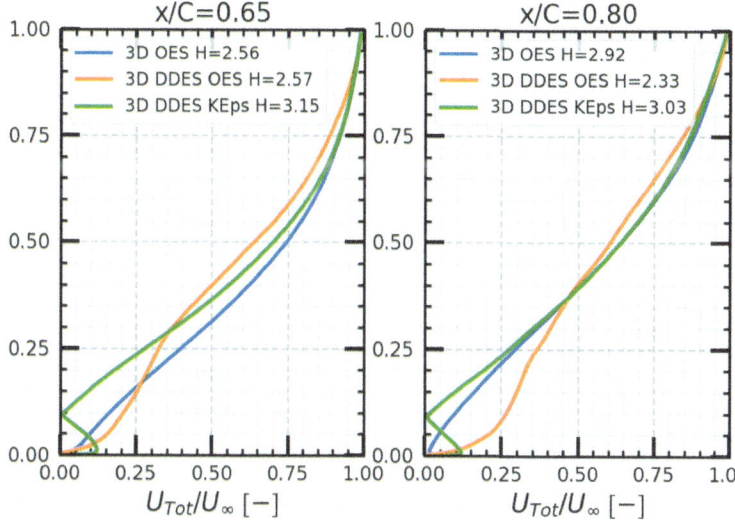

Fig. 7 Time-averaged surface-parallel boundary layer velocity profiles at two chord-wise locations over the wing's suction side. The shape factor, H, is indicated in the legend for each case

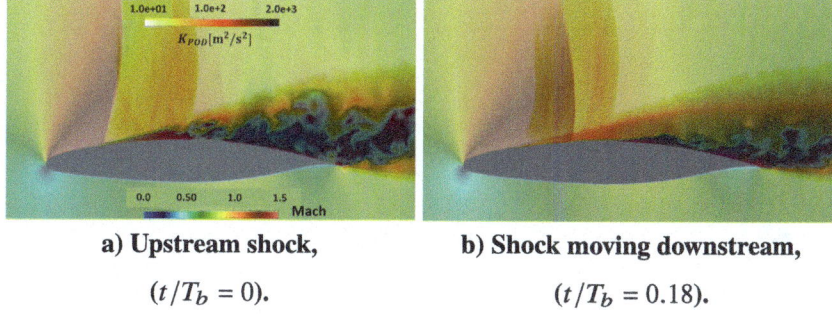

a) **Upstream shock,**

$(t/T_b = 0)$.

b) **Shock moving downstream,**

$(t/T_b = 0.18)$.

Fig. 8 Mach number contours overlapped with the reconstructed k_{POD} field in the DDES $k - \varepsilon$ model, showing the reduction of the suction area on the downstream motion of the shock

The results have been compared with the experimental data and numerical simulations from the TFAST project [8]. The mean pressure distribution and fluctuation levels are displayed in Fig. 11 and the mean force values and buffet frequencies are presented in Table 1. Only the tripped simulation with the DDES $k - \varepsilon$ model from Dassault Aviation predict the shock's mean location with acceptable accuracy but failed on predicting the amplitude of the motion and unsteady levels. The tripped DNS from the Sapienza University of Rome can accurately predict the mean forces with some misagreements on the region of appearance of the shock and the motion characteristics. The 2D $k - \varepsilon$ model shows the worst predictions, as it is expected. No significant effects on the forces or flow modification were found in the cases

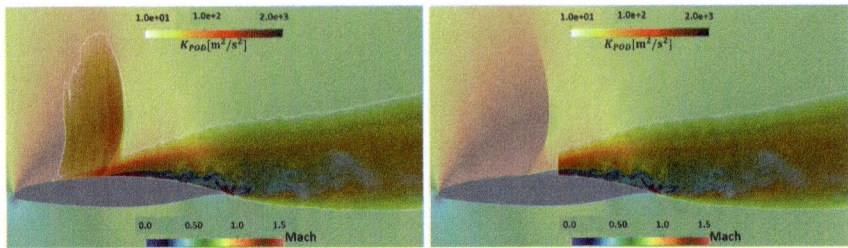

Fig. 9 Mach number contours overlapped with the reconstructed k_{POD} field in the DDES $k-\varepsilon$ model when the shock wave it is at its most downstream location with (left) the original k_{POD} field and (right)) a trimmed k_{POD} field where $k_{POD} = 0$ for $x/C < 0.55$

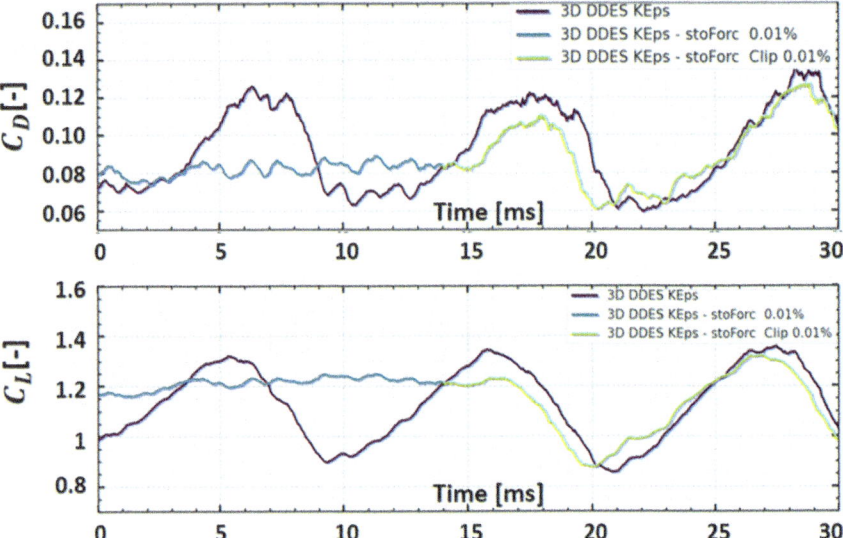

Fig. 10 Effects on the lift coefficient when substituting the original k_{POD} field with the trimmed k_{POD} field where $k_{POD} = 0$ for $x/C < 0.55$

where the stochastic forcing was applied due to the previous discussion, with the exception of an increase of the buffet's amplitude. The results from the stochastic simulations by Simiriotis [23] have been added at the end of the Table 1. They successfully showed the benefits of this kind of modelling where the expected thinning of the sheared regions was achieved. Many parameters differ from the simulations presented in this study, mainly the grid, the modified turbulence model and the selected modes for the k_{POD} generation. Nevertheless, the present work constitutes a learning step towards the correct use of the stochastic forcing in hybrid simulations. The time-averaged force coefficients have been compared with previous research works in Table 1. An increase of the lift force is obtained in the simulations with stochastic forcing thanks to the insertion of energy from smaller structures that help to promote

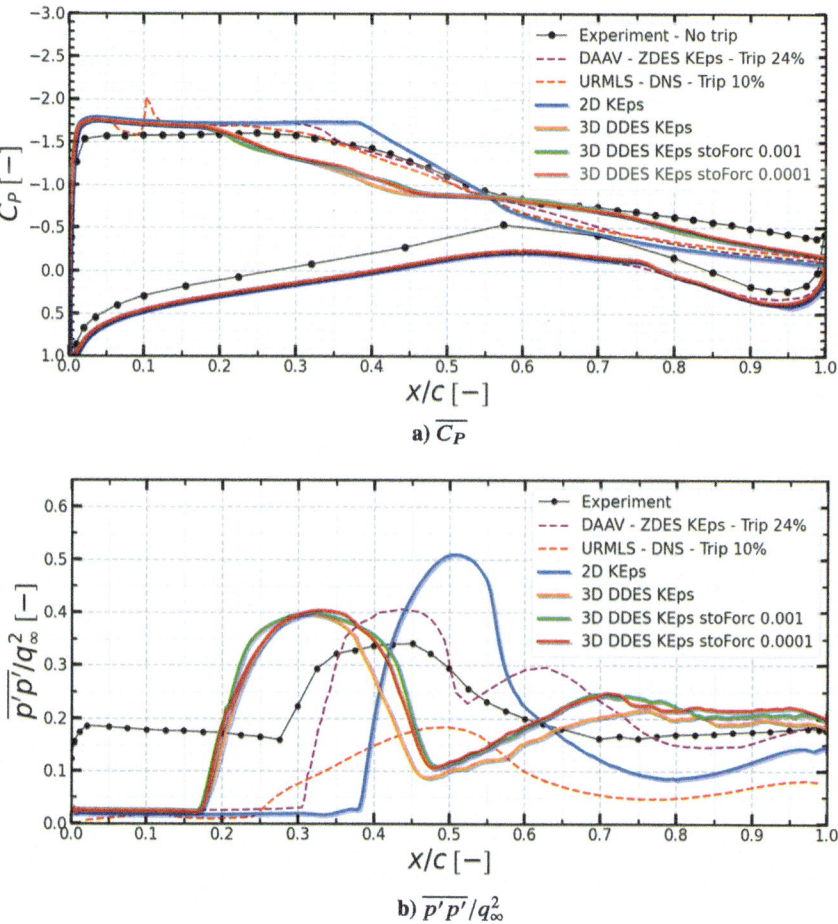

Fig. 11 Time-averaged pressure coefficient (left) and pressure fluctuations product (right) compared with the experimental data and numerical simulations from the TFAST project [8]

the circulation on the suction side of the wing as well as a decrease of drag with the 0.001% rate of the k_{POD}. The stochastic forcing through this kind of turbulence reinjection in the shearing regions is a general approach that can be used in other modelling approaches, statistical, LES or hybrid ones. It is worth noting that morphing concepts inject small vortices and are able to manipulate coherent and chaotic turbulent structures and turbulent interfaces through the actuation. The effects of the forcing can therefore be materialized through electroactive morphing strategies since morphing concepts can add kinetic energy at desired frequencies and amplitudes to enhance upscale effects and improve the aerodynamic performance.

Table 1 Comparison of force prediction and buffet frequencies. The Strouhal number is based on the wing's chord

	$\overline{C_L}$	$\overline{C_D}$	f_b (Hz)	f_b (St)
EXP V2 clean	1.0077	0.0748	82	0.09
Dassault aviation DDES $k - \varepsilon$—Trip 24%	–	–	119	0.13
Sapienza University of Roma DNS—Trip 10%	1.1131	0.0791	120	0.13
IMFT 2D $k - \varepsilon$ (Chien's damping)	1.2000	0.0749	66	0.07
IMFT DDES $k - \varepsilon$ (Chien's damping)	1.0906	0.0868	91	0.10
IMFT DDES Sto. Forc. $1 \cdot 10^{-3}$%	1.1130	0.0893	91	0.10
IMFT DDES Sto. Forc. $1 \cdot 10^{-4}$%	1.1116	0.0898	91	0.10
IMFT DDES Sto. Forc. $1 \cdot 10^{-3}$% (Trimmed)	1.1043	0.0883	91	0.10
IMFT DOES from Simiriotis (2020) [23]	1.0360	0.0884	95	0.10
IMFT DOES Sto. Forc. from Simiriotis [23]	1.0570	0.0795	95	0.10

3 Electroactive Morphing of the tRS—"transonic Reduced Scale A320 Prototype" of the H2020 SMS EU Project

This section aims at studying advanced morphing concept able to increase the aerodynamic performance. The study is based on the tRS prototype of the SMS project. A detailed experimental study was done by IMP-PAN. In the TEAMAERO project, a detailed numerical study has been undertaken to analyse the modification of the transonic interaction in cruise conditions by means of near-trailing-edge region actuation and slight deformation in two and three dimensions. The 2D study allowed evaluation of optimal vibration frequencies and amplitude ranges and a reduced number of selected parameters from these ranges allowed performing a 3D study, that evaluated the increase of the aerodynamic performances and reduction of the noise sources produced by the trailing edge. As mentioned in the previous section, the idea of using different morphing concepts in the rear part of a wing was expected to produce significant feedback effects on the overall area of the wing and particularly, over a key distance below the SBLI and upstream of it. This kind of morphing, that is a materialization of the mechanisms of the stochastic forcing discussed in the previous section, is able to drastically modify the forces. The work presented in this section goes beyond the state of art by studying small trailing edge deformations at relatively higher frequencies, in the range of 100–800 Hz, that are possible thanks to the usage of new generation electroactive materials.

3.1 Flow Parameters and Numerical Method

The A320 tRS wing prototype has a chord of 15 cm (Fig. 12). The boundary conditions selected for the numerical simulations have been taken in accordance with the wind tunnel experimental parameters of IMP-PAN, using an angle of attack of $\alpha = 1.8°$, a free-stream Mach number of 0.78 with the values of static pressure and static temperature set to 67574 Pa and 261.22 K respectively. This leads to a chord-based Reynolds number of 2.06×10^6. The upstream (inlet) freestream turbulence intensity is $3.75 \cdot 10^{-5}\,\mathrm{m^2\,s^{-2}}$. The code NSMB has been used. An implicit dual time- stepping second-order accurate method was used with an outer time step of $dt = 5 \cdot 10^{-6}$ and the number of inner iterations set to $N_{inner} = 150$. These values were determined after detailed tests to be the most suitable in terms of accuracy and convergence [17]. Regarding the discretization of the convective fluxes, Roe's upwind scheme was utilized together with the 3rd order Monotonic Upwind Scheme for Conservation Laws (MUSCL). The diffusion terms have been discretized by second-order central scheme of Jameson. For the 2D computations, the domain and the geometry have been discretized in a 2D structured multi-block C type mesh with 64 blocks and 194000 elements. For the 3D simulations, this mesh has been extruded over a spanwise length of 7 cm. The values of y^+ have been in the order of 10^{-1} with around 70 points in the boundary layer. The turbulence modelling is the the Organized Eddy Simulation (OES) approach [22, 24, 30] sensitized for the physically correct development of the instabilities and coherent structures. This model allows respecting the dual character of turbulence, organized and chaotic, where the chaotic turbulence is embedded in the coherent structures. By using the ensemble average, a turbulence spectrum splitting is operated, where the coherent part is the resolved turbulence and the chaotic turbulence is modeled by means of a modified closure for the turbulence stresses to consider the energy subtracted by the coherent eddies. The OES approach has been largely used in previous research activities along federative EU projects DESIDER, ATAAC, UFAST, TFAST and showed remarkable efficiency in the prediction of high Reynolds number highly detached flows, by providing a physically correct development of the flow instabilities and related coherent structures, even in two dimensions, because the approach is not intrinsically 3D. In the present study, due to main physical characteristics of the coherent structures that can be assessed by

 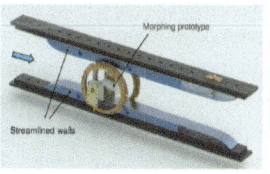 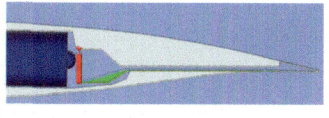

Fig. 12 The tRS A320 prototype in the IMP-PAN's transonic wind tunnel (left), (middle) wind tunnel streamlined walls and (right) internal schematics of the tRS prototype constructed by LAPLACE with the embedded piezoelectric actuator (blue) and the force transmission chain (red and green)

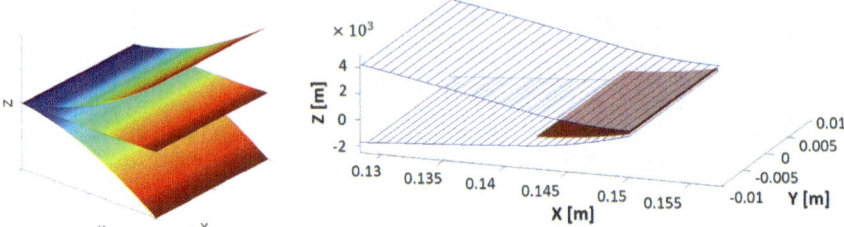

Fig. 13 Schematic description of the displacement by means of the second order polynomial deformation in two and three-dimension. Left: Patch deformation in different stages; right: Instantaneous upward mesh deflection

a 2D approximation, it was possible to perform a large parametric study in 2D concerning the vibration effects and to detect optimal intervals in which a narrower space of these parameters has been used in a 3D study afterwards. The statistics have been performed over 1 s of physical time where at least 100 buffet cycles were captured. Since morphing needs a moving grid, the cells associated to the moving trailing edge region and the cells in the nearby region are moved and deformed by the Arbitrary Lagrangian Eulerian (ALE) method in NSMB with a Transfinite 2D interpolation (TFI2D). The displacement of the trailing edge is modelled by using a quadratic function given by equation $z(x, t) = \left(\frac{2h_p}{3L_p^2}(x - x_0)^2 + \frac{h_p}{3L_p}(x - x_0) \right) sin(\omega t)$ where h_p is the patch semi-amplitude, L_p is the patch length, x_p is the initial coordinate of the patch in the x direction and ω is the angular frequency ($\omega = \frac{2\pi}{f_a}$) obtained as a function of the actuation frequency (f_a). Figure 13 is provided for giving the reader a better understanding of the numerical implementation.

3.2 Physical Analysis of the Flow Structure in the Static Case

Figure 14 is a 3D numerical simulation of the flow by using the 2D grid extruded over 7cm along the span. The λ-shock structure can be seen as well as the separated region and different classes of vortices due to the shear layers and wake unsteadiness. These vortices undergo specific spanwise undulations, due to development of secondary instabilities. The larger undulated vortices are the 3D von Kármán vortices. The Kelvin-Helmholtz (KH) vortices in the upper shear layer are formed at the beginning of the shear layer created between the separated area and the outer flow region, due to the intense shear between the high- speed flow coming from the supersonic region and the low-speed recirculating flow. The lower shear layer is created from the trailing edge and forms the lower TNT (Turbulent-Non Turbulent) Interface. The von Kármán vortices are created after interaction of the upper and lower shear layers due to the shearing of the velocity profiles much downstream of the trailing edge in the wake which creates the alternating vortices, due to the presence of two inflexion

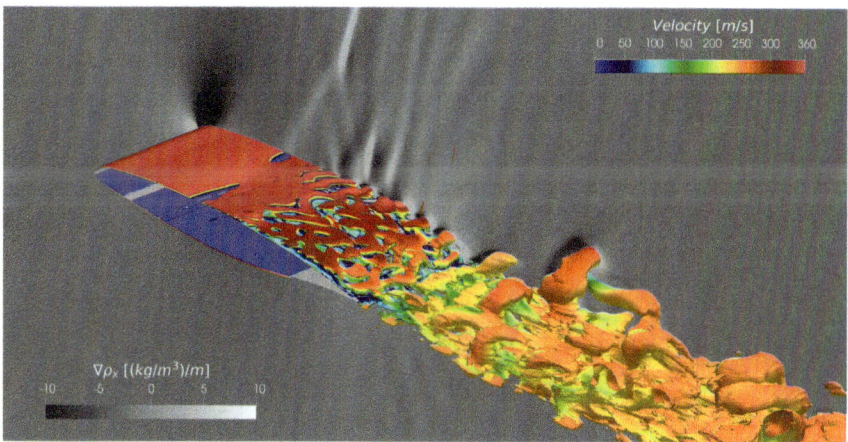

Fig. 14 Flow over the A320 tRs prototype: Q-criterion contours coloured with velocity magnitude and density gradients displayed at the symmetry plane

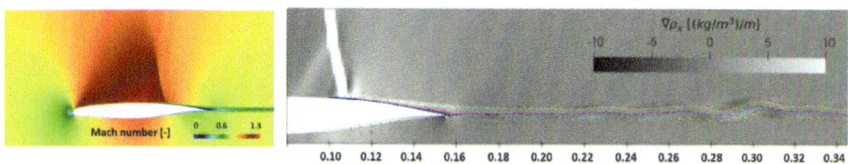

Fig. 15 Left: Time-averaged Mach number contours for the unactuated A320 airfoil. Right: Instantaneous density gradient contours with streaklines for the unactuated A320 airfoil

points in the velocity profiles in that area. The development of these instability modes has significant effects on the aerodynamic performance. They can cause drag and lift fluctuations, which can lead to structural vibrations that affect the stability and control of the aircraft and in addition, noise generation.

The flow conditions considered lead to a weak form of buffet where the mean position of the shock is $\frac{\bar{x}}{c} = 0.659$ and its (*rms*) 0.6×10^{-3}. In Fig. 15 (left), the averaged flow is illustrated through the Mach number iso-contours, showing the mean shock location. The wake structures have been visualized by means of streaklines, Fig. 15 (right). The buffet dynamics have been studied by means of Power Spectral Density (PSD), Fig. 16. This figure shows a a prominent peak corresponding to the buffet frequency ($f_b = 106$ Hz, $St = 0.063$), for the static (unactuated) case. In Fig. 16 (right) the higher frequency predominant peaks or bumps are associated to the shedding of the KH vortices around 20 kHz. The VK frequency is in the order of 6.8 kHz, whose frequency bump is present in the PSD from Fig. 16 (left) and was also confirmed by tracking the von Kármán vortices from flow visualizations. Therefore, these two figures highlight that the three main modes that play an important role in the evolution of the forces are the buffet, the von Kármán, and Kelvin-Helmholtz modes.

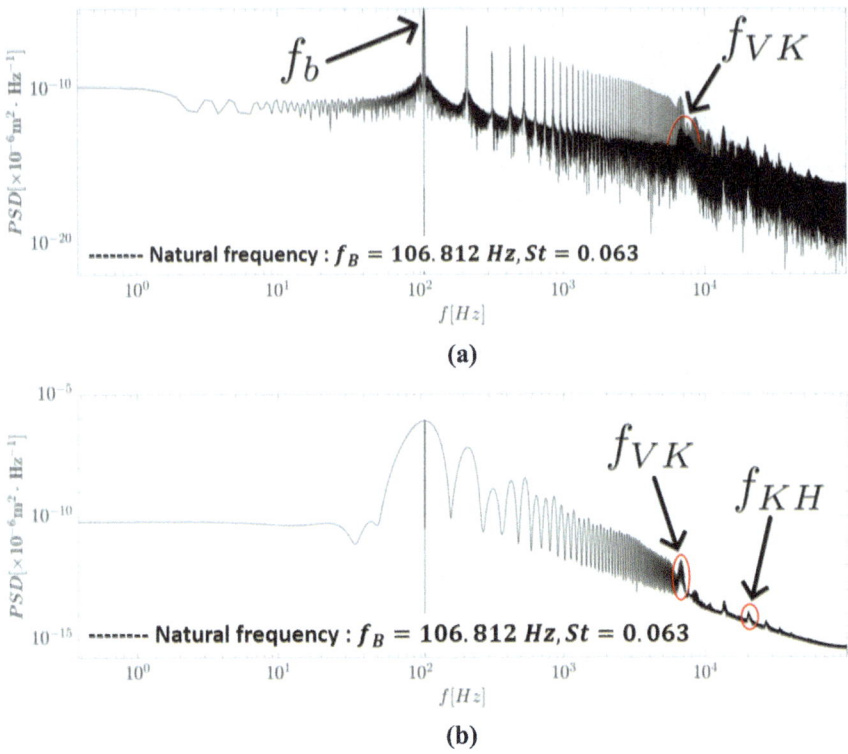

Fig. 16 PSD on the lift coefficient showing **a** the buffet frequency and its harmonics and **b** a zoom on the buffet frequency, the von Kármán (VK) and Kelvin Helmoltz (KH) modes

PSDs were computed for the lift coefficient signal for the numerical simulations and for the shock wave's location evolution for the experimental data. The results are displayed in Fig. 17.

3.3 Parametric Study on the Vibration Frequency with Amplitude Set to $\theta = \pm 1°$

The actuator's length was considered as a constant parameter in all the simulations according to the experiments with a value of $7.5 \cdot 10^{-3}$ m (i.e. 5% of the airfoil chord). The frequencies considered in this study are contained in the range $f_a = [100{-}2000]$ Hz. A detailed study on the amplitude effects in the range of $\theta = \pm 1°, \pm 2°, \pm 3°, \pm 5°$ can be found in the Ph.D. thesis of César Jimenez Navarro [17], where amplitudes higher than 2° increase the drag. The force coefficients are summarized in Fig. 18. In general terms, the trailing edge vibration leads to an

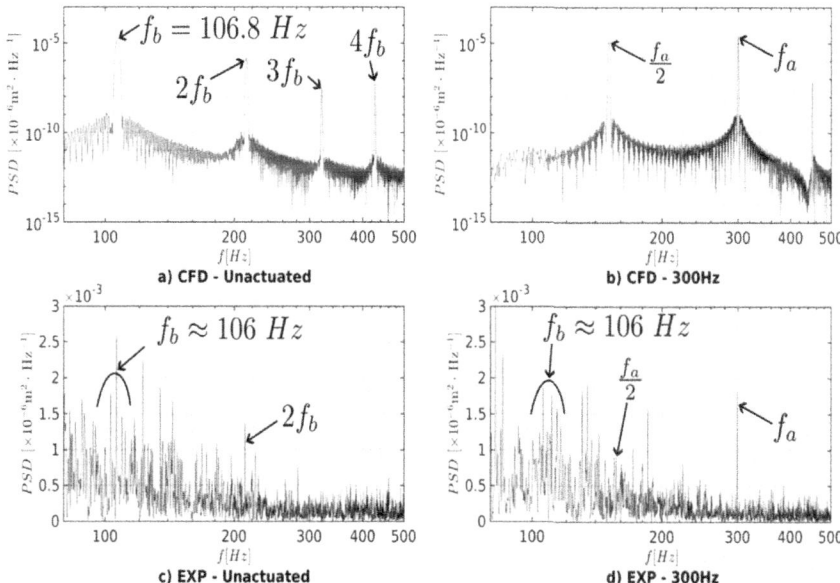

Fig. 17 PSDs of C_L **a** static case, **b** actuated at 300 Hz, **c** shock from experimental Schlieren results of IMP-PAN static case and **d** actuated at 300 Hz

increase in lift and drag but not always in the same proportion, as shown in the lift to drag ratio evolution. The red areas indicate optimal ranges. The frequency ranges (250–390) Hz is a promising area in terms of lift increase and simultaneous *rms* reduction comparing to the static case. A synchronization of the shock motion and the actuation frequency was observed between $f_a = [365-510]$ Hz, where a significant increase of C_L/C_D with frequency is perceived and at $f_a = 720$ Hz which is the most optimal frequency in terms of lift to drag ratio. This phenomenon referred to as *lock-in* effect [10] has been investigated through the PSD of the pressure fluctuations in Fig. 19, where a modified buffet frequency appears exactly at the actuation frequency. To study this phenomenon, the shock wave's position has been tracked during the simulation. This was done by means of density gradients recorded on a line placed across the shock wave, as shown in Fig. 20. Mean and *rms* of the signals are very meaningful because they highlight that by acting at the right frequencies it is possible to push the shock downstream and reduce its fluctuating level, leading to a significant buffet attenuation. The frequencies that lead to an optimal buffet attenuation are between the range (365, 510) Hz and around $f_a = 720$ Hz which are the frequencies where the lock-in effect occurs.

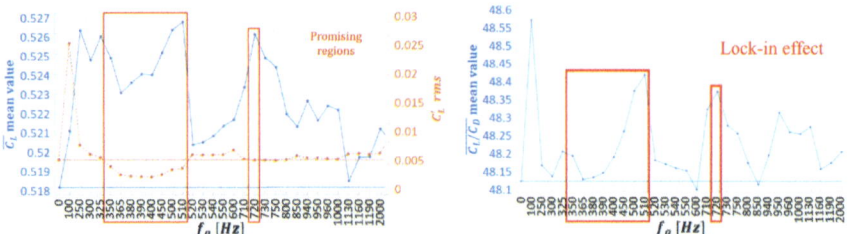

Fig. 18 Left: Mean (blue) and *rms* (orange) drag coefficient evolution for the studied frequencies. The two horizontal lines show the relative evolution with respect to the static case ($f_a = 0$ Hz). Right: Mean (blue) and *rms* (orange) lift to drag ratio evolution with $\theta = \pm 1°$

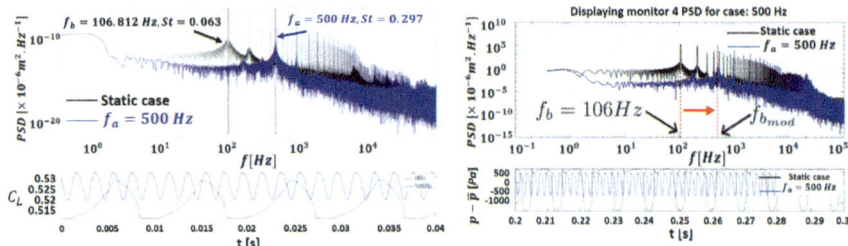

Fig. 19 Left: PSD analysis of the lift coefficient signal for the case $f_a = 500$ Hz compared with the unactuated case $f_a = 0$ Hz. Right: PSD of the pressure signal at the monitor point 4 for the case $f_a = 500$ Hz compared with the unactuated case $f_a = 0$ Hz

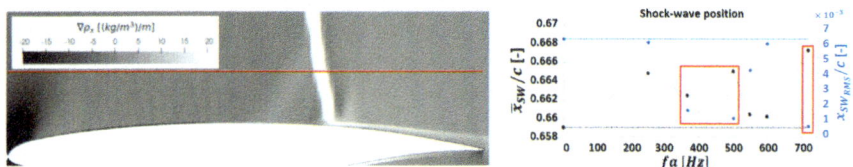

Fig. 20 Left: Instantaneous density gradient contours in X direction along the red horizontal line used to track the shock's position. Right: Mean (black) and *rms* (orange) shock's position evolution for interesting frequencies with $\theta = \pm 1°$. Two horizontal lines are presented for each set of results to show the relative evolution with respect to the static case ($f_a = 0$ Hz)

3.4 Three-Dimensional Parametric Study

After the 2D parametric study has been completed, the most promising frequencies found have been used for 3D simulations. These frequencies are $f_a = 500$ Hz and $f_a = 700$ Hz due to the increase of the aerodynamic performance and reduction of the *rms* values that they provide. The actuation amplitude is $\theta = \pm 1°$. The grid has been extruded in the spanwise direction with 7 cm span. It contains 9.7 million elements and 50 cells uniformly distributed in the spanwise direction. Figure 14 shows the flow structure discussed in the beginning of this section.

Numerical Study and Physical Analysis of the Transonic Interaction...

The relative change of the aerodynamic coefficients is presented in Fig. 21. As has been initially predicted in the 2D simulations, the wing experiences a gain in aerodynamic performance for the considered frequencies but the *improvements are significantly higher* in the 3D computations (around 2–3% of gain). The same trend is confirmed for the lift coefficient. The drag shows a different behaviour from the 2D computations *with a decrease in the order of 1.2%*. The mean shock location can be found by displaying the time-averaged pressure fluctuations product (pressure *rms*). For the morphing cases, the front of maximum *rms* is "pushed" downstream and this is associated to the obtained benefits, Fig. 21.

As previously discussed, the importance of the downstream unsteadiness on the SBLI is a key factor to optimize the performances. A similar effect as the one produced by the introduction of a splitter plate [14, 16] can be induced if this trailing edge plate is replaced by another element or interface that reduces the interactions between the upper and lower shear layers. In the present case, this element is the piezoactuator patch that introduces new small scale vortices through the vibrations. The mechanism is illustrated in Fig. 3. The shear layers are the boundaries between the "inviscid" and rotational regions, forming the two Turbulent-Non-Turbulent (TNT) interfaces and a Turbulent-Turbulent (TT) inner interface. The latest is the one manipulated by the trailing edge vibrations. The injection of energy from the fluid-structure interactions creates new eddies at specific frequency ranges whose effect is to block the interactions between the upper and lower interfaces in the near-wake region and to reinforce the vortex structures in these shear layers, constricting them under shear sheltering effect. This action "pushes" the VK mode much farther downstream and therefore, the related interaction of this mode with the SBLI becomes very weak, thus attenuating the buffet and the *rms* of the pressure and therefore of the forces, leading to the improved aerodynamic performances. The most promising parametric configurations take place when the lock-in occurs at $f_a = [365-510]$ Hz and $f_a = 720$ Hz, acting with 0.2 mm of amplitude ($\theta = \pm 1°$). An increase of the aerodynamic efficiency of 1% in 2D and of 3% in 3D has been obtained, together with a drag reduction in the order of 1.2%, by actuating in optimal frequency ranges in accordance to those detected by the first step of 2D simulations.

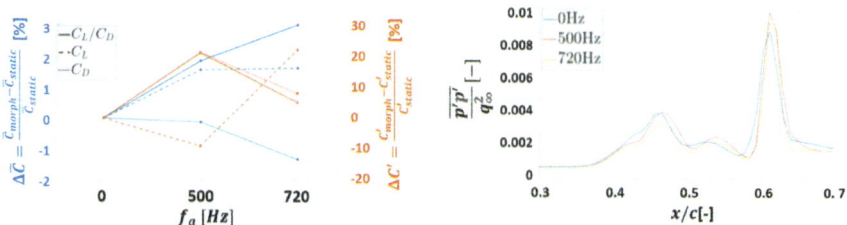

Fig. 21 Left: Morphing effects on the mean (blue) and *rms* (orange) for the lift to drag ratio (solid line), the lift coefficient (dashed line) and drag coefficient (dotted line). Right: Time-averaged pressure fluctuations product, $\overline{p'p'}$

4 Morphing of an Intermediate Scale A320 Airfoil of 70 Cm Chord in Transonic Conditions

In this section, two different morphing concepts are investigated on the 70 cm chord RS- Reduced scale A320 prototype of the SMS project, to provide realisability of the morphing in higher Reynolds number transonic regime. The present chord dimension is almost half of the near-tip chord of the swept A320 wing. The domain and the geometry have been discretized in a structured multi-block C type mesh with 83 blocks and $461 \cdot 10^3$ elements. The values of y^+ are of the order of 10^{-3} with around 60 mesh points in the boundary layer. The boundaries of the computational domain are placed 75 chords away from the studied geometry. The flight parameters used in the numerical simulations were chosen to match the flight conditions provided by Airbus, using an angle of attack of $\alpha = 1.8°$ and a free-stream Mach number of 0.78. Regarding the values of pressure and temperature, they were selected to match the atmospheric conditions for a flight that connects Paris and Toulouse simulated by ONERA-Toulouse [26].

4.1 Analysis of the Unactuated (Static) Case

For the given flight conditions, as shown in the lift coefficient signal (Fig. 22), there is an intermittent behavior of high-lift 'plateaus' followed by shorter durations of low lift intervals. This indicates that the dynamic system is near the limit of a reduced buffet appearance. The frequency of this intermittent phenomenon has been evaluated around 6.9 Hz corresponding to a Strouhal number based on the chord of 0.02 and is considered as a slow buffet mode. The shock's excursion takes place between the chordwise locations $\frac{x_{SW}}{c} = [0.68 - 0.72]$ and the amplitude of the oscillation is around 4% of the chord. A PSD of the lift coefficient is presented in Fig. 23 (left). The von Kármán mode is assessed to have a broadband frequency peak centered in $f_{VK} \approx 3.5$ KHz ($St \approx 9.94$). The reason for this broadband frequency range is linked to the chaotic turbulence motion that affects the VK mode and produces smearing of the alternating vortices. The origin of this modification of the von Kármán vortices is a periodic premature breakdown of the lower shear layer. The last found frequency belongs to the lower shear layer mode which interacts with the von Kármán vortices. Thanks to a frequency analysis that will be later presented, it was possible to assess the frequency of the lower shear layer mode which induces high frequency oscillations in the last part of the pressure side of the wing at a frequency around $f_{LSL} \approx 4$ KHz ($St \approx 12.05$). While peaks have been observed for the frequencies f_R and f_M, the available evidence falls short of providing conclusive proof to support these claims, leading us to consider them as conjecture. A POD decomposition of the system allows detecting the spatial distribution of the von Kármán mode, Fig. 24. The PSDs of the pressure fluctuations on selected monitors are presented in Fig. 25a. The PSDs of the POD temporal coefficients from modes 1 and 10 are presented in Fig. 25b. The

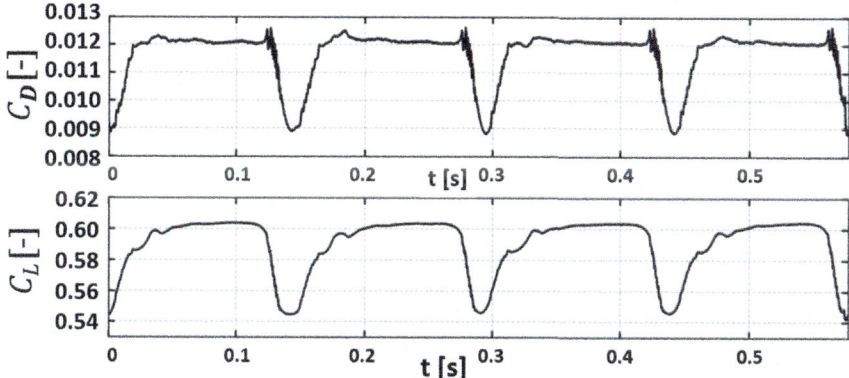

Fig. 22 Time evolution of the drag and lift coefficients for the unactuated case

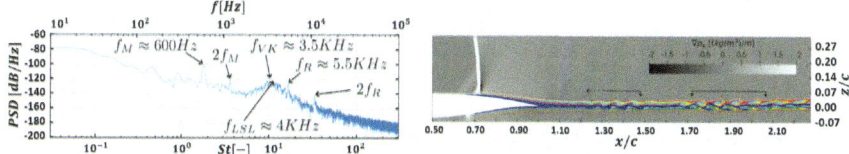

Fig. 23 Left: PSD on the lift coefficients for the unactuated case. Right: Spatial modulation of the von Kármán mode

von Kármán mode is evaluated in the order of 3.5 KHz (St = 10.54), by tracking of the coherent vortices and identification in the PSDs.

Two key phases of the shock's excursion are displayed in Fig. 26. On the left, the shock is initially located at its most downstream location. Then, a thickening of the shear layers occurs due to the growth of the wake instabilities. These structures are illustrated by means of streaklines in Fig. 26 (right) where the KH and VK modes are clearly visible. The feedback effects coming from these instabilities cause the shock to move forward and produce stronger flow recirculation underneath the shock which can be appreciated due to the disappearance of the blue particles in the lower region of the suction side that are being accumulated at the recirculation bubble. During this upstream excursion, the wing loses gradually lift until it reaches the minimum lift value when the shock is located at its most upstream location. The loss in lift is approximately 10% when this occurs. Beyond this point, the wake instabilities decrease their strength and the shock can move downstream. During this motion it is possible to observe the recovery of the blue particles flow since the recirculation bubble moves with the shock. This kind of lift variations have a negative impact on the aircraft's performance. The levels of improvement when morphing actuation is applied have been evaluated comparing the unactuated (static case) results.

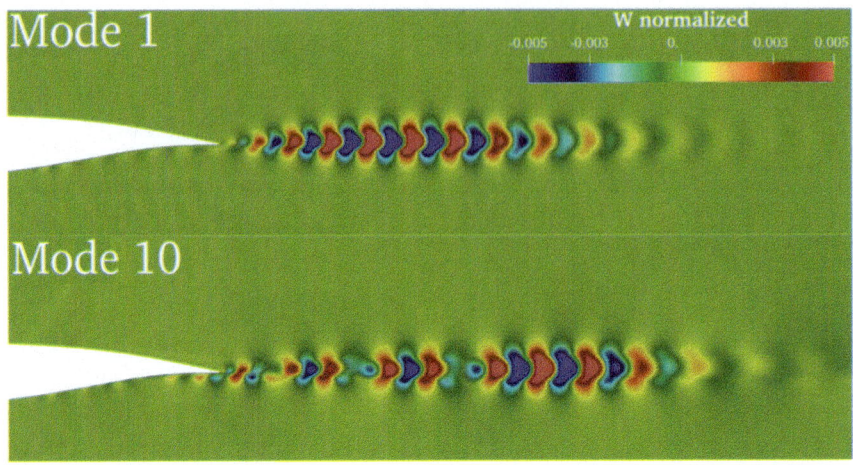

Fig. 24 Contours of vertical velocity component of spatial POD mode 1, associated to von Kármán mode, and mode 10, associated to the spatial modulation of the von Kármán mode

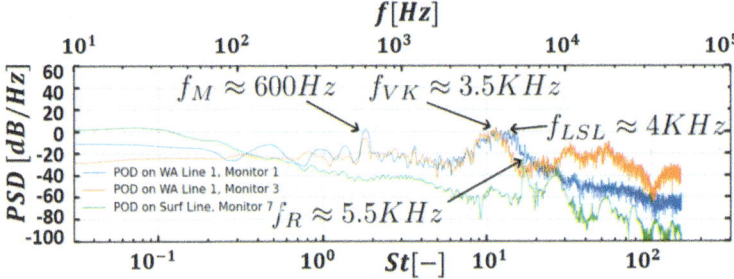

a) **PSD on pressure fluctuations at monitor points.**

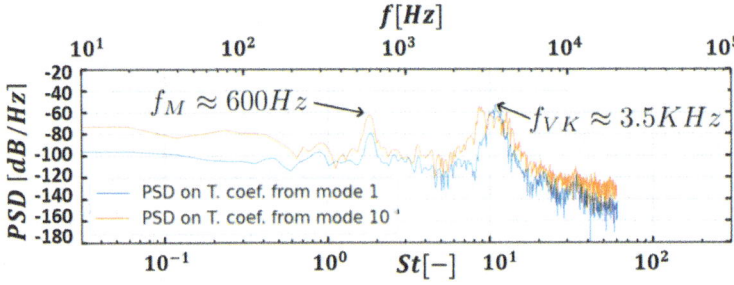

b) **PSD on temporal coefficients associated to von Kármán (mode 1)**

Fig. 25 Comparison of the frequency peaks found in PSDs for the monitor points and temporal coefficients signals for the unactuated case

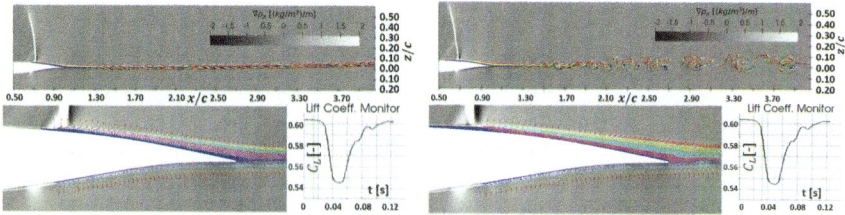

Fig. 26 Instantaneous density gradients in the streamwise direction contours, streaklines and lift coefficient monitored in 2 relevant phases of the shock's motion. Left: Shock at its most downstream location. Right: Shock moving upstream

4.2 Simulations with Travelling Wave Actuations

The morphing actuation considered in this study are the travelling waves which are modelled numerically with a piecewise periodic travelling wave function (Eq. 2).

$$z(x,t) = \begin{cases} \left(\frac{x-x_0}{x_1-x_0}\right) h_{tw} \sin(x\kappa - \omega t) & \text{if } x_0 \leq x < x_1 \\ h_{tw} \sin(x\kappa - \omega t) & \text{if } x_1 \leq x \leq x_2 \\ \left(\frac{x_f-x}{x_f-x_2}\right) h_{tw} \sin(x\kappa - \omega t) & \text{if } x_2 \leq x < x_f \end{cases} \quad (2)$$

The grid is displaced by using the ALE method. Figure 27 shows a 2D and 3D examples of the motion of the travelling waves.

Where h_{tw} is the travelling wave amplitude, κ is the wave number ($\kappa = \frac{2\pi}{\lambda}$) calculated from the wave length (λ), ω is the angular frequency ($\omega = \frac{2\pi}{f_{tw}}$) calculated from the wave frequency (f_{tw}), x_0 is the initial coordinate of the patch in the x direction, x_1 defines the end of the linear amplitude increase zone, x_2 defines the beginning of the linear amplitude decrease zone and x_f defines the end of the travelling waves patch.

The following parametric values were chosen for the first exploration of travelling waves actuations in transonic regime.

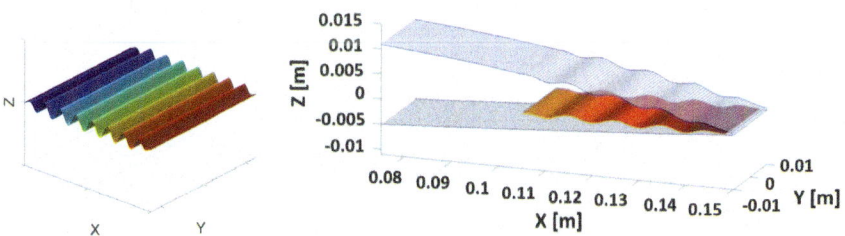

Fig. 27 Schematic description of the displacement by means of the travelling waves function. Left: Illustation of the ondulated travelling wave. Right: Instantaneous grid deformation

4.3 Morphing Effects on the Aerodynamic Coefficients

The variation of the mean and *rms* values of the aerodynamic coefficients with respect the unactuated case from the numerical simulations performed with the parameters presented in Table 2 are displayed in Fig. 28. The mean values of the lift and drag forces are decreased but not in the same proportion as it is shown in the aerodynamic efficiency graph, due to the fact that the drag reduction is higher than the reduction in lift. The reason for these significant variations in the aerodynamic coefficients is a more upstream mean location of the shock which makes it weaker and the suction region is reduced. For the case of the large amplitudes one can observe opposite evolutions with the increase of frequency for each wavelength. Regarding the levels of fluctuations of these coefficients, the lowest values of fluctuations are found for the lowest amplitudes which is consistent with the previous findings in the studies with trailing edge vibrations. The lift force variations are in the order of 3–4% for most of the cases but there is one case that provides a slight increase of lift (0.14%). The highest reductions in the fluctuations are also found for a specific set of parameters ($f_{tw} = 750$ Hz, $\lambda = 6$ cm, $h_{tw} = 1.5$ mm) which prevent the shock to move from its initial position. This result is quite noticeable in the parametric study because it shows that travelling wave actuations can fix the shock and provide a more stable flow behaviour. It seems that there is only a very specific set of parameters that can stabilize the shock for a given set of flow conditions and changing any of them leads to an undesirable increase of the fluctuations.

Figure 29 shows the effect of the TW on the aerodynamic coefficients and on the shock's motion. The TW increases the velocity in the inner region of the upper shear layer, thus decreasing the global shearing rate in this layer comparing to the static case and renders this rate under-critical in respect of the shear-layer instability and disappearance of the KH vortices. This results in moving the shock's position farther downstream and in reducing the shock wave amplitude to 0.0033 C. This produces a reduction in the oscillations of the forces and a reduction of the mean drag. It has been found that an *equal reduction* in the lift oscillations as the one of the shock's amplitude is obtained.

Table 2 Morphing parameters used in the present numerical campaign

Parameter	Values	Units
Frequency (f_{tw})	[250, 500, 750]	Hz
Wavelength (λ)	[3, 6]	cm
Amplitude (h_{tw})	[0.3, 0.6, 1.2, 1.5]	mm
Actuation length from trailing edge (p_l), $\frac{x_f - x_0}{c}$	30	%

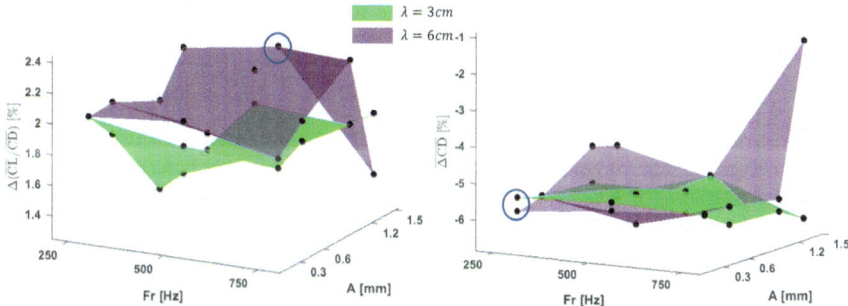

Fig. 28 Percentage differences of lift-to-drag ratio (left) and mean drag (right) comparing to the static case. AoA = 1.8°, M = 0.78, Re = 4.5 Million; amplitude $A \in \{0.2; 0.6; 1.2; 1.5\}$. Violet colour corresponds to $\lambda = 6$ cm, green to $\lambda = 3$ cm. Actuation zone length: 30% of the chord, starting from $x/c = 0.99$ up to $x/c = 0.69$ (that is slightly upstream of the most downstream position of the shock in the static case). The blue circles indicate the most promising cases

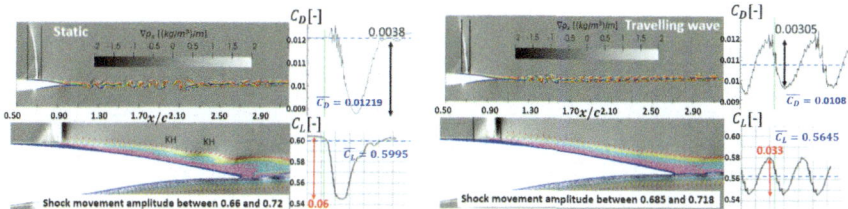

Fig. 29 Drag reduction effect for $f_a = 250$ Hz, A = 0.3 mm and $\lambda = 3$ cm. A decrease of 45% (0.06–0.033) in the drag oscillation amplitude is obtained with a decrease of the lift coefficient amplitude of 0.033, *equal to the decrease of the shock's excursion amplitude*

5 Conclusions

The intermediate scale supercritical A320 wing profile of 70 cm chord used in the EU SMS project in subsonic conditions has been investigated numerically in the transonic regime corresponding to cruise conditions. A simplified electroactive skin has been implemented in the form of travelling waves over the suction side of the wing. A large parametric space study based on the vibration frequency, amplitude and wavelength has been accomplished at Re = 4.5×10^6, upstream Mach number of 0.78 and angle of incidence 1.8°. These parameters had been indicated by AIRBUS. For selected values of these parameters, a drag reduction in the order of **7%** has been obtained and a lift-to-drag ratio increase in the order of **2.2%**. The "live-skin" morphing concept is under thorough realisation and investigation with a high number of DoF through experiments and simulations in the French ANR project EMBIA, https://anr.fr/Project-ANR-21-CE05-0006 concerning low subsonic conditions on the RS prototype and in the ongoing since December 2023 EU project HORIZON-2023-2027-PATHFINDER $N°$ 101129952-BEALIVE-"Bioinspired Electroactive multi-scale Aeronautical Live skin", regarding transonic regimes as well as a large high-lift system equipped of the live-skin for take-off and landing. In this project, a strong collaboration with IMP-PAN for these regimes is carried out.

Acknowledgements The authors are grateful to the French supercomputing Centres TGCC of CEA, CINES and CALMIP for the important CPU allocation provided.

References

1. Seegmiller, H.L., Marvin, J.G., Levy, L.L.: Steady and unsteady transonic flow. AIAA J. **16**(12), 1262–1270 (1978). https://doi.org/10.2514/3.61042
2. Levy, L.L.: Experimental and computational steady and unsteady transonic flows about a thick airfoil. AIAA J. **16**(6), 564–572 (1978). https://doi.org/10.2514/3.60935
3. Jacquin, L., Molton, P., Deck, S., Maury, B., Soulevant, D.: Experimental study of shock oscillation over a transonic supercritical profile. AIAA J. **47**(9), 1985–1994 (2009). https://doi.org/10.2514/1.30190
4. Deck, S.: Numerical simulation of transonic buffet over a supercritical airfoil. AIAA J. **43**(7), 1556–1566 (2005). https://doi.org/10.2514/1.9885. Viscosity model in Strain-Adaptive formulation. AIAA J. **41**(7), 1396–1399. https://doi.org/10.2514/2.2089
5. Grossi, F., Braza, M., Hoarau, Y.: Prediction of transonic buffet by delayed detached-Eddy simulation. AIAA J. **52**(10), 2300–2312 (2014). https://doi.org/10.2514/1.J052873
6. Dandois, J., Mary, I., Brion, V.: Large-Eddy simulation of laminar transonic buffet. J. Fluid Mech. **850**, 156–178 (2018). https://doi.org/10.1017/jfm.2018.470
7. Paladini, E., Beneddine, S., Dandois, J., Sipp, D., Robinet, J.C.: Transonic buffet instability: from two-dimensional airfoils to three-dimensional swept wings. Phys. Rev. Fluids **4**(10), 103906 (2019). https://doi.org/10.1103/PhysRevFluids.4.103906
8. Doerffer, P., Flaszynski, P., Dussauge, J.P., Babinsky, H., Grothe, P., Petersen, A. and Billard, F.: Transition location effect on shock wave boundary layer interaction. In: Experimental and Numerical Findings from the TFAST Project. Notes on Numerical Fluid Mechanics and Multidisciplinary Design, vol. 144, Science Eds. Springer (2021). https://link.springer.com/book/10.1007/978-3-030-47461-4
9. Duncan, W.J., Ellis, L., Scruton, C.: First report on the general investigation of tail buffeting. Aeronaut. Research. Com. R. and M. 1457, Part I (1932). https://doi.org/10.1017/S0368393100110156
10. Tô, J.B., Simiriotis, N., Marouf, A., Szubert, D., Asproulias, I., Zilli, D.M., Hoarau, Y., Hunt, J., Braza, M.: Effects of vibrating and deformed trailing edge of a morphing supercritical airfoil in transonic regime by numerical simulation at high reynolds number. J. Fluids Struct. **91**, 102595 (2019). https://doi.org/10.1016/j.jfluidstructs.2019.02.011
11. Bouhadji, A., Braza, M.: Organised modes and shock-vortex interaction in unsteady viscous transonic flows around an aerofoil: Part I: Mach number effect. Comput. Fluids **32**(9), 1233–1260 (2003)
12. Bouhadji, A., Braza, M.: Organised modes and shock-vortex interaction in unsteady viscous transonic flows around an aerofoil: Part II: Reynolds number effect. Comput. Fluids **32**(9), 1261–1281 (2003)
13. Bourdet, S., Bouhadji, A., Braza, M., Thiele, F.: Direct numerical simulation of the three-dimensional transition to turbulence in the transonic flow around a wing. Flow Turbul. Combust. **71**, 203–220 (2003). https://doi.org/10.1023/B:APPL.0000014932.28421.9e

14. Jimenez-Garcia, A.: Etude de l'interaction tremblement transsonique-instabilité de von Kármán à l'aide d'une plaque de bord de fuite par approche de modélisation de la turbulence statistique avancée (Diploma dissertation, ENSEEIHT) (2012)
15. Doerffer, P., Hirsch, C., Dussauge, J.P., Babinsky, H. Barakos, G.N.: Unsteady Effects of Shock Wave induced Separation. Notes on Numerical Fluid Mechanics and Multidisciplinary Design, vol. 114, Science Eds. Springer (2011). https://doi.org/10.1007/978-3-642-03004-8
16. Szubert, D., Grossi, F., Jimenez-Garcia, A., Guibert, V., Hoarau, Y., Saintlos, S., Hunt, J., Braza, M.: Feedback effects and stochastic forcing response of the transonic buffet around an airfoil including trailing-edge-plate at high Reynolds. In: ECCOMAS Conference, Vienna (2012)
17. Jimenez-Navarro, C.: Numerical study and physical analysis of the transonic interaction and morphing around supercritical wings at high Reynolds number. Ph.D thesis, Institut National Polytechnique de Toulouse-INPT (2023). https://theses.fr/s256504
18. Hunt, J., Eames, I., Da Silva, C.B., Westerweel, J.: Interfaces and inhomogeneous turbulence. Philos. Trans. Royal Soc. A: Math. Phys. Eng. Sci. **369**(1937), 811–832 (2011). https://doi.org/10.1098/rsta.2010.0325
19. Scheller, J., Chinaud, M., Rouchon, J.F., Duhayon, E., Cazin, S., Marchal, M., Braza, M.: Trailing-edge dynamics of a morphing NACA0012 aileron at high Reynolds number by high-speed PIV. J. Fluids Struct. **55**, 42–51 (2015). https://doi.org/10.1016/j.jfluidstructs.2014.12.012
20. Scheller, J.: Electroactive morphing for the aerodynamic performance improvement of next generation airvehicles. Doctoral dissertation, Institut National Polytechnique de Toulouse-INPT (2015). https://theses.fr/2015INPT0105
21. Jodin, G., Motta, V., Scheller, J., Duhayon, E., Döll, C., Rouchon, J.F., Braza, M.: Dynamics of a hybrid morphing wing with active open loop vibrating trailing edge by time-resolved PIV and force measures. J. Fluids Struct. **74**, 263–290 (2017). https://doi.org/10.1016/j.jfluidstructs.2017.06.015
22. Szubert, D., Grossi, F., Garcia, A.J., Hoarau, Y., Hunt, J.C., Braza, M.: Shock-vortex shear-layer interaction in the transonic flow around a supercritical airfoil at high Reynolds number in buffet conditions. J. Fluids Struct. **55**, 276–302 (2015). https://doi.org/10.1016/j.jfluidstructs.2015.03.005
23. Simiriotis, N.: Numerical study and physical analysis of electroactive morphing wings and hydrodynamic profiles at high Reynolds number turbulent flows. Doctoral dissertation, Toulouse, INPT (2020). https://theses.fr/2020INPT0041
24. Bourguet, R., Braza, M., Harran, G., El Akoury, R.: Anisotropic organised Eddy simulation for the prediction of non-equilibrium turbulent flows around bodies. J. Fluids Struct. **24**(8), 1240–1251 (2008). https://doi.org/10.1016/j.jfluidstructs.2008.07.004
25. Marouf, A.: Physical analysis of electroactive morphing concepts for the aerodynamic performance increase of future wing design through high-fidelity numerical simulation and turbulence modelling in high Reynols number. Doctoral dissertation, Strasbourg University (2020). https://theses.fr/2020STRAD025
26. Braza, M., Rouchon, J. F., Tzabiras, G., Auteri, F., Flaszynski, P. (Eds.): Smart Morphing and Sensing for Aeronautical Configurations: Prototypes, Experimental and Numerical Findings from the H2020 N° 723402 SMS EU Project, vol. 153. Springer Nature (2023). https://doi.org/10.1007/978-3-031-22580-2
27. Simiriotis, N., Jodin, G., Marouf, A., Elyakime, P., Hoarau, Y., Hunt, J.C., Rouchon, J.F., Braza, M.: Morphing of a supercritical wing by means of trailing edge deformation and vibration at high Reynolds numbers: experimental and numerical investigation. J. Fluids Struct. **91**, 102676 (2019). https://doi.org/10.1016/j.jfluidstructs.2019.06.016
28. Hoarau, Y., Pena, D., Vos, J. B., Charbonier, D., Gehri, A., Braza, M., Deloze, T., Laurendeau, E.: Recent developments of the Navier Stokes multi block (NSMB) CFD solver. In: 54th AIAA Aerospace Sciences Meeting, p. 2056 (2016). https://doi.org/10.2514/6.2016-2056
29. Berkooz, G., Holmes, P., Lumley, J.L.: The proper orthogonal decomposition in the analysis of turbulent flows. Ann. Rev. Fluid Mech. **25**, 539–575 (1993). https://doi.org/10.1146/annurev.fl.25.010193.002543

30. Braza, M., Perrin, R., Hoarau, Y.: Turbulence properties in the cylinder wake at high Reynolds numbers. J. Fluids Struct. **22**(6–7), 757–771 (2006). https://doi.org/10.1016/j.jfluidstructs.2006.04.021

Open Access This chapter is licensed under the terms of the Creative Commons Attribution 4.0 International License (http://creativecommons.org/licenses/by/4.0/), which permits use, sharing, adaptation, distribution and reproduction in any medium or format, as long as you give appropriate credit to the original author(s) and the source, provide a link to the Creative Commons license and indicate if changes were made.

The images or other third party material in this chapter are included in the chapter's Creative Commons license, unless indicated otherwise in a credit line to the material. If material is not included in the chapter's Creative Commons license and your intended use is not permitted by statutory regulation or exceeds the permitted use, you will need to obtain permission directly from the copyright holder.

Transonic Compressor

Numerical Investigations of Transitional SBLI on a Highly Loaded-Transonic Compressor

Selin Kahraman and Ilias Vasilopoulos

Abstract The complex flow physics of transonic compressors is linked with strong adverse pressure gradients, shock-wave boundary layer interactions (SBLI) and high level of unsteadiness. These are exacerbated by transitional effects coming along with altitude excitation. This chapter presents implications towards designing more efficient and compact turbomachineries for aeronautics, putting the mechanisms shock wave-boundary layer interactions in the center. Within this scope, a new design solution is proposed, which aims to mitigate limiting effects of transonic speeds at altitude, still maintaining the performance across the operating envelope with a holistic approach. The laminar boundary layer at altitude incurs a multi-shock pattern in the flow with a larger separated region accompanied with high losses and low-frequency unsteadiness. High fidelity numerical approaches are employed to capture the physics of the phenomenon accurately. It is followed by a numerical validation study carried out on a canonical test case, namely the Sandia axisymmetric transonic bump. The captured physics over the transonic, spherical bump is applicable to the flow in transonic compressor passages. Therefore, it is aimed to evaluate the strength and weakness of various numerical methods and different modelling approaches in respect to investigations of shock-boundary layer interaction.

Keywords Transonic flow · Shock wave—boundary layer interaction · Shock induced separation · Boundary layer transition · Multi-point optimization

---

S. Kahraman (✉) · I. Vasilopoulos
Rolls-Royce Deutschland Ltd. & Co. KG, Blankenfelde-Mahlow, Germany
e-mail: selin.kahraman@rolls-royce.com

I. Vasilopoulos
e-mail: ilias.vasilopoulos@rolls-royce.com

1 Introduction

The trend towards more compact engine designs leads to higher engine speeds while maintaining high performance targets such as low specific fuel consumption and longer range and concerning the environmental effects. Higher overall pressure ratios are typically required for highly efficient thermodynamic cycles, leading to higher loaded designs, wherein the stator vanes become more susceptible to the adverse effects caused by SBLI. In this study, several performance aspects of highly loaded stator vanes are discussed.

High loading in transonic stator passages can lead to additional losses due to strong shock wave and SBLI in the presence of adverse pressure gradients [1]. The effects become more pronounced when vane boundary layers remain laminar while the shock interaction occurs, and, therefore, tend to be more severe at low Reynolds numbers, present at high altitude [2]. The behavior of transitional and turbulent SBLI was studied experimentally in Swoboda and Nitsche [3]. It was demonstrated that, for the transitional case, the shock wave is positioned further downstream while the pre-shock Mach number is lower, and flow separates earlier than in the fully turbulent case. In addition, in the present study, it is found, that the operating range of the compressor is also significantly reduced at low Reynolds numbers. Unlike external flows, in addition to Re and Mach numbers, the confinement parameter, which is defined as the ratio of undisturbed boundary layer displacement thickness, δ_u^* to the duct half-height, h is one of the primary parameters affecting the confined interaction and the shock system [4–8] in internal flows. It was shown that, under proper conditions which are driven by the combined effects of Re number, Mach number and δ_u^*/h, multiple normal/oblique shock wave boundary layer interaction, also so called shock-train may form, and causes a highly unsteady flow downstream. As the confinement level increases, a series of repeated shock waves may be observed, and with an increase in Mach number, the interaction length and the tendency to the multiple oblique shock system increases. An increase in Re number may cause a shift from a repeated to a single shock system. In an experimental study, while, at nearly same -undisturbed Mach and Re numbers, the lower confinement parameter, (2.27%) causes a single shock interaction, increasing confinement (5.15%) promoted multiple shock interactions and longer overall length of the interaction [9]. It is noted that unsteadiness from SBLI at high operating speeds may impact the structural integrity of the VSVs.

The detrimental effects at transonic speeds can be addressed by appropriate design of the compressor aerofoils. With the introduction of controlled diffusion aerofoils (CDA) in the 1980s, a landmark improvement was achieved for the operation at moderately transonic speeds [10]. The shock-free flow around controlled diffusion aerofoils results in strongly reduced losses, which is the reason for their continuing appeal. The inflow Mach numbers however, where CDA has a performance advantage to double circular arc (DCA) and NACA 65 didn't exceed 0.8 [11, 12]. Similarly, using a cascade of controlled diffusion aerofoils designed for M = 0.85, showed reduced losses for inlet Mach-numbers below 0.8 [13]. Several optimization algorithms have been employed, notably summarized in John et al. [14]. While most optimization studies focused on the inflow Mach numbers range between 0.7 and 0.8, a cascade design optimized using a multi-objective genetic algorithm [15], demonstrated significant loss reduction at Mach numbers up to M=0.83 where the losses increase rapidly beyond this Mach number [16].

This type of flow poses a particular challenge to engineering-level methods such as RANS for the prediction of engineering quantities of interest. Scale resolving methods such as LES are considered to improve the accuracy for capturing the critical aspects of the flow at the expense of higher computational costs, the accuracy of which also depends on the resolution. To have confidence with both RANS methods and LES approaches requires a well defined experimental database aligning with industrial needs.

This chapter briefly summarizes the content of the work performed within the context of a doctoral study for the TEAMAero research consortium. The readers are referred to the existing and future publications for further detail. The first step was to optimize a robust compressor cascade at multiple operating points for seeing to what extent the detrimental effects could be alleviated, and then to use it for further investigations beyond the conventional operating limits with high fidelity methods, which are validated with a canonical experimental case. This chapter then begins with a brief overview of the design optimization of a highly loaded compressor cascade, then followed by the further investigations of the new design via high fidelity numerical approaches and finalized with the highlights of the numerical validation study performed with the Sandia experimental results.

2 Shape Impact on Mitigating Adverse Effects of Transitional SBLI

A new design solution is proposed, which mitigates the limiting effects of transitional SBLI at altitude condition and transonic speeds. An optimization study has been carried out at multiple mission points of a representative engine. The optimization is mainly aimed at improving the performance parameters of the VSV section, specifically considering the laminar-transitional SBLI at altitude, and maintaining performance at part-speed conditions. The new section design alleviates the detrimental effects by decreasing the acceleration in the front portion of the suction surface, controlling the shock location. The losses are significantly reduced, and the choking limit is notably improved at over-speed. In this condition, a multiple-shock structure is introduced with high loading observed in the passage. The unsteadiness of this phenomenon is investigated via high fidelity methods in the Sect. 3. This section provides a brief overview to the multi-point optimization methodology and the improvement achieved with the new design. For the details of the optimization work and more analysis on the optimized design, the reader is referred to Kahraman et al. [17].

2.1 Overview of the Multi-point Optimization Process

The Rolls-Royce *Section Optimizer* [18] was used for designing the new 2D cascade geometry. This tool is a platform in which geometry creation, optimization and analysis tasks can be performed interactively within a well-constructed process chain. Within this scope, *Parablading* [19] is used for parametric airfoil geometry creation, *Mises* coupled with the boundary layer equations is used for blade-to-blade calculations, and *Isight* is used for process chain and optimization management. Further details of the optimization workflow can be found in the studies by Flassig [18], Gräsel et al. [20], Dutta [21].

Multiple optimization points at an altitude Reynolds number of 200 k while considering three Mach numbers corresponding to design speed, over speed, and part speed, are evaluated first in the presence of transitional conditions, and second with fully turbulent conditions at a Re number of 750 k. The constraints (see Fig. 1) related to the boundary layer control and the position of the peak Mach number then enable the designer to optimize for laminar SBLI as well.

The tool, *Section Optimizer* carries out single objective optimization, minimizing loss at a chosen design point. Therefore, the design space and the constraints need to be prescribed for each operating point individually. In order to optimize the geometry at multiple conditions, first the optimizer is run to optimize the performance at each individual operating condition, and then each of these designs are evaluated at the other relevant operating conditions. The optimization process yields six best candidates for each operating point individually. For the evaluation of the candidates,

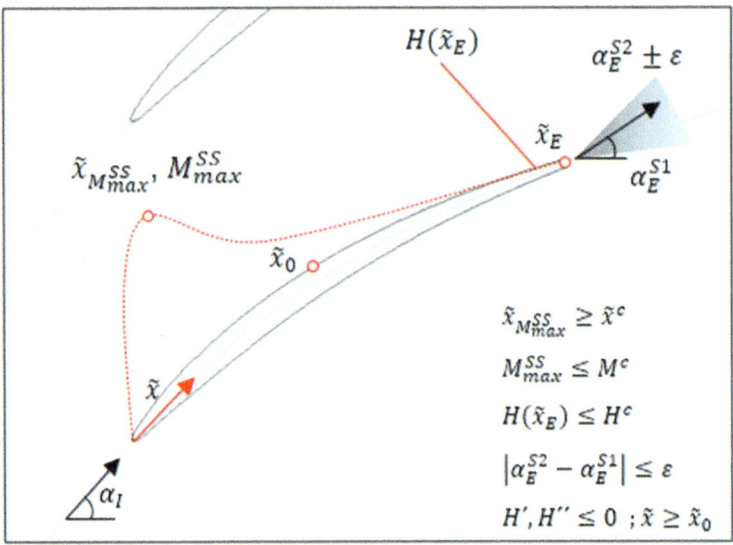

Fig. 1 Optimization constraints [17]

two objective functions are defined which combine the performance of the operating points. The objective functions incorporate terms that account for the total loss (ω) and the working range (WR) from choke to stall. The weighting coefficients have been chosen here in a way that yields a reasonable final design for a representative mission profile, but for other cases, further tuning of the weights may be required. Once the f_1 *(loss)* is evaluated against $1/f_2$ *(increasing WR)*, the direction towards the overall best design in terms of smallest loss, highest working range forms a pareto front. With this approach, the candidate which was optimized for the design speed, and at altitude condition (Re = 200 k) gives the overall best performance for the given mission profile. It would be recommended that in the future, profile optimization should be carried out at the highest altitude Reynolds numbers to achieve a robustness of the design.

Several RANS computations have been carried out with the $\gamma - Re_\Theta$, correlation-based transition model [22] using the Rolls-Royce CFD code, *Hydra*, with the $k - \omega$ SST turbulence model [23] displaying the improved performance of the optimized design, and providing flow topology in more detail.

2.2 Overall Performance Improvement Optimized Versus Baseline Design

The performance gains with the new design in terms of total loss reduction and working range improvement are shown in Fig. 2. The bar chart illustrates the delta difference of the new design over the baseline design as a percentage. The loss is reduced by 12.6 and 13.1% at the over-speed condition, respectively for sea level and altitude. The same condition is found to have 12.5 and 26.2% larger working range. The deltas at design speed case at altitude are also noteworthy where 7.7 and 6.9% in terms of loss reduction and working range increase, respectively. Overall significant improvement at design and over-speed conditions is achieved at both sea level and altitude, although small deficits (5% loss in working range for sea level and 0.5% loss increase for altitude) are observed at part-speed. This is considered acceptable since the altitude, over-speed condition is where the most improvements are needed to alleviate the adverse effects of transitional SBLI.

As was observed with blade-to-blade calculations of Mises, Quasi-3d CFD computations predicts loss reduction to be most significant at high Mach number. The choking limit of the baseline design is found to be $M_\infty = 0.882$ while the optimized design pushes this limit to $M_\infty = 0.903$. Whilst the loss coefficient at part-speed is nearly identical, a 16% improvement at the altitude over-speed condition is found (slightly higher than predicted by Mises).

Figure 3a and b show the optimized design compared to the baseline design. Turning with the new design is seen to increase slightly. There is a notable difference in thickness and camber distributions. Although both are front-loaded designs, the entrance to the passage of the optimized design is more flat considering the local

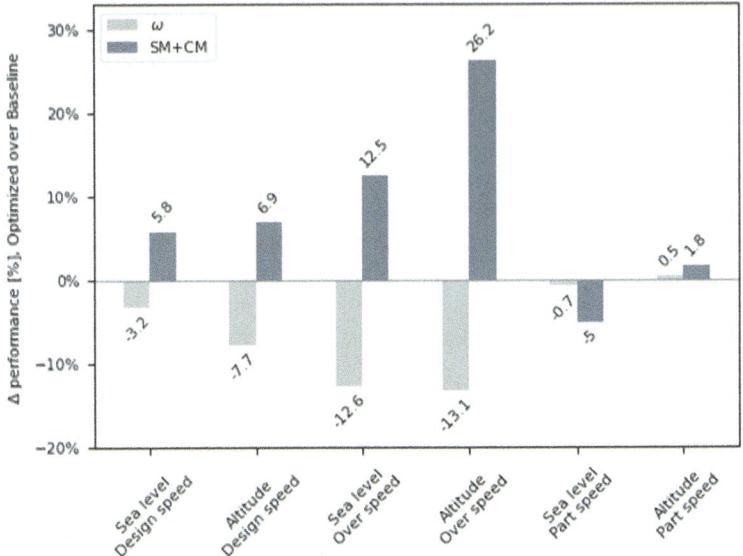

Fig. 2 Performance Improvement at each mission point [17]

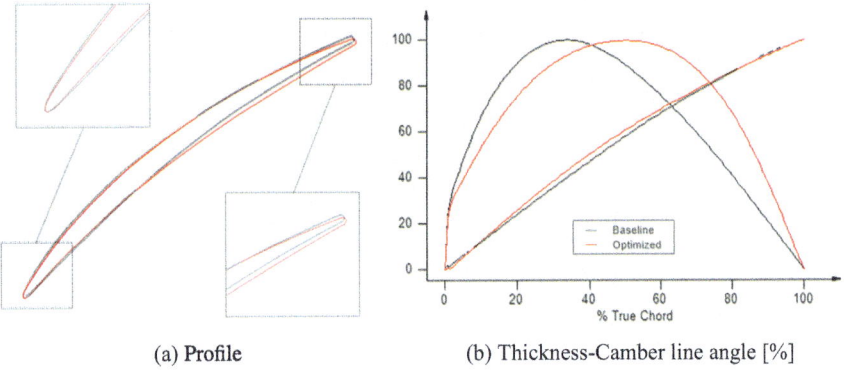

(a) Profile (b) Thickness-Camber line angle [%]

Fig. 3 Baseline (black) versus Optimized design (red)

camber at the leading edge. This is considered to be a leading factor in controlling the acceleration in the vicinity of leading edge. The position of the maximum thickness is shifted to the mid chord whilst the maximum thickness remains constant. It can be noted that the new section design was also analyzed in terms of stress and no concerns were raised over the new design.

3 Wall-Resolved Large Eddy Simulations of the New Design

The new designed geometry under investigation in this section is introduced in Sect. 2.2.

3.1 Numerical Setup and Approach

The Rolls-Royce in-house solver, Hydra, is used for all simulations presented. For the RANS computations, the scheme is set to non-linear and steady, using a semi-implicit Runge-Kutta scheme with 4 multi-grid levels for convergence acceleration. The steady solver of Hydra was used with the $k - \omega$ SST turbulence model [23] coupled with the $\gamma - Re_\Theta$ correlation-based transition model [22].

WRLES computations were performed with the unsteady, non-linear, second order, in-house WRLES solver of Hydra, coupled with the implicit dual time-stepping numerical scheme where small eddies below a spatial resolution are modelled with the Sigma sub-grid scaling (SGS) model. The turbulent fluctuations imposed with the inflow are generated with an isotropic synthetic turbulence generator and it is based on the model proposed by Davidson [35].

The CFD performed in this study is done in a quasi-3D computational domain (see Fig. 4b) wherein an annular extension of the new design is carried out in the spanwise direction by as much as 15% of the whole span. Since the CFD is focused on the sectional design, it does not include end-walls. However, the contraction of the flow (caused by endwall blockage or contraction of the annulus) is taken into account by taking the full 3D CFD results of the relevant blade (see Fig. 4a) and using the streamlines of the circumferentially averaged flow as the upper and lower inviscid walls of the domain. This naturally will mimic the axial velocity-density-ratio (AVDR) of the baseline design (=1.14 locally around the section). Figure 4b shows the representative CFD domain used for both LES and RANS computations. The lateral walls are modelled as periodic, while the upper and lower walls are modelled as inviscid. Total pressure and total temperature are imposed as the inlet boundary condition. By doing so, the chord-based inlet Reynolds number was set to 200 k which features the laminar incoming boundary layer at altitude condition along with a turbulence intensity of 4%. For the convenience of the back-to-back comparisons, the same static pressure was imposed at the outlet which results in slightly different inlet Mach numbers for LES and RANS simulations.

The numerical grid was generated by the *Autogrid* meshing utility. Meshes created are hexahedral with a multi-block O-H grid topology. For the RANS computations, grid independency was achieved with 3×10^6 grid cells with y^+ value around 1 over the blade wall. The spatial resolution achieved with our WRLES simulations is given in Fig. 5 in both streamwise and spanwise directions as well as the wall normal

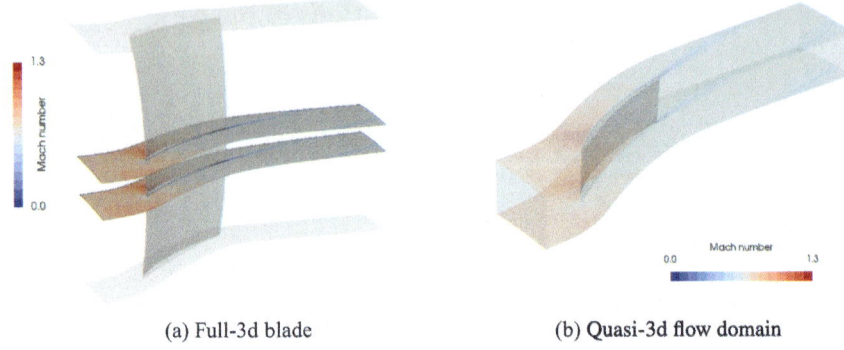

(a) Full-3d blade (b) **Quasi-3d flow domain**

Fig. 4 Illustration of generation of the quasi-3d flow domain

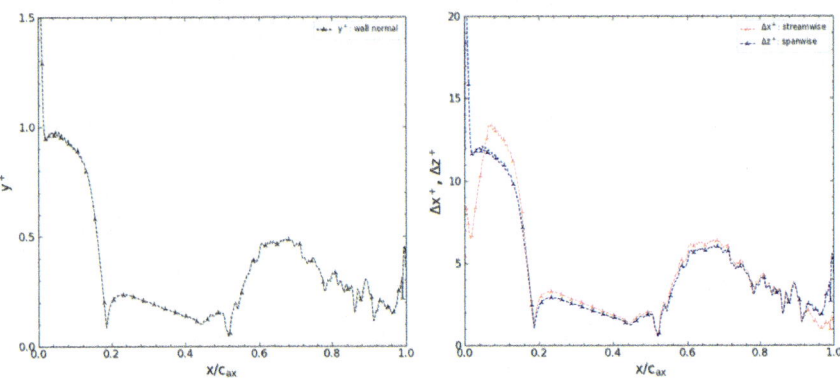

Fig. 5 Spatial resolution over the suction surface—wall normal distribution (*left*), streamwise and spanwise distributions (*right*)

direction. In the front part, flow accelerating over the wall is resolved with $\Delta x^+ = 14$, $\Delta \theta^+ = 12$, $\Delta y^+ = 1$ as maximum values on the wall. The quasi-3d domain is discretized with a total of 69×10^6 grid points.

The time step size, Δt, is set to 1×10^{-7}s which corresponds to a CFL number of around 10. The simulation is initialized with a RANS solution and monitored until the quantities such as total pressure loss calculated from outlet to inlet, reaches steady-state before averaging starts. A statistically converged solution has been averaged through 10 convective time units, which is defined as time through one chord flow. Numerical probes are placed throughout the computational domain and used for collecting statistics during the averaging time.

3.2 Results and Discussion

In this section, we discuss the WRLES and RANS results with respect to the mean flow field, and with engineering quantities of interest. Unsteadiness predicted by WRLES is explicitly presented in the second part of the section.

3.2.1 Time Averaged Flow Properties

The time averaged flow field contour of the Mach number at mid-span and the magnitude of density gradient, $|\nabla \rho|$, over the blade wall are consolidated in Figs. 6 and 7 for the WRLES and RANS, respectively. Inlet Mach numbers corresponding to the same exit back pressure are computed as 0.90 and 0.91 respectively for the WRLES and RANS. Whilst the entrance to the passage seems similar for both, size of the supersonic region and the shock structure are found to be significantly different. The RANS results predicts a large, stronger main shock wave followed by a weak secondary shocklet. The WRLES predicts a shock train-like structure with an early occuring main shock wave followed by secondary shocks in the confined region of the passage by the adjacent blade. This topology is likely to occur when the confinement parameter, which is defined as the ratio of displacement thickness, δ_u^*, to the duct half-height, h of the passage is relatively high as mentioned earlier. The ratio is calculated as around 4.5% and higher in the confined region in our case which aligns with the literature in which the multiple-shock system is observed. Isentropic Mach number distribution over the blade suction surface in Fig. 7 quantifies the accelaration behaviour with a higher pre-shock Mach number (1.276) for RANS, whilst

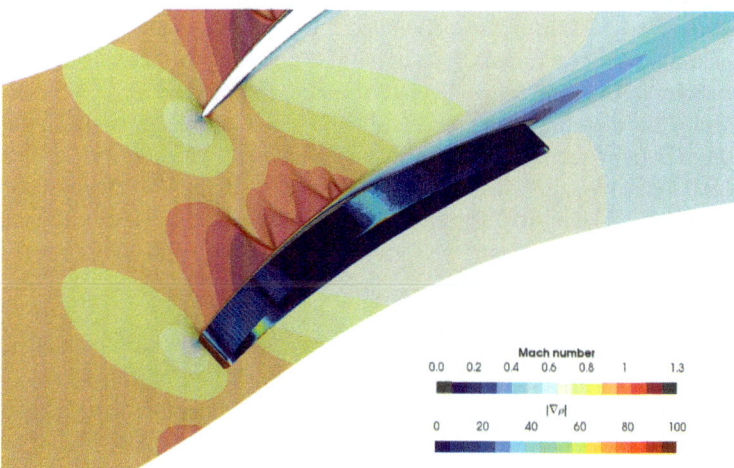

Fig. 6 Time averaged flow field—WRLES

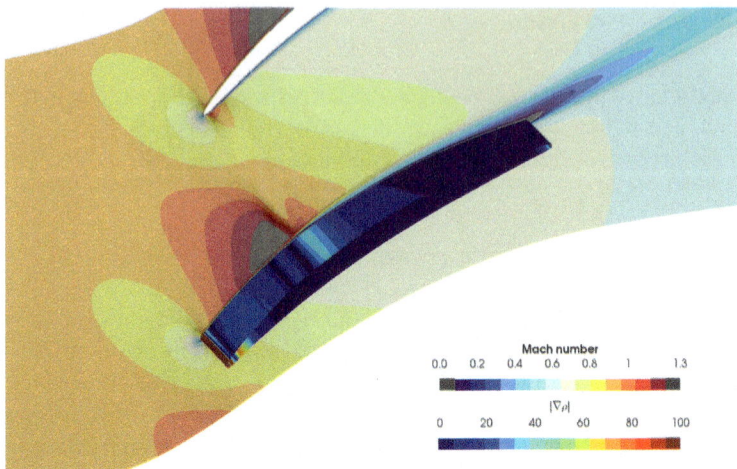

Fig. 7 Flow field—RANS

this peak value is around 1.2 for WRLES. In the same figure, the shock position is seen at 36% of the chord for RANS, and 24% of the chord for WRLES.

Figure 8 shows the variation of average incompressible BL parameters over the suction surface for WRLES in terms of boundary layer thickness, δ, displacement thickness, δ^*, momentum thickness, θ and shape factor, H. In the front portion up to 15% of the chord where the laminar boundary layer exists, δ^* and θ follows a steady trend and H is seen to be below 2.5. Once the upstream interaction begins between the shock wave and the laminar boundary layer, the trend goes up rapidly for each parameter where the boundary layer is smeared by the upstream interaction with oblique waves. Skin friction coefficient, C_f, in Fig. 9 indicates the formation of a separation bubble at around 19% of the chord. Due to the fact that the boundary layer thickness grows significantly with the laminar interaction, flow is deflected outward and this directional change leads to the formation of secondary shocks. The multi-shock pattern is attributed to the post-shock flow re-acceleration in the confined region, and to the passage which contracts ever increasingly with growing boundary layer through the blade wall. The rising trend in the boundary layer parameters, in particular H, keeps growing until the 35% of the chord where the shape factor reaches around the value of 10. This position corresponds to the reattachment point in the C_f plot (see Fig. 9).

Despite the boundary layer going into the healthier state with the reattaching flow (decreasing H), flow separates once again at around 42% of the chord (see Fig. 9). This is attributed to the sustained interaction between the secondary shocklets and the boundary layer. It shows itself as a small spike in the H plot (see Fig. 8) which is normally in a downtrend. This phenomenon concludes and flow reattaches at 53% of the chord, while H goes to below 2.5 critical limit at around the same position.

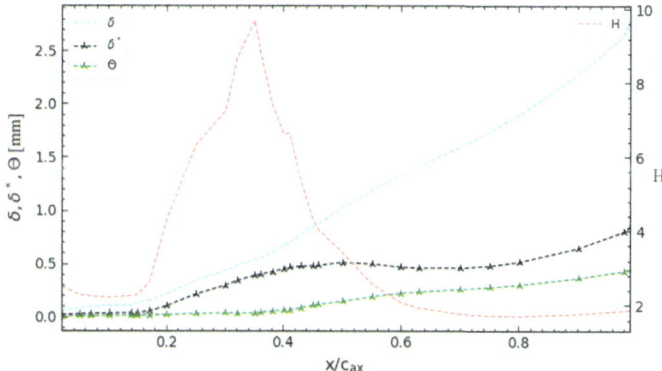

Fig. 8 Incompressible boundary layer parameters over the suction surface

Pressure distribution given in the bottom of Fig. 9 for WRLES shows a two-step rise, the first of which is quite small and occurs at the laminar separation, and is followed by a large plateau region where the transition takes place. The second rise is significantly higher, and occurs at the reattachment. The RANS predicts a smeared pressure rise with a very little plateau region before the main pressure rise occurs.

The overall loss has been calculated across the entire span as $(PT_1 - PT_2)/PT_1$: where PT_1 is the reference total pressure taken upstream of the leading edge and PT_2 is the mass-averaged total pressure at the wake location just downstream of the trailing edge. Based on this, total loss for LES is found to be approximately 15% higher than for RANS.

3.2.2 Unsteadiness

In order to investigate the wall pressure over the blade, the numerical probes have been placed throughout the suction surface. Figure 10 shows space-time plots of the suction side normalized static pressure at mid span. The data has been sampled at a frequency of $f_s = 10^7$ Hz. The imprint of the main shock is visible as the relatively sharp increase in the contour at $x^*/c \approx 0.22$ and it fluctuates with an amplitude of $\pm 3\%$ of the axial chord length.

Figure 11 shows a sequence of the snapshots which are uniformly spaced in time with $\Delta T = 6 \times 10^{-5}$ s. The snapshots shows the magnitude of density gradient, and are contoured with span-wise velocity, V_z. It is visible that the laminar separated shear layer and its transition to turbulence are highly unsteady. The markers are placed in the figures with equal intervals of 1% of the axial chord length at around the main shock foot and where the shear layer breaks down to turbulence. In Fig. 11a–c, downstream phase of the unsteadiness is illustrated where the main shock wave and transition to turbulence moves increasingly far downstream. In Fig. 11d upstream

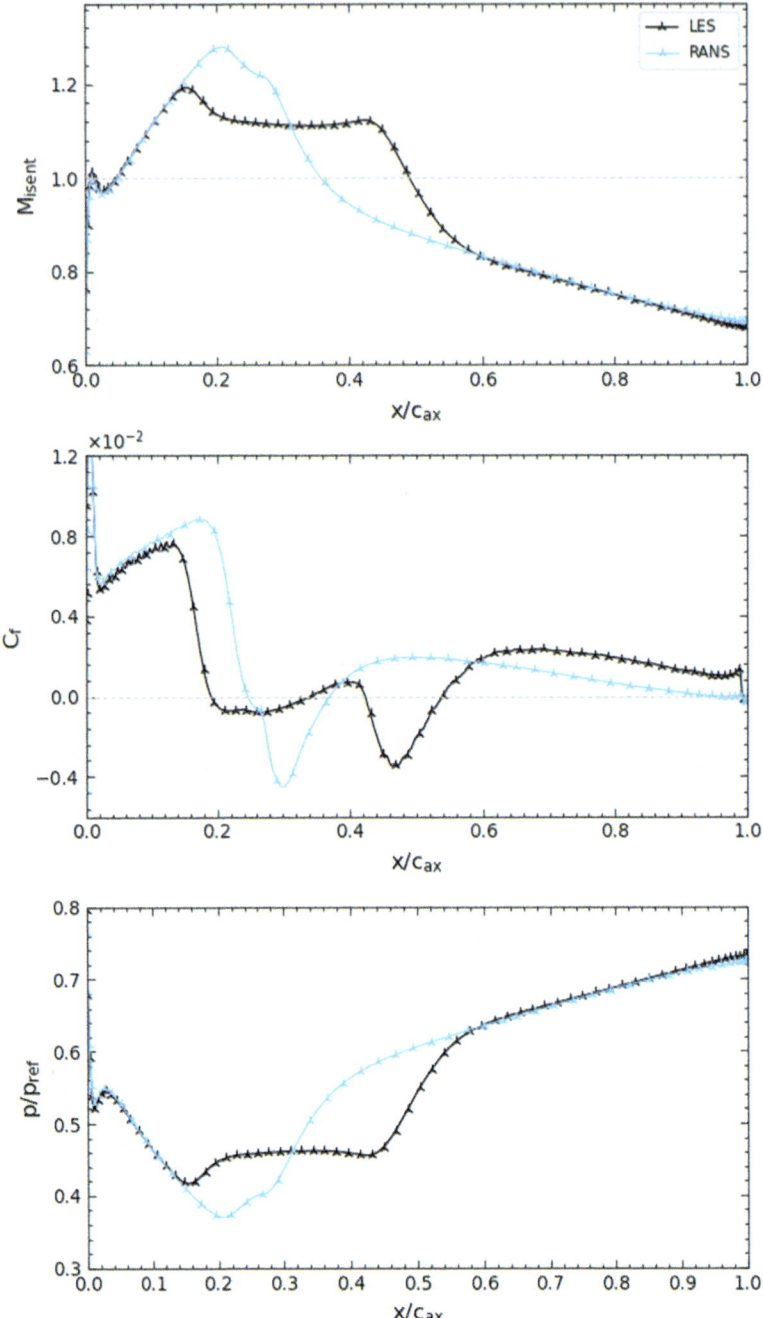

Fig. 9 Comparison of LES and RANS results—Isentropic Mach number (*top*), Skin friction coefficient (*middle*), Pressure coefficient (*bottom*)

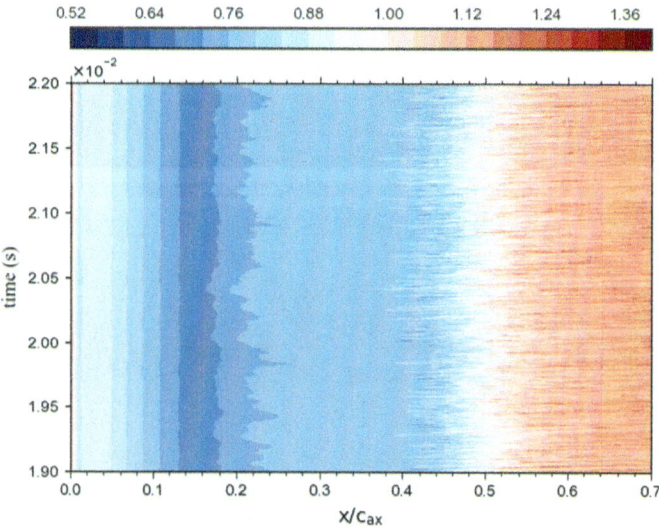

Fig. 10 Space-time plot of pressure over the suction surface

phase starts, and continues in Fig. 11e. Early transition also affects the boundary layer thickness downstream of the interaction which seems to be thicker in Fig. 11d.

In order to further investigate unsteadiness, Fig. 12 shows the normalized pre-multiplied Power Spectral Densities (PSD) obtained from a series of probes recording the wall pressure at mid span. The PSDs are obtained by using the Welch's method [24] with a 75% overlap. Figure 12a demonstrates the streamwise evolution of the PSDs as a map. The frequency, f is normalized with the Strouhal number ($St_L = f \times L_{sep}/U_\infty$) where L_{sep} represents the length of the separation region, and U_∞ is the free-stream velocity. A low frequency region in a narrow band is observed at $x^*/c \approx 0.19$ which corresponds to the laminar separation in the C_f plot in Fig. 9. Along with the interaction between secondary shocks and shear layer, a secondary separation region is observed in the C_f plot as mentioned earlier. This shows itself as a second low frequency region centered at $x^*/c \approx 0.42$. This is followed by the interaction region wherein intermediate frequencies develop in the detached shear layer. It is represented by broad band energetic regions which occurs between $x^*/c \approx 0.44$ and 0.60 with two centers, one of which is at $x^*/c \approx 0.44$ corresponding to $St_L \approx 0.03$ and the second at $x^*/c \approx 0.49$ with the highest energy content across the map. Markers on the map identify the five stations which are individually investigated in Fig. 12b, respectively corresponding to the first and second negatives in the C_f plot, center of the low frequency, high energetic region, highest energy content in the spectrum and the relaxation zone. The $f*$PSDs obtained in Fig. 12b are normalized with the global maximum. In order to recognize the peaks at low amplitude regions, zoomed-in view is shown explicitly at the top left. The low frequency content of the primary laminar separation clearly stands out with a peak at $St_L = 0.02$–0.03 (dark blue). The widespread range of the second negative shows a peak at $St_L \approx 0.05$

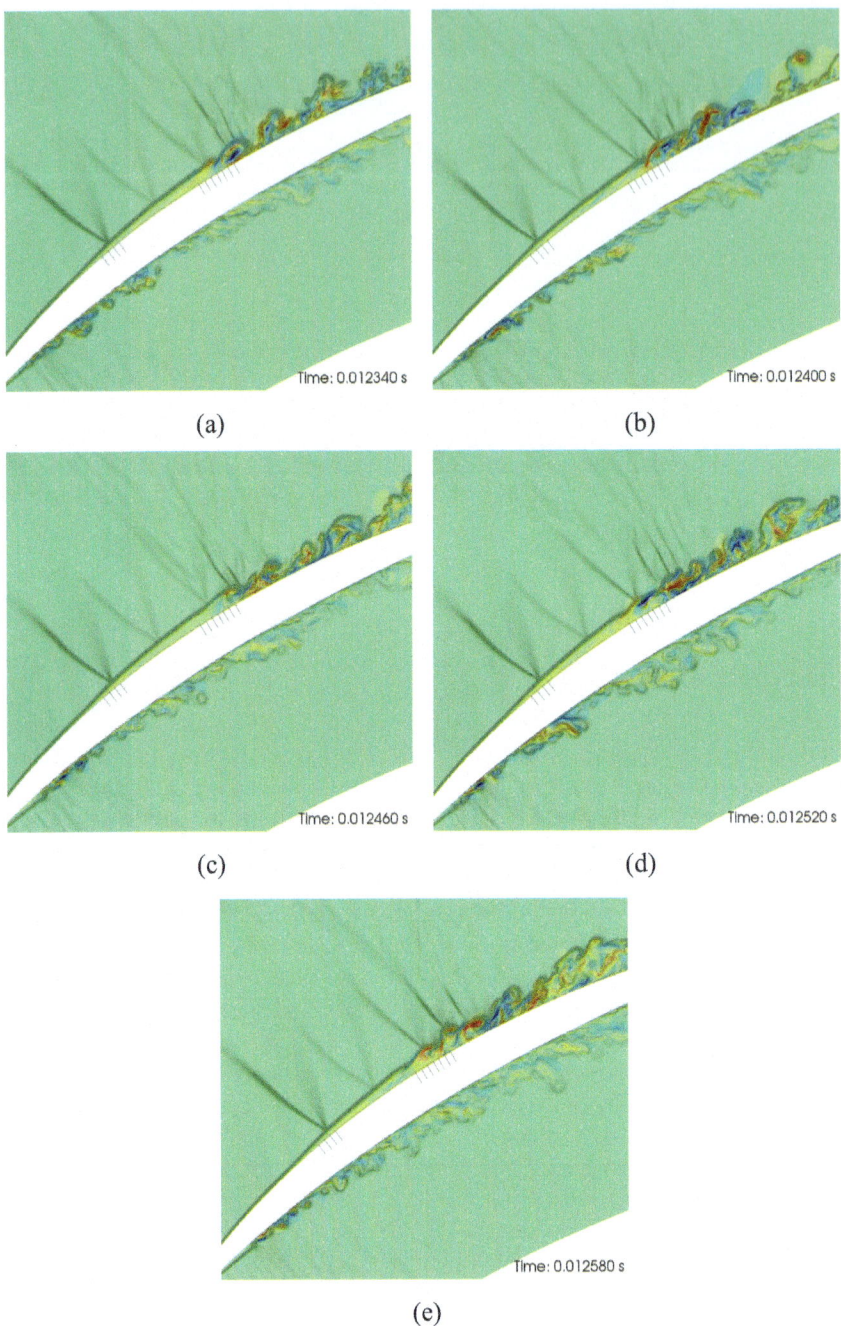

Fig. 11 Instantaneous flow field, contoured with the span-wise velocity

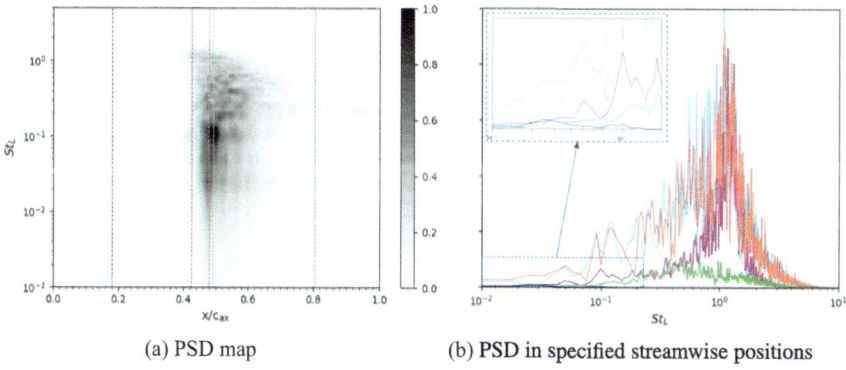

(a) PSD map   (b) PSD in specified streamwise positions

Fig. 12 Normalized premultiplied power spectral densities of blade wall pressure

(magenta). The station with high energy content shows peaks at both $St_L \approx 0.05$ and $St_L \approx 1$ (orange and lightblue). In the relaxation region, frequencies are widespread and at low energy content.

4 Numerical Validation Study—Sandia Test Case

In this section, wall-resolved large eddy simulations (WRLES) and Reynolds averaged Navier-Stokes (RANS) computations with different turbulent closures are carried out for the flow over the Sandia axisymmetric transonic bump (ATB) at a Reynolds number of 971 k and Mach number of 0.875 based on the chord length of the bump. Flow approaching the bump accelerates to the supersonic state locally near the apex and forms a λ shock structure resulting in an unsteady separation bubble and then reattaches. The captured physics over the transonic, spherical bump is applicable to the flow in transonic compressor passages. Therefore, it is aimed to to evaluate the strength and weakness of various numerical methods and different modelling approaches in respect to investigations of shock-boundary layer interaction. The results of both LES and RANS are presented comparatively with the test results in terms of the wall pressure coefficient (C_p), skin friction coefficient (C_f), boundary layer profiles and flow topology.

4.1 Background of the Test Campaign

An experimental characterization of the separated flow over an axisymmetric, transonic bump had been carried out by Bachalo and Johnson [25] first in 1986, at a Reynolds number of 3 million. This effort addressed the request of the CFD

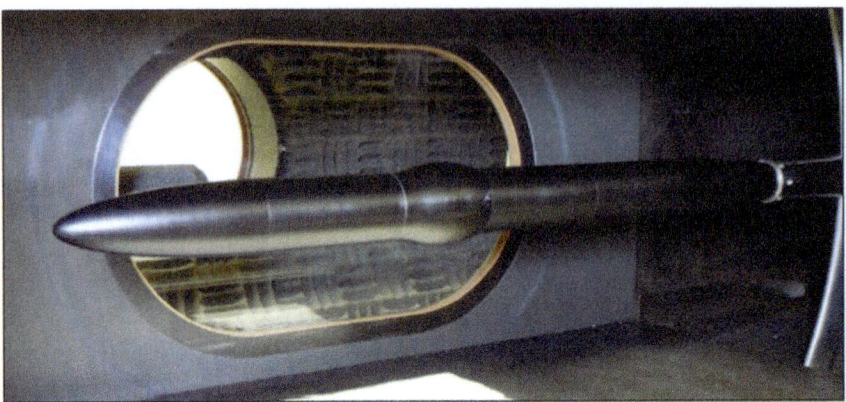

Fig. 13 Sandia trisonic wind tunnel [29]

community to a large extent for additional validation cases including SBLI. As proposed by Bachalo and Johnson, this case has been used for several numerical studies including RANS model calibrations. Menter [26] utilized it as a benchmark of two-equation eddy-viscosity turbulence models. The results of his work suggested that the Shear-Stress Transport (SST) model predicted better results due to its improved transport equations. The study of Spalart et al. [27], which was performed to simulate Bachalo and Johnson test case [25], found that WMLES could not predict the shock location accurately while hybrid calculations, into which a partial DNS domain was embedded, were able to give better agreement with the test results despite the difficulties of a smaller domain in the azimuthal and streamwise directions compared to WMLES. In Uzun and Malik's work [28], WRLES calculations were performed where the impact of grid resolution and azimuthal span were investigated. They found that simulating 20° of span was adequate to provide a grid independent reattachment location, however, the Reynold stresses differed from 20° to 120° azimuthal extent particularly in the separated region. The shock location was also observed to shift upstream as the grid was refined.

The Bachalo and Johnson experiment was redesigned by Lynch et al. [29–31] at a lower Reynolds number of 971 k. This makes it more affordable to the community for high fidelity methods such as LES/DNS. The new Sandia ATB test campaign eliminated the unquantified boundary condition effects such as inflow turbulence intensity, the lack of skin-friction data measurements and the uncertainty of permeable tunnel walls. The experiment was conducted at the same transonic Mach number, 0.875, with a downscaled axisymmetric bump model in a small size test cell. The metrics for the evaluation of the capabilitiy of the various numerical methods and modelling approaches included the mean velocity and Reynolds stress profiles, and the shock characterization in terms of the mean shock location, separation/reattachment positions as well as the mean surface pressure and wall shear-stress distributions.

The early numerical studies which were submitted as part of the original blind-CFD-challenge of the Sandia ATB experiment (only the inputs to the CFD works were provided with key experimental outcomes held-back until after the results were submitted) exhibit that modelling the inflow properties accurately, such as the freestream turbulence and the profile of the boundary layer, plays a prominent role for the accurate prediction of the shock location, size and, the strength of the recirculation zone. RANS methods applying the Spalart-Allmaras turbulence closure predicted the separation and reattachment locations reasonably well but overestimated the skin friction along the wall. A WMLES computation predicted reattachment accurately whilst the skin friction in the upstream wall was mispredicted and the location of the shock was incorrect [32]. The explicit LES calculation of Gupta et al. with a finer grid captured the separation and reattachment points well enough, however, the skin-friction coefficient was mispredicted over the majority of the model wall [33]. Finally, an implicit LES underpredicted the skin-friction coefficient and the location of the separation was inaccurate [34]. Below, we present the findings of our numerical simulations of the Sandia ATB experiment with WRLES and RANS.

4.2 Numerical Methodology

4.2.1 Flow Solver

The Rolls-Royce in-house solver, Hydra, is used for all simulations presented. For the RANS computations, the scheme is set to non-linear and steady, using a semi-implicit Runge-Kutta scheme with 4 multi-grid levels for convergence acceleration. The Spalart-Allmaras and k-ω SST turbulence models have been used for the comparisons. WRLES computations were performed with the unsteady, non- linear, second order, in-house LES solver, coupled with the implicit dual time-stepping numerical scheme where small eddies below a spatial resolution are modelled with the Sigma sub-grid scaling (SGS) model. The turbulent fluctuations imposed with the mean inflow profile are generated with an isotropic synthetic turbulence generator and it is based on the model proposed by Davidson [35].

4.2.2 Computational Setup and Boundary Conditions

The geometry of the test cell and the details of the bump model are provided in Lynch et al. [30, 36]. It was stated by Lynch et al. that the use of a solid-wall test section results in a significant non-uniform pressure distribution along the wall compared to a porous-wall test section. Due to model blockage and boundary layer growth on the tunnel wall and model, the Mach number increases between 5 and 10% at the streamwise stations corresponding to the model [29]. Since the non-uniformity makes the definition of freestream flow conditions ambiguous, the freestream pressure and Mach number were defined as the condition at the tunnel wall at a streamwise station

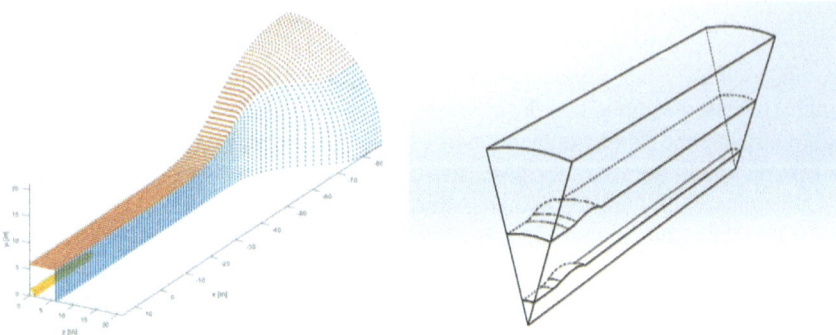

Fig. 14 Test setup versus Computational setup

corresponding to x/c = 0. The target condition of M_∞ = 0.875 is therefore obtained by adjusting the back-pressure in the simulation at the tunnel wall at x/c = 0 as given in the experiments. The experimental boundary layer measurements are taken at x/c = −0.78 and this profile is therefore imposed at this location as an inlet boundary to ensure the incoming boundary layer conditions are modelled correctly in the forebody. The experimental velocity profile is converted into total pressure and total temperature profiles in the simulation. This aspect of the modelling is to reduce the computational domain into a smaller size in both streamwise and azimuthal extent to make the WRLES computations affordable for the available computational resources. For the azimuthal direction, the square-section tunnel walls are transformed into an equivalent effective radius to be used as an axisymmetric inviscid (slip-wall) boundary. The contour of the effective radius $R^*(x)$ has been applied as reported with the test database by Miller et al. This contour is meant to produce nearly the same mass flow rate, streamwise "free stream" Mach number distribution, wall surface pressure distribution along the model, and wall shear stress magnitues along the model as were produced with the tunnel-resolved simulation. A contraction of $R^*(x)$ along the streamwise coordinate naturally mimics the blockage due to the solid test walls.

The axisymmetric computational domain starts with x/c = −0.78 with an azimuthal extent of 30° with the effective radius of $R^*(x)$. The RANS solution was used as an initial average flow and the isotropic synthetic turbulence generator was introduced at the inlet for the velocity fluctuations of the turbulence intensity of 0.23% (Fig. 14).

4.2.3 Spatial Resolution

The spatial resolution achieved with our LES simulations is given in Fig. 15 in both streamwise and azimuthal directions as well as the wall normal direction. In the near wall region, flow approaching the bump is resolved with $\Delta x^+ = 36$, $\Delta \theta^+ = 32$, Δy^+

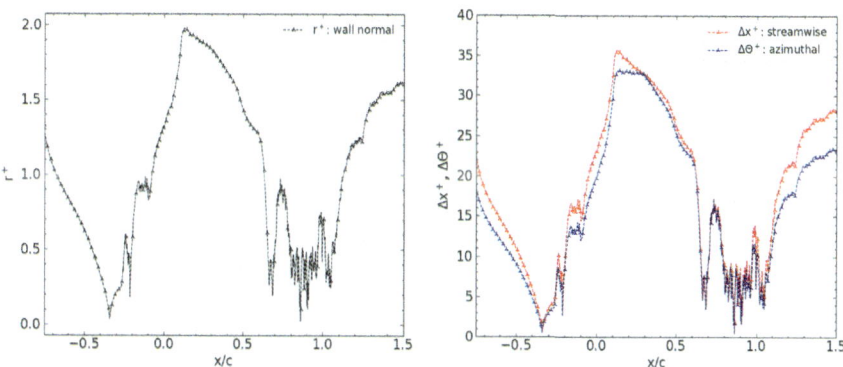

Fig. 15 Near-wall grid spacings in wall units—wall normal distribution (*left*), streamwise and azimuthal distributions (*right*)

= 2 as maximum values on the wall. The study of Gupta et al. reports the streamwise grid resolution of the EFLES on the Sandia ATB as 370 as the maximum value over the wall. The resolution reported by Uzun et al. is $\Delta x^+ = 25$, $\Delta \theta^+ = 12.5$, $\Delta y^+ = 0.8$ which was carried out on the Bachalo and Johnson test case. In either case, the values could be an average or a maximum over the wall, it is not stated explicitly.

4.3 Results and Discussion

In this section, we discuss the WRLES and RANS results with respect to the experiments.

4.3.1 Flow Topology

The instantaneous flow field is given that illustrates the state of the boundary layer and the shock structure over the bump (Fig. 16). It shows a good agreement with the flow topology observed in the experiments qualitatively.

4.3.2 The Incoming Boundary Layer Profile

As pointed out in previous numerical studies [32–34], properly modelling the experimental inflow boundary layer profile is crucial for accurate prediction of the wall shear stresses, and therewith the downstream flow physics. To this end, a number of RANS simulations were carried out with an upstream extended domain, from which total pressure and temperature profiles of the fully turbulent boundary layer

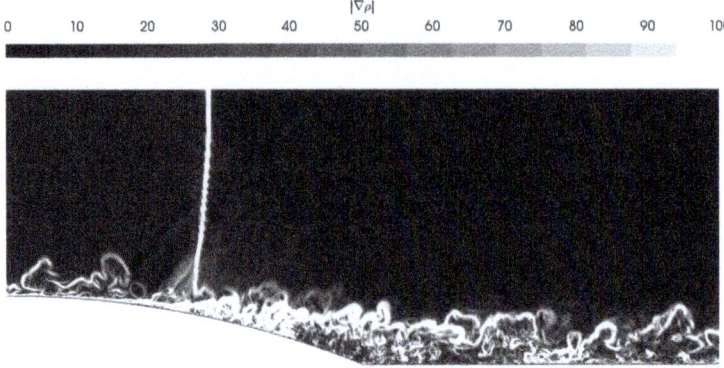

Fig. 16 Instantaneous flow field

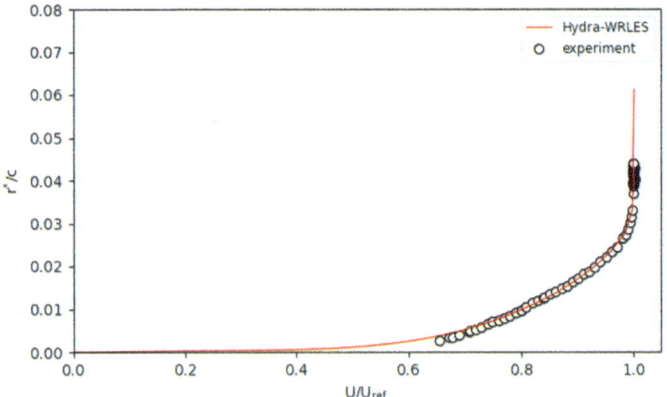

Fig. 17 Incoming boundary layer profile

was extracted at a certain position providing the experimental boundary layer profile. Then, they were imposed as the inlet boundary condition on the short-inlet LES domain (Fig. 16). This is not only to save the total number of the grid points but also to ensure the experimental boundary layer at $x/c = -0.78$. The inlet profile computed in this way can be seen in Fig. 17 to agree quite well with the experiment.

4.3.3 Engineering Quantities of Interest

In Fig. 18, we compare the skin friction coefficient of our numerical simulations with respect to the experiments. The Spalart-Allmaras turbulence model predicts the separation point at around $0.67c$ and the reattachment point at $1.19c$ whereas the measurements reported are 0.69 and $1.18c$ respectively [31]. The characteristic can be seen in Fig. 18. While the k-ω SST turbulence model predicts the separation point at $0.67c$, the reattachment point is predicted at $x/c = 1.10$. Despite the fact that upstream shear stresses are significantly overpredicted for both turbulent

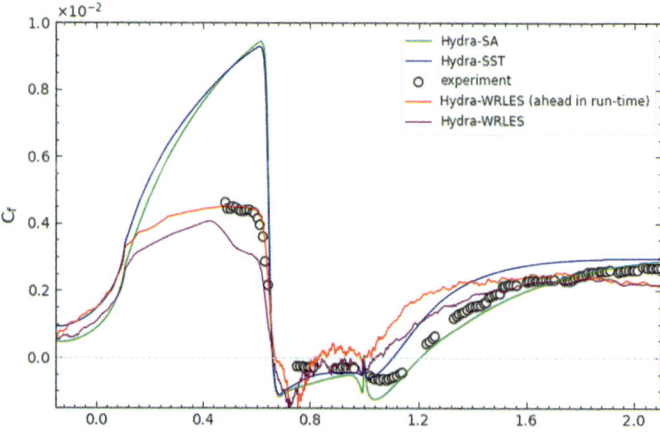

Fig. 18 Skin friction coefficient, C_f

models, the separation and reattachment locations agree well with the experiments. According to the interim averaging, unlike the RANS methods, WRLES predicts the upstream shear stresses reasonably well and thus the separation point accurately as the computation approaches to the statistical convergenge. The modelling of the incoming turbulent boundary layer profile accurately poses a significant role in this agreement. The reattachment is estimated to occur earlier at $x/c = 1.04$ compared to the experiments.

The recent results of WRLES on Sandia ATB test case has been presented in this section. It is noted that the statistically converged solution was not achieved yet because of the high computational resources and lengthy run time required. After statistically converged solution is ensured with finer time resolution, further data analysis and further back to back comparison with the measurements, will be performed on the time-averaged solution. The present WRLES predictions are foreseen to improve with final results of statically converges solution, and so the readers are referred to the future publications for further detail.

Once the recent results are evaluated overall, it is worth stating that the shear stress distribution predicted by WRLES in the forebody shows a good compromise with the experiments which can not be captured accurately with RANS methods. The modelling of the incoming turbulent boundary layer profile accurately poses a significant role in this agreement.

5 Conclusion

The shock boundary layer interaction in transonic compressor flows is a complex phenomenon in which several factors and mechanisms interplay one another. This type of flow poses a particular challenge to engineering-level methods such as RANS for the prediction of engineering quantities of interest. Scale resolving methods such as LES are considered to improve the accuracy for capturing the critical aspects of the flow at the expense of higher computational costs, the accuracy of which also depends on the resolution. To have confidence with both RANS methods and LES approaches requires a well defined experimental database aligning with industrial needs.

This chapter presents implications towards designing more efficient and compact turbomachineries for aeronautics, putting the mechanisms shock wave-boundary layer interactions in the center. Within this scope, a multi-point re-design study was performed and a new design solution was proposed, which aimed to mitigate limiting effects of transonic speeds at altitude, still maintaining the performance across the operating envelope with a holistic approach. The methodology followed was briefly introduced in this chapter and further details were adressed. The performance improvement of the new design over the baseline design was compared back to back and presented. While the quantitative results of the Euler-based methods with integral boundary layer equations and transitional RANS methods shows differences, the delta improvement with the new design was found to be consistent for both methods.

The new design, then, was further investigated with high fidelity methods under the altitude excitation and high loading conditions. The laminar boundary layer at altitude was observed to incur a multi-shock pattern in the flow with a larger separated region accompanied with high losses and low frequency unsteadiness. This pattern was found to result in a broad band, low-frequency energetic oscillations shown in the PSD map. The main shock foot was also seen to oscillate with a significant amplitude over the suction surface. The fluctuation of the break-down of the separated shear layer into turbulent state was illustrated in order to address the high-level of unsteadiness.

Following that, our efforts towards numerical modelling and validation of the Sandia ATB case was presented. Both RANS methods and ongoing WRLES methods were briefly compared back to back with test results with the key assessments quantities. Once the recent results are evaluated overall, as the computation progressed in runtime, the shear stress distribution predicted by WRLES in the forebody shows a good compromise with the experiments which can not be captured accurately with RANS methods. The modelling of the incoming turbulent boundary layer profile accurately poses a significant role in this agreement. RANS methods were found to significantly overpredict the shear stresses over the wall in the forebody, however, the Spalart-Allmaras turbulent closure shows a good agreement with the experiment in terms of separation and reattachment points. The present WRLES predictions are foreseen to improve with final results of statically converged solution, and so the readers are referred to the future publications for further detail.

Acknowledgements This work is part of a project that has received funding from the European Union's Horizon 2020 research and innovation program under grant agreement no. 860909 (TEAMAero-Towards Effective Flow Control and Mitigation of Shock Effects In Aeronautical Applications).

References

1. Kamin, M. and Mathew, J.: Prediction of transitional and separated boundary layers in a compressor cascade. In: Turbo Expo: Power for Land, Sea, and Air, vol. 45615, p. V02BT39A038. American Society of Mechanical Engineers (2014). https://doi.org/10.1115/GT2014-26892
2. Bode, C., Kožulović, D., Stark, U., Hoheisel, H.: Performance and boundary layer development of a high turning compressor cascade at sub-and supercritical flow conditions. In: Turbo Expo: Power for Land, Sea, and Air, vol. 44748, pp. 49–61. American Society of Mechanical Engineers (2012). https://doi.org/10.1115/GT2012-68382
3. Swoboda, M., Nitsche, W.: Shock boundary-layer inter- action on transonic airfoils for laminar and turbulent flow. J. Aircr. **33**(1), 100–108 (1996). https://doi.org/10.2514/3.46909
4. Merkli, P.E.: Pressure recovery in rectangular constant area supersonic diffusers. AIAA J. **14**(2), 168–172 (1976)
5. Mateer, G., Viegas, J.: Effect of Mach number and Reynolds number on a normal shock-wave/turbulent boundary-layer interaction. In: 12th Fluid and Plasma Dynamics Conference, p. 1502 (1979)
6. Carroll, B.F., Dutton, J.C.: Multiple normal shock wave/turbulent boundary-layer interactions. J. Propul. Power **8**(2), 441–448 (1992)
7. Om, D., Viegas, J.R., Childs, M.E.: Transonic shock-wave/turbulent boundary-layer interactions in a circular duct. AIAA J. **23**(5), 707–714 (1985)
8. Carroll, B.F., Dutton, J.C.: Characteristics of multiple shock wave/turbulent boundary-layer interactions in rectangular ducts. J. Propul. Power **6**(2), 186–193 (1990)
9. Om, D., Childs, M.: An experimental investigation of multiple shock wave/turbulent boundary layer interactions in a circular duct. In: 16th Fluid and Plasmadynamics Conference, p. 1744 (1983)
10. Hobbs, D.E., Weingold, H.D.: Development of controlled diffusion airfoils for multistage compressor application. In: Turbo Expo: Power for Land, Sea, and Air, vol. 79511, p. V001T01A058. American Society of Mechanical Engineers (1983). https://doi.org/10.1115/1.3239559
11. Steinert, W., Eisenberg, B., Starken, H.: Design and testing of a controlled diffusion airfoil cascade for industrial axial flow compressor application. In: Turbo Expo: Power for Land, Sea, and Air, vol. 79047, p. V001T01A044. American Society of Mechanical Engineers (1990). https://doi.org/10.1115/1.2929119
12. Steinert, W., Starken, H.: Off-design transition and separation behavior of a CDA cascade (1996). https://doi.org/10.1115/1.2836627
13. Hoheisel, H., Seyb, N.J.: The boundary layer behaviour of highly loaded compressor cascade at transonic flow conditions. Technical report no. (1987)
14. John, A., Shahpar, S. and Qin, N.: Alleviation of shockwave effects on a highly loaded axial compressor through novel blade shaping. In: Turbo Expo: Power for Land, Sea, and Air, vol. 49699, p. V02AT37A040. American Society of Mechanical Engineers (2016). https://doi.org/10.1115/GT2016-57550
15. Yamaguchi, Y., Arima, T.: Multi-objective optimization for the transonic compressor stator blade. In: 8th Symposium on Multidisciplinary Analysis and Optimization, p. 4909 (2000). https://doi.org/10.2514/6.2000-4909
16. Song, B., Ng, W.F.: Performance and flow characteristics of an optimized supercritical compressor stator cascade (2006). https://doi.org/10.1115/1.2183316
17. Kahraman, S., Bourgeois, J.A., Janke, C., Swoboda, M.: Numerical investigations of transitional SBLI on a highly loaded transonic compressor passage. In: Turbo Expo: Power for Land, Sea, and Air, vol. 87080, p. V13AT29A024. American Society of Mechanical Engineers (2023)
18. Flassig, P.M.: Unterstützende Optimierungsstrategien zur robusten aerodynamischen Verdichterschaufelauslegung. Doctoral thesis (2016)
19. Gräsel, J., Keskin, A., Swoboda, M., Przewozny, H., Saxer, A.: A full parametric model for turbomachinery blade design and optimisation. In: International Design Engineering Technical

Conferences and Computers and Information in Engineering Conference, vol. 46946, pp. 907–914 (2004). https://doi.org/10.1115/DETC2004-57467
20. Keskin, A.: Process integration and automated multi-objective optimization supporting aerodynamic compressor design. Doctoral thesis, BTU Cottbus, Senftenberg (2007)
21. Dutta, A.K.: An automated multi-objective optimization approach for aerodynamic 3D compressor blade design. Doctoral thesis (2016)
22. Langtry, R.B., Menter, F.R.: Correlation-based transition modeling for unstructured parallelized computational fluid dynamics codes. AIAA J. **47**(12), 2894–2906 (2009)
23. Menter, F.R., Kuntz, M., Langtry, R.: Ten years of industrial experience with the SST turbulence model. Turbul. Heat Mass Transf. **4**(1), 625–632 (2003)
24. Welch, P.D.: IEEE Trans. Audio Electroacoust. **15**, 70 (1967)
25. Bachalo, W., Johnson, D.: Transonic, turbulent boundary-layer separation generated on an axisymmetric flow model. AIAA J. **24**(3), 437–443 (1986)
26. Menter, F.R.: Two-equation eddy-viscosity turbulence models for engineering applications. AIAA J. **32**(8), 1598–1605 (1994)
27. Spalart, P.R., Belyaev, K.V., Garbaruk, A.V., Shur, M.L., Strelets, M.K., Travin, A.K.: Large-eddy and direct numerical simulations of the Bachalo-Johnson flow with shock-induced separation. Flow Turbul. Combust. **99**(3), 865–885 (2017)
28. Uzun, A., Malik, M.R.: Wall-resolved large-eddy simulations of transonic shock-induced flow separation. AIAA J. **57**(5), 1955–1972 (2019)
29. Lynch, K.P., Barone, M.F., Beresh, S.J., Spillers, R., Henfling, J., Soehnel, M.: Revisiting Bachalo-Johnson: the Sandia axisymmetric transonic hump and CFD challenge. In: AIAA Aviation 2019 Forum, p. 2848 (2019)
30. Lynch, K.P., Lance, B., Lee, G.S., Naughton, J.W., Miller, N.E., Barone, M.F., Beresh, S.J., Spillers, R., Soehnel, M.: A CFD validation challenge for transonic, shock-induced separated flow: experimental characterization. In: AIAA Scitech 2020 Forum, p. 1309 (2020)
31. Lynch, K.P., Lance, B.W., Miller, N.E., Barone, M.F., Beresh, S.J.: Experimental characterization of an axisymmetric transonic separated flow for computational fluid dynamics validation. AIAA J. **61**(4), 1623–1638 (2023)
32. Riley, L.P., Adler, M.: RANS and wall-modeled LES predictions for the SANDIA challenge on transonic, separated flow. In: AIAA Aviation 2021 Forum, p. 2757 (2021)
33. Gupta, M., Datta, A., Mathew, J., Hemchandra, S.: Shock induced separation in a transonic flow past an axi-symmetric hump. In AIAA Aviation 2021 Forum, p. 2756 (2021)
34. Rahmani, S.K., Wang, Z.J.: Large Eddy simulation of the sandia axisymmetric transonic hump using a high-order method. In: AIAA SCITECH 2022 Forum, p. 1534 (2022)
35. Davidson, L.: Using Isotropic Synthetic Fluctuations as Inlet Boundary Conditions for Unsteady Simulations. Advances and Applications in Fluid Mechanics. Citeseer (2007)
36. Lynch, K.P., Lance, B., Lee, G.S., Naughton, J.W., Miller, N.E., Barone, M.F., Beresh, S.J., Spillers, R., Soehnel, M.: Correction: A CFD Validation Challenge for Transonic, Experimental Characterization, Shock-Induced Separated Flow, p. 1309 (2020)

Open Access This chapter is licensed under the terms of the Creative Commons Attribution 4.0 International License (http://creativecommons.org/licenses/by/4.0/), which permits use, sharing, adaptation, distribution and reproduction in any medium or format, as long as you give appropriate credit to the original author(s) and the source, provide a link to the Creative Commons license and indicate if changes were made.

The images or other third party material in this chapter are included in the chapter's Creative Commons license, unless indicated otherwise in a credit line to the material. If material is not included in the chapter's Creative Commons license and your intended use is not permitted by statutory regulation or exceeds the permitted use, you will need to obtain permission directly from the copyright holder.

Reynolds Number Effects on Shock Wave Boundary Layer Interaction in Highly Loaded Compressor Stator

Arun Joseph, Pawel Flaszynski, Michal Piotrowicz, Piotr Doerffer, and Marcin Kurowski

Abstract When an aircraft engine operates at a transonic regime, the Shock Wave Boundary Layer Interaction (SBLI) plays a significant role in aerodynamic performance and its effects are detrimental while operating at low Reynolds conditions. To investigate SBLI effects numerically and experimentally, a linear cascade 3-profile test section has been designed for the IMP-PAN transonic blow-down wind tunnel facility. A novel technique of inlet valves with perforated plates setup has been installed upstream of the test section to reduce the Reynolds number in the test section. Three different Reynolds cases (7.4×10^5, 5.8×10^5, and 2.8×10^5) have been chosen for detailed flow structure comparison. The main focus of the research is on the suction side of the middle profile where the shock wave interacts with the boundary layer resulting in boundary layer separation. A detailed boundary layer investigation and the location of the separation bubble beneath the shock foot have been compared. The total pressure losses have been compared based on wake measurements downstream of the middle profile.

Keywords Transonic flow · Shock induced separation · Boundary layer transition · Numerical simulations

---

A. Joseph (✉) · P. Flaszynski · M. Piotrowicz · P. Doerffer · M. Kurowski
Institute of Fluid-Flow Machinery, Polish Academy of Sciences (IMP PAN), Gdańsk, Poland
e-mail: ajoseph@imp.gda.pl

P. Flaszynski
e-mail: pflaszyn@imp.gda.pl

M. Piotrowicz
e-mail: mpiotrowicz@imp.gda.pl

P. Doerffer
e-mail: Doerffer@imp.gda.pl

M. Kurowski
e-mail: marcin.kurowski@imp.gda.pl

1 Introduction

Transonic flow effects in air-breathing engines have been researched for decades due to their detrimental impact on performance. Aerodynamic performance improvement in engines has pushed design engineers to achieve high-performance goals like compact engine designs with reduced length and weight. Reduced blade count in axial compressors helps in achieving these performance goals but in return results in higher blade loading which causes performance penalties at transonic flows [1]. The increase of root tip velocity of the compressor is limited by centrifugal stresses whereas the large blade turn angle leads to an adverse pressure gradient resulting in flow separation and severe passage blockage [2]. These effects get aggravated when shock waves interact with the boundary layer ensuing higher losses [3].

Normal shocks and SBLIs are natural occurrences that have a major role in limiting the aerodynamic performance of axial compressors. SBLI depends on the nature of the boundary layer developed upstream of shock wave interaction [4]. A strong interaction submits the boundary layer into an adverse pressure gradient resulting in boundary layer separation [5]. The boundary layer is mostly laminar upstream of the shock where the laminar boundary layer is more sensitive to adverse pressure gradient [6]. Shock-induced separation most often results in large unsteadiness which limits the engine performance. Unsteady pressure fluctuation caused by normal SBLI results in unsteady aerodynamic loading [7]. Laminar SBLI can lead to buffeting at airfoils and shock oscillation in internal flows [8]. The transition from laminar to a turbulent boundary layer has a serious impact on the overall drag and performance of the blade [9].

At higher altitudes lower density causes a reduction in the Reynolds number. Reynolds number drops to a factor of 4 compared to sea level conditions [10]. As a result, the transition happens further downstream on the blade, resulting in an extended laminar boundary layer length. At low Reynolds numbers due to the increased viscosity effect the boundary layer developed on the blade surface becomes thicker compared to high Reynolds conditions. When shock waves interact with a thicker boundary layer the resulting separation would be larger than high Reynolds conditions. The shear stress on the blade surface produced by the boundary layer is also lower, with a reduction in the Reynolds number. As a result, a small disturbance could trigger separation of the boundary layer at low Reynolds conditions exhibiting larger shock unsteadiness in the blade passage [11].

Understanding the fundamental flow mechanics of axial turbomachinery blade rows has been, and continues to be, dependent on cascade testing. Even with the significant advancements in computational prediction methods, understanding the complex flow interactions within blade rows still requires experimental research on cascade flows conducted with defined inflow boundary conditions. Generally, a turbomachinery flow through the blade row is three-dimensional and unsteady. Over time, numerous experimental setups and measurement methods have been created to determine flow numbers more exactly and realistically. Literature reviews, such as those by Sieverding [12], Baines et al. [13], Hirsch [14], Sonoda et al. [15], Hergt et al. [16] cover several types of cascade tunnels, including low-speed, high-speed, intermittent blow-down, and suction. A rectilinear test section with three blades has been designed for the IMP-PAN transonic wind tunnel to investigate the SBLI effects on stator cascade operating at different Reynolds number conditions. The objective of this design was to model the flow over the blade suction side. It can not be done with a single blade but two other neighboring blades are necessary to create a realistic flow structure.

Few research could be found in the literature focusing on the low Reynolds effect in compressor cascades. Some of the prominent researches from DLR includes [15–18]. Similar research on SBLI effects on compressor profile from IMP-PAN transonic wind tunnel facility can be found in Joseph et al. [19], Flaszynski et al.[20], Doerffer et al. [21]. The main challenge in investigating the low Reynolds number in the wind tunnel is to achieve low density by reducing the total pressure in the test section. Therefore most of the researches are carried out in a closed-loop loop wind tunnel, where inlet conditions could be controlled according to the Reynolds number. A novel technique of inlet valve has been introduced upstream of the test section in IMP-PAN transonic wind tunnel to achieve a low Reynolds number. A detailed numerical and experimental comparison of flow structure has been carried out for three chosen Reynolds number conditions.

2 Test Section Design Approach

The primary goal of this research is to explore the effects of SBLI numerically and experimentally on a highly loaded compressor stator cascade operating at various Reynolds number conditions. The geometrical and flow parameters for the compressor stator profile were defined by the project partner Rolls Royce Deutschland [22]. A rectilinear 3-profile linear cascade test section has been designed for the IMP-PAN transonic blow-down wind tunnel facility. The main objective of the test section is to reproduce the flow structure on the suction side of the stator profile as in cascade. To achieve better optical access in experiments the chord length of the delivered stator profile has been up-scaled with a scale factor of 1.72. A comparison of geometrical parameters of the delivered profile and up-scaled profile has been defined in Table 1.

The span of the stator profile has been defined as 100 mm with an aspect ratio of 2 for the up-scaled profile based on the IMP-PAN wind tunnel dimensions. 2D and 3D cascade simulations were carried out for the up-scaled profiles to have a direct comparison with the test section flow structure and experiment data. Mainly the 2D simulations were carried out to extract the streamlines to design the preliminary shape of the upper and lower walls of the test section as shown in Fig. 1.

The streamlines, depicted in Fig. 1, were extracted from 2D simulations for cascade simulations. The initial geometry of the upper and lower walls was designed from the streamlines extracted whereas the final geometry has been optimized based on the boundary layer developed at end-walls causing blockage effects. The test section configuration has been chosen with 3 profiles where the main focus of the

Table 1 Geometrical parameters of cascade profile mounted in the test section

Parameters	Units	Test section profile
Chord length	mm	50
Stagger angle	deg	30.31
Inlet Mach number	–	0.90
Blade thickness	mm	2.75
Blade inlet angle	deg	48.14
AVDR	–	1.129

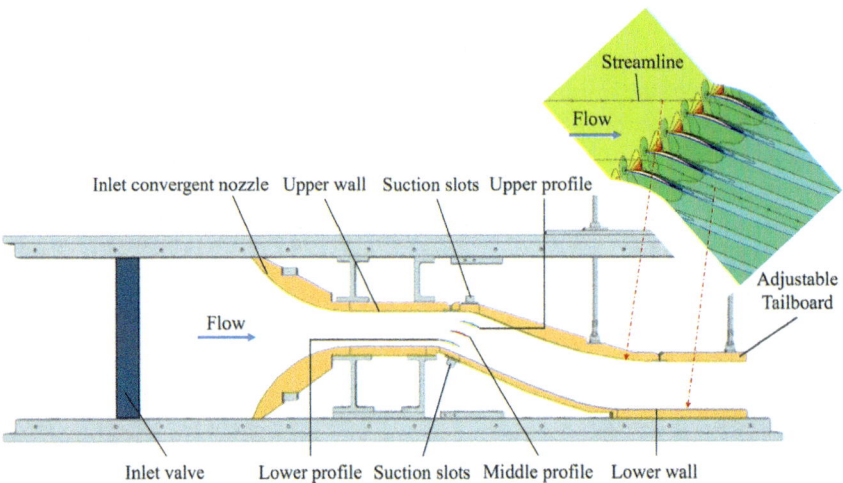

Fig. 1 CAD model of the designed test section based on the streamlines extracted from 2D cascade simulation

research is on the suction side of the middle profile. There is no possibility of modifying the inflow angle in this test section configuration since the blades are fixed. The inlet of the test section is designed as a convergent nozzle to achieve the required inlet Mach number upstream of the profiles. The inlet Mach number of 0.9 has been considered one of the main design criteria for Re investigation. The end-wall boundary layer plays a huge role in flow structure in the test section an active flow control method of boundary layer suction has been implemented at the upper and lower walls to reduce the blockage effects caused by the boundary layer. An adjustable tailboard has been used downstream of the test section to adjust the opening of the outlet cross-section to control the static pressure and adjust the inflow Mach number.

2.1 Introduction of Suction Slots in Test Section

It is challenging to obtain a similar flow structure in the test section as in cascade configurations. To reduce the adverse effect on flow structure caused by the development of boundary layers at end-walls, suction slots have been designed on the upper and lower walls close to the blade passage. Suction slots help in achieving flow periodicity in the blade passages. The final position of the suction slots at the upper and lower walls has been determined based on flow uniformity predicted numerically in each blade passage. The suction slots are connected to the outlet vacuum tank and the mass flow through them is controlled with a valve. The suction slots also influence the inflow conditions upstream of the blade passage. Figure 2 illustrates the difference between the Mach number contour plot and the isentropic Mach number plotted at all three blades in the test section with and without suction slots for the same inflow conditions.

According to Fig. 2, the suction side of the upper and lower profile benefit the most from suction slots, which also aid in producing a similar shock distribution over all three blades in comparison to the test section without suction slots. The distribution over the suction side of the upper and lower profile improves with the help of active suction at end-walls. Whereas the isentropic Mach number distribution over the middle profile is least influenced by suction in the test section. The numerical and experimental investigation for the Reynolds number includes the effect of active suction at both end-walls.

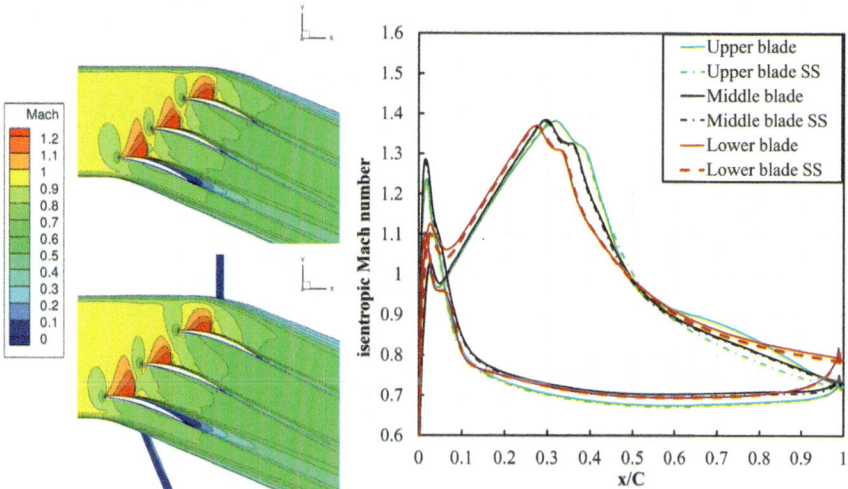

Fig. 2 Comparison of isentropic Mach number plotted on each blades in the test section with (SS) and without suction slots

2.2 Introduction of Blade Supports in Test Section

To have the numerical model defined as similar as possible to the experimental setup the blade holders for the upper and lower profiles have been also modeled in the numerical model of the test section. The blade holders are positioned in the upper and lower passages the fluid domain for the mesh has been adapted accordingly to include the effect of the blade holder. A comparison of flow structure with and without blade supporting holders for the same boundary conditions has been shown in Fig. 3.

The main focus of this research is on the middle profile, therefore the shock distribution across the suction side has been compared for configuration with and without blade support and is compared with the experiment. From Fig. 3, it is visible that the shock moves slightly downstream on both the suction and pressure side and has a good agreement with measurements. Even though the holders are placed in the uppermost and lowermost passages the effect plays a huge role in the flow structure on the suction side of the middle profile. Due to the installation of blade holders at both side walls of the bottom and top passage of the test section the mass flow through individual blade channels has been modified resulting slightly higher Mach number compared to the configuration without holders as depicted in Fig. 3. The passage shock wave generated between blade passages is comparatively stronger with holders compared to the test section without holders.

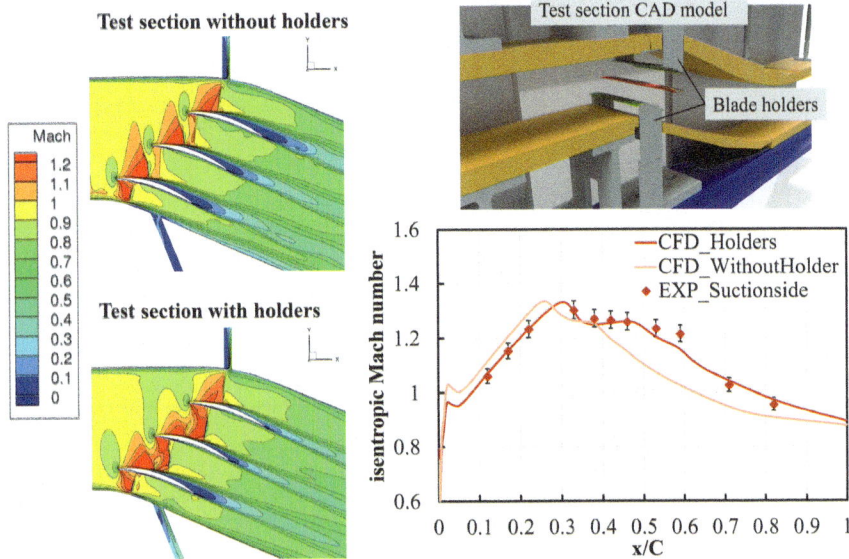

Fig. 3 Comparison of (Left) Mach number contour plot, (Right) isentropic Mach number plotted on the middle profile for test section with and without blade holders

3 Numerical Model Description

The commercial code from $FINE^{TM}$ Turbo Numeca (Cadence) has been used to perform numerical computations. On a structured multi-block grid configuration, the steady-state 3D Reynolds-averaged Navier-Stokes equations (RANS) are solved. For the numerical simulations spatial discretization using a second-order central difference scheme with scalar artificial dissipation formulated by Jameson [23] was applied. The Baseline Explicit Algebraic Reynolds Stress Model (BSL-EASRM), proposed by Menter [24], is a two-equation nonlinear eddy viscosity turbulence model that extends the kω-SST turbulence model as shown in Eqs. (1) and (2).

$$\frac{D\rho k}{Dt} = \tau_{ij}\frac{\partial u_i}{\partial x_j} - \beta^* \rho\omega k + \frac{\partial}{\partial x_j}\left[(\mu + \sigma_k\mu_t)\frac{\partial k}{\partial x_j}\right] \quad (1)$$

$$\frac{D\rho\omega}{Dt} = \frac{\gamma}{\nu_t}\tau_{ij}\frac{\partial u_i}{\partial x_j} - \beta\rho\omega^2 + \frac{\partial}{\partial x_j}\left[(\mu + \sigma_\omega\mu_t)\frac{\partial \omega}{\partial x_j}\right] + 2\rho(1-F_1)\sigma_{\omega 2}\frac{1}{\omega}\frac{\partial k}{\partial x_j}\frac{\partial \omega}{\partial x_j} \quad (2)$$

where k is the turbulent kinetic energy and ω is the specific dissipation rate. The coefficients $\beta = 3/40$, $\beta^* = 9/100$, $\sigma_k = 0.85$ and $\sigma_\omega = 0.50$ are the model constants. A generalized transition model [25] has been used for predicting transition effects. This model is defined on two transport equations, Eqs. (3) and (4), for the intermittency (γ) and the transition momentum thickness Reynolds number ($\tilde{Re}_{\theta t}$).

$$\frac{\partial \rho \gamma}{\partial t} + \frac{\partial \rho u_j \gamma}{\partial x_j} = P_\gamma - E_\gamma + \frac{\partial}{\partial x_j}\left(\left(\mu + \frac{\mu_t}{\sigma_f}\right)\frac{\partial \gamma}{\partial x_j}\right) \quad (3)$$

$$\frac{\partial \rho \tilde{Re}_{\theta t}}{\partial t} + \frac{\partial \rho u_j \tilde{Re}_{\theta t}}{\partial x_j} = P_{\theta t} + \frac{\partial}{\partial x_j}\left(\sigma_{\theta t}(\mu + \mu_t)\frac{\partial \tilde{Re}_{\theta t}}{\partial x_j}\right) \quad (4)$$

where P_γ and E_γ are the transition sources for intermittency whereas $P_\theta t$ is a source term for momentum thickness Reynolds number ($\tilde{Re}_{\theta t}$). The structured grid for the computation was generated using Numeca/Cadence IGG (Interactive Geometry Modeler and Multi Block Structured Grid Generator). To achieve y^+ of 1, the mesh resolution near the walls has been defined as 5×10^{-6} (m). The test section mesh domain consists of 130 blocks with 41×10^6 cells as shown in Fig. 4. The mesh domain shown in Fig. 4 includes the position of the blade supporting holders for upper and lower blades where the mesh block has been adapted according to the shape and size of real geometry in the experimental setup. The mesh topology close to the leading and trailing edge of the middle profile has been depicted in Fig. 4. The convergence study procedure was performed as described by Celik [26]. Firstly, three mesh resolutions of roughly (332×145, 280×101 and 196×85) cells (N) on the suction side of the middle blade keeping $y^+ = 1$ close to the wall were created, and the representative cell size was measured (h).

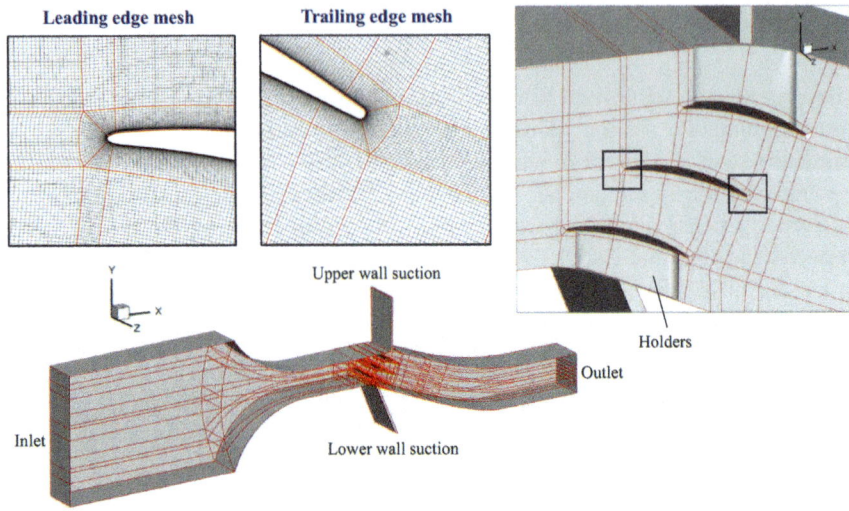

Fig. 4 Structured grid topology and mesh domain of the test section created using Numeca/IGG

Table 2 Mesh convergence study on test section model

Parameters	Unit	Values
N_1, N_2, N_3	–	48140, 28280, 16660
h_1, h_2, h_3	mm	0.3, 0.4, 0.5
ϕ_1, ϕ_2, ϕ_3	kPa	59360, 59327, 59157
r^{21}, r^{32}	–	1.3047, 1.3029
GCI^{21}, GCI^{32}	–	0.0166, 0.0866

Secondly, RANS simulations were performed with the numerical setup described in the next section. Finally, the static pressure at the suction side of the middle profile was picked as the critical variable (ϕ), and the discretization error was estimated. The results obtained are summarized in Table 2. From these results, the mid-refinement level shown in Fig. 4 was chosen for the numerical investigation. This level provides the best balance between accuracy in predicting boundary layer effects in SBLI for test section simulations, with a GCI of 0.0866, and calculation time within the scope of limits.

4 Experimental Setup Description

To carry out an experimental investigation of SBLI effects the designed test section has been manufactured and mounted at IMP-PAN transonic blow-down wind tunnel facility with the ambient condition at inlet. The close-up image of the 3-blade profiles mounted at the test section is depicted in Fig. 5.

The detailed description of the designed test section can be found in the previous Sect. 2. The upper and lower profiles are mounted in the test section using an adjustable supporting holder, whereas the middle profile has been fixed to the Plexiglas side walls. Both side-wall regions close to the profiles have been equipped with Plexiglases to have a large field of optical access in experiments. The test section is mounted to a blow-down wind tunnel facility where the inlet is defined as ambient condition and the outlet is connected to a vacuum tank. The measurement time for the experiment is estimated based on the time required to refill the vacuum tanks from the empty state. The suction slots have been connected to a separate vacuum tank. The mass flow through the suction slots has been adjusted using a control valve to set the inflow conditions in the test section. The static pressure has been measured with pressure taps at the suction side of the middle profile. Also, the static pressure is measured at the upper and lower walls upstream of the blade passage to estimate the inlet Mach number.

A novel approach of inlet valve setup has been introduced at IMP-PAN transonic blow-down wind tunnel facility to investigate the Reynolds number effect as shown in Fig. 1. The inlet valve is a set of 3 plates attached to a rectangular frame that has been mounted upstream of the test section as shown in Fig. 6. Two sets of plates are positioned upstream of the frame: one stationary plate fixed to the frame and another sliding plate is a bit shorter in total length so it can slide over the stationary plate. The two plates are defined with 15 rectangular slots with 8.3 mm slot openings. The length of the upper and lower slots are slightly smaller to accommodate the mounting bolts. The main purpose of these slots is to control the mass flow through the rectangular slots which results in a reduction of ambient total pressure and density, which helps

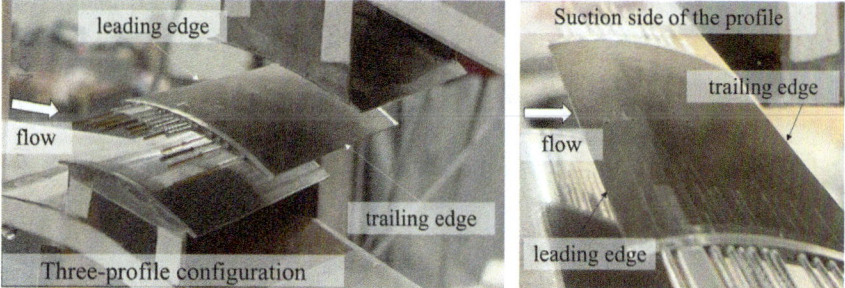

Fig. 5 (Left) The designed test section mounted at IMP-PAN transonic blow down wind tunnel facility, (Right) zoomed image of the suction side of the middle profile

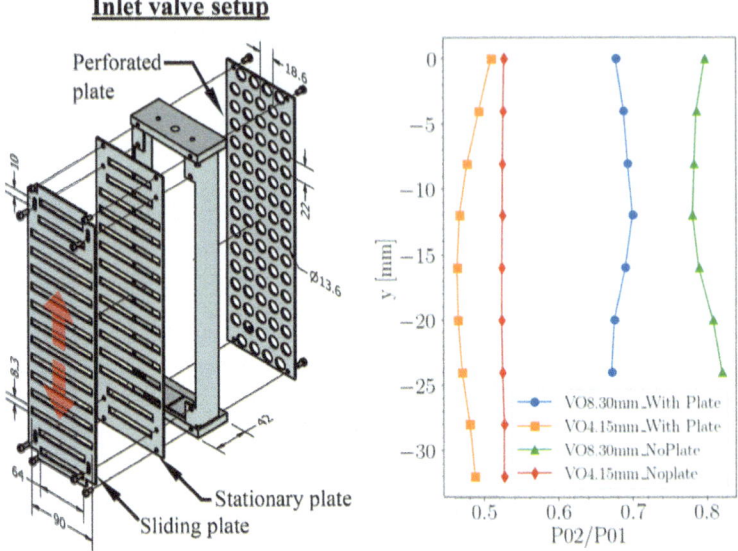

Fig. 6 (Left) CAD model of the inlet valve setup, (Right) total pressure measurements downstream the inlet valve for two valve openings with and without perforated plate included in the setup

in achieving the required pressure drop in the test section resulting in a reduction of Reynolds number. The flow through rectangular slots from the inlet valve creates turbulent vortex structures, resulting in increased turbulent intensity. A third plate with a circular hole with a slightly lower perforation rate than concerning other two plates has been defined as shown in Fig. 6.

The third plate is attached to the frame downstream of the first set of plates to achieve flow uniformity downstream of the inlet valve setup. These perforated plates diffuse the large vortices coming from the rectangular slot openings. The perforation of this third plate is kept lower to avoid choking conditions in the perforated plate. Another added advantage was that further pressure drop was also achieved using the perforated plate which added up to achieving even lower Reynolds conditions. A Pitot probe was placed downstream of the inlet valve at a distance of 45 mm at a midspan location to estimate the total pressure drop for different Reynolds conditions in the test section. The inlet convergent nozzle also helps in making the flow uniform close to the blade passage therefore the novel approach of the inlet valve could be an optimal method for Reynolds number investigation in blow down wind tunnel. The vertical traverse measurement of total pressure at different inlet valve slot opening configurations has been plotted in Fig. 6. The plot comprises two slot openings of 8.3 and 4.15 mm with and without perforated plate included. The y-axis corresponds to the vertical travel of the total pressure probe. An additional pressure drop of 10% could be achieved for the same slot opening with the addition of a perforated plate as shown in Fig. 6.

5 Numerical and Experimental Investigation of Reynolds Number Effect on SBLI

The detailed flow structure investigation has been carried out numerically and experimentally for the chosen three Reynolds number cases. The Reynolds number in the test section has been estimated based on the total pressure measurements from the Pitot tube located upstream of the blade passage. The main goal of this research is to understand the fundamental effect of SBLI at three different Reynolds conditions (corresponding to different altitudes). To have similar inflow conditions in comparison the chosen Reynolds cases have been defined with an inlet Mach number as 0.9. The required inflow Mach number is achieved by adjusting the outlet static pressure in the test section. The static pressure in the suction slots also plays a huge role in getting the required inflow conditions in the test section. The boundary conditions in the test section for chosen Reynolds number cases have been described in Table 3.

The total temperature range for corresponding Re cases has been defined in Table 3 and in experiments the temperature varies with ambient conditions in the range of $\pm 6\,°C$. The turbulent intensity described in Table 3, corresponds to the effect of the inlet valve in the test section. The turbulent intensity for the *High* Re case has been estimated without an inlet valve in the test section. Whereas the *Mid* Re case and *Low* Re cases have been defined with an inlet valve (with different valve openings) in the test section resulting in higher turbulent intensity compared to *High* Re case. The estimated high level of turbulent intensity is normal for rotor downstream planes in aircraft engines. Based on the boundary conditions defined in Table 3, the shock structure on the middle profile in the test section has been visualized using the numerical schlieren method as depicted in Fig. 7.

The numerical schlieren method is presented using the density gradient magnitude. A normal shock wave (passage shock) is generated in the blade passage. A lambda foot is visible close to the boundary layer upstream of the passage shock. For *Low* Re the lambda foot is significantly larger compared to *High* Re. The separation bubble beneath the shock wave is visible as a bump in the numerical schlieren. A set of compression waves and expansion fans are visible upstream and downstream of the passage shock wave. Secondary shocklets consisting of alternating waves of

Table 3 Boundary conditions at inlet and outlets defined in test section based on Reynolds number estimations

Configuration	Reynolds number	Inlet total pressure (Pa)	Outlet static pressure (Pa)	Suction static pressure (Pa)	Inlet total temperature (K)	Turbulent intensity (%)
High Reynolds	7.4×10^5	101,000	77,000	55,000	293	2.5
Mid Reynolds	5.8×10^5	67,000	51,000	36,000	293	5.2
Low Reynolds	2.8×10^5	38,400	28,000	21,000	293	4.8

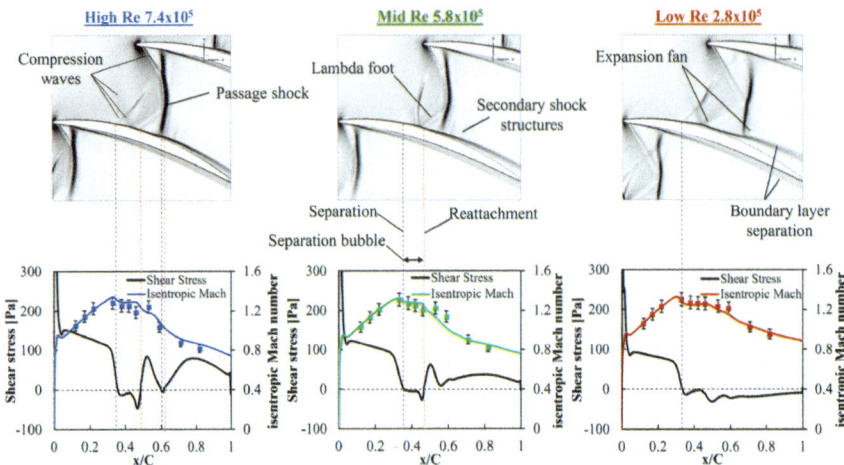

Fig. 7 Comparison of (above) numerical schilieren, (below) isentropic Mach number and wall shear stress plotted on the suction side of middle profile for three different Reynolds conditions

Table 4 Comparison of AVDR in test section

Configuration	AVDR
High Reynolds	1.231
Mid Reynolds	1.238
Low Reynolds	1.258

compression and expansion are more visible at *High* Re and *Mid* Re cases, whereas they are less visible in *Low* Re case. This may be due to the contraction of the main flow and corner flows as evidenced by the higher AVDR coefficient relative to the high Reynolds number from Table 4 in addition, the position of the shock wave is different, which may also be relevant. From Fig. 7, it is noticeable that boundary layer separation is more pronounced when the Reynolds number is reduced which is indicated by the negative value of shear stress. To visualize the shock wave distribution at the mid-span location along the chord length over the suction side of the middle profile, the isentropic Mach number has been estimated based on the static pressure measurements from the pressure taps as shown in Fig. 7.

The position of the shock wave on the suction side of the middle profile for *High* Re is at (x/C = 0.57), *Mid* Re at (x/C = 0.54) and *Low* Re at (x/C = 0.63). The chosen turbulence model has a good agreement with the isentropic Mach number estimation from experimental data. The strength of the shock wave decreases with the reduction of Reynolds number in the test section (depicted in isentropic Mach number plot), which represents a less intense density gradient in Fig. 7. To visualize the separation bubble generated on the suction side of the middle profile beneath the shock wave, wall shear stresses have been plotted at the mid-span location in Fig. 7. The shear stress value zero indicates the separation location and the negative value of shear stresses represents the reverse flow and the size of the separation bubble. It is apparent that the separation starts earlier for *Low* Re (x/C = 0.32) compared to *Mid* Re (x/C = 0.35) and *High* Re (x/C = 0.37). The separation bubble reattaches at (x/C = 0.49) for *High* Re and at (x/C = 0.48) for *Mid* Re. Whereas for *Low* Re separation bubble is extended till the trailing edge. This indicates that the boundary layer is volatile at *Low* Re conditions and tends to separate for even weak shock interactions. The passage shock due to the upper blade also has an effect on separation which is visible in the shear stress plot. The boundary layer separates and reattaches in *High* Re case and *Mid* Re case whereas for *Low* Re cases boundary layer continues to be separated till the trailing edge. For further visualization of separation zones generated due to SBLI on the suction and pressure side of the middle profile in the test section, the negative value of axial velocity (V_x) component has been plotted using iso-surfaces marked as blue in Fig. 8.

The location and shape of the separation bubble along the span and chord length on the middle profile can be visualized in the Fig. 8a. The size of the separation bubble gets larger with a decrease in Reynolds number. Also the separation bubble generated at pressure side leading edge has a similar trend as that on the suction side. The Fig. 8a illustrates the corner flows that occur as a result of the end-wall, causing a contraction of the blade passage area towards the trailing edge on the suction side. To quantify the secondary flows developed at different Reynolds conditions Axial Velocity Density Ratio (AVDR) has been calculated using Eq. (5).

$$AVDR = (\rho_2 v_2 \sin \beta_2) / (\rho_1 v_1 \sin \beta_1) \tag{5}$$

To investigate the AVDR at the middle blade passage the traverse has been extracted 10% upstream and downstream the middle profile as shown in Fig. 8b. AVDR is an important parameter to check the two-dimensionality of cascade flows.

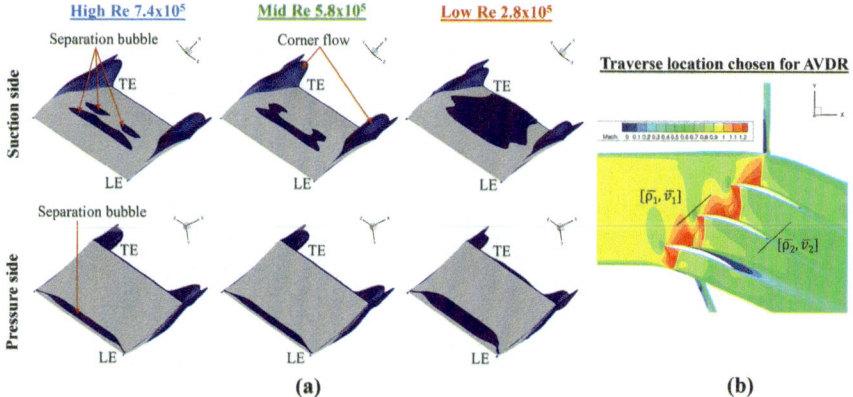

Fig. 8 Comparison of **a** Flow structure on suction and pressure side of middle profile. **b** Traverse location for calculating AVDR for different Reynolds number conditions

Based on Eq. (5), AVDR has been estimated for chosen three Reynolds conditions as shown in Table 4.

The AVDR has been estimated using mass averaged velocity component and density at the traverse location shown in Fig. 8b. From the comparison of AVDR estimated in Table 4, the passage contraction due to secondary flows increases when the Reynolds number is reduced for the chosen traverse location marked in Fig. 8b. A larger value for AVDR translates to a reduced passage area resulting in flow acceleration towards the trailing edge. The chosen traverse location for estimating AVDR takes into account the effects of upper and lower passages which gives a better estimation of the blockage effect of separation at the suction and pressure side of the middle profile. To visualize the flow structure above the suction side of the middle profile oil flow visualization has been compared for *Mid* Re and and are compared with velocity streamlines plotted above the blade surface as shown in Fig. 9.

The chosen numerical model could accurately predict the size and location of the separation bubble generated beneath the shock wave as marked in Fig. 9. The corner flow separation due to the side walls is visible towards the trailing edge of the blade. It is visible from Fig. 9, that the corner separations are similar in size at both sides of the blade passage. This is evident as well in the iso-surface generated for separation region in Fig. 8a. A detailed analysis of the boundary layer on suction side of middle profile has been investigated numerically and compared for three different Re numbers. Therefore velocity profiles at 9 traverse locations have been plotted and they are normalized to compare with theoretical flow on a flat plate as shown in Fig. 10.

The theoretical Blasius profile of the laminar boundary layer has been defined by Blasius [27] and the Prandtl turbulent boundary layer is defined with power law defined with factor (n)power exponent equal 7 [28]. According to Fig. 10, the velocity profile appears to be laminar at traverse (x/C = 0.1, x/C = 0.2 and x/C = 0.3). For both the *High* 7.4×10^5 and *Mid* 5.8×10^5 Reynolds cases, the boundary layer thickness is too thin and comparable as depicted in the Fig. 11. Whereas in the *Low* Reynolds case, the boundary layer is thicker from leading edge to trailing edge compared to high and mid-Reynolds numbers. At traverse x/C = 0.4, the traverse is already in a separation bubble for all three Re cases which is depicted as a negative value close to the boundary layer. The evidence for elongated separation bubble for *Low* Re case

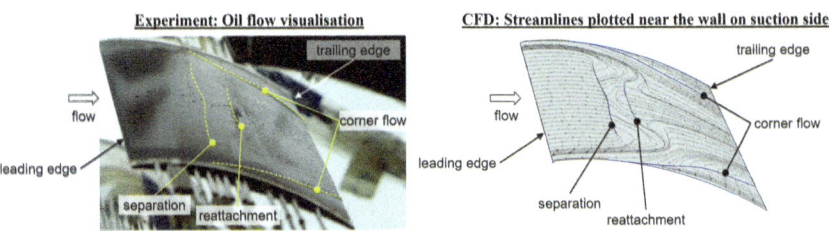

Fig. 9 Comparison of oil flow visualization (Left) and velocity streamlines plotted above the suction side ($i = 2$) (Right) for the *Mid* Reynolds case

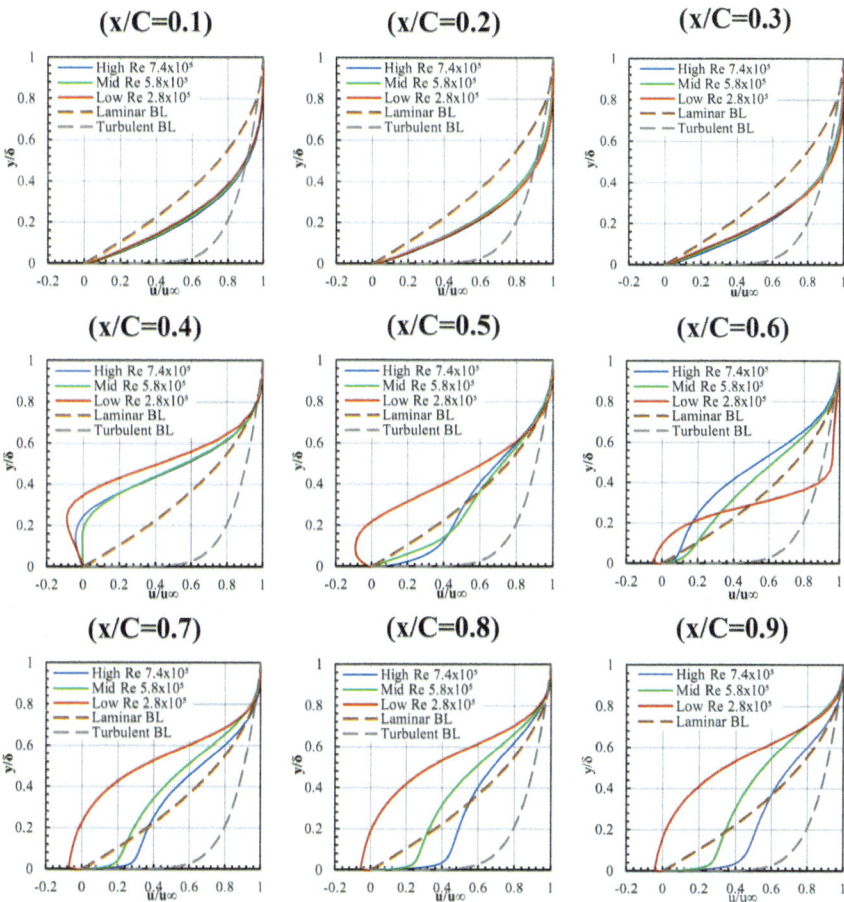

Fig. 10 Comparison of normalised velocity profiles at nine traverse location on the suction side of middle profile

could be visible at all traverses from (x/C = 0.4 to x/C = 0.8). Boundary layer shape is very similar for *Mid* and *High* Re but the differences are more prominent from x/C = 0.4. From the traverse x/C = 0.5, the boundary layer tends to be turbulent for both *Mid* and *High* Re cases. To further analyze the boundary layer, the traverses have been plotted at nine locations across the suction side of the middle profile a detailed boundary layer thickness and integral parameters of the boundary layer have been plotted in Fig. 11.

In the boundary layer thickness plot, it is evident that the thickness of the boundary layer increases when the Reynolds number is reduced. In the *High* and *Mid* Reynolds cases the thickness of the boundary layer is comparable to the shock interaction zone and differences are visible downstream of the reattachment zone. It is clear from the thickness of the boundary layer that laminar SBLI is unfavorable for conditions

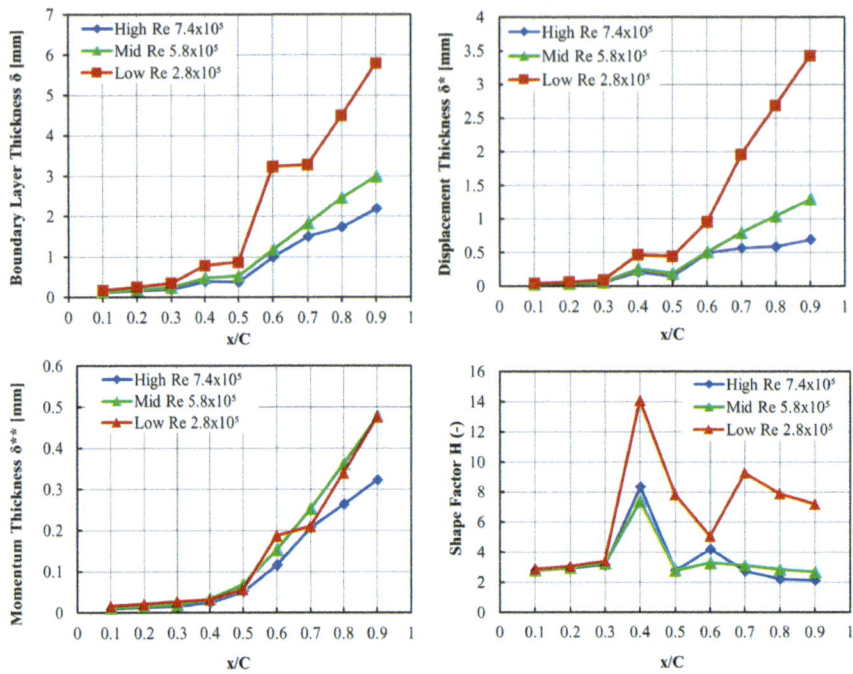

Fig. 11 A detailed comparison of boundary layer thickness and integral parameters plotted on the middle profile

with *Low* Reynolds numbers because of its enhanced susceptibility for separation. Looking at the displacement thickness the spike in the plot corresponds to the location separation of the boundary layer. The *Mid* and *High* Re have a similar range till x/C = 0.6 and differences are more pronounced downstream of x/C = 0.6. Interesting is the plot for momentum thickness. The momentum thickness is comparable for all three Reynolds cases till x/C = 0.5 and differences are more pronounced downstream. Towards the trailing edge, the momentum thickness predictions are similar for *Mid* and *Low* Re cases. Based on the displacement thickness and momentum thickness, the shape factor has been plotted along the axial chord as shown in Fig. 11. Theoretically, H=2.59 for laminar flows Blasius boundary layer [27] and H = 1.3–1.4 for turbulent flows. From the shape factor plot it is evident that in all three cases, the flow upstream of the shock wave has a laminar boundary layer. Strong shock-induced boundary layer separation at x/C = 0.5 with reattachment close to x/C = 0.5 is distinguished by a steep increase in shape factor in the interaction zone and a steep decrease further downstream for *High* and *Mid* Re. Whereas for the *Low* Re case the boundary layer separates at (x/C = 0.3) and tries to reattach at (x/C = 0.6), due to the passage shock interaction the separation happens again as the second spike arises further downstream and continues detached towards trailing edge. The separation bubble due to passage shock is stronger for *High* Re compared to *Mid* Re which is visible in the shape factor plot. Numerical estimation of the boundary layer separation on the suction side of the middle profile has been estimated by the wake downstream the middle blade passage at 10 and 50% traverse location as depicted in Fig. 12.

The inlet velocity has been compared upstream of the three Reynolds cases at traverse extracted at one chord length upstream of the blade passage at the mid-span location. The inlet velocity upstream for all three cases has been kept similar since the inlet Mach number and temperature have been kept in a similar range to have a direct flow structure comparison at different Re numbers. The wake width is the extension of the boundary layer separation downstream of the blade passage. The two traverses have been plotted close to the blade and one further downstream of the blade passage. It is visible that the wake gets thicker when the Reynolds number is decreased. The trend of thicker wake is visible even at further downstream traverse (III). One could also visualize higher velocity for reduced Reynolds number cases outside the boundary layer due to higher contraction shown by AVDR. Laser Doppler Anemometry (LDA) measurements have been carried out along the traverse (I) and traverse (II) marked in Fig. 13 for the *Mid* Re case and are compared with numerical predictions.

The magnitude of the velocities at the traverse has been normalised to have a direct comparison with the LDA predictions. The LDA measurements at upstream traverse (I) have a good agreement with the numerical estimations. This implies the designed test section could achieve the required inflow conditions in experiments as predicted in CFD. The wake has been estimated at 10% downstream the middle profile and has been compared for *Mid* Re case as shown in Fig. 13. The shape and width of the

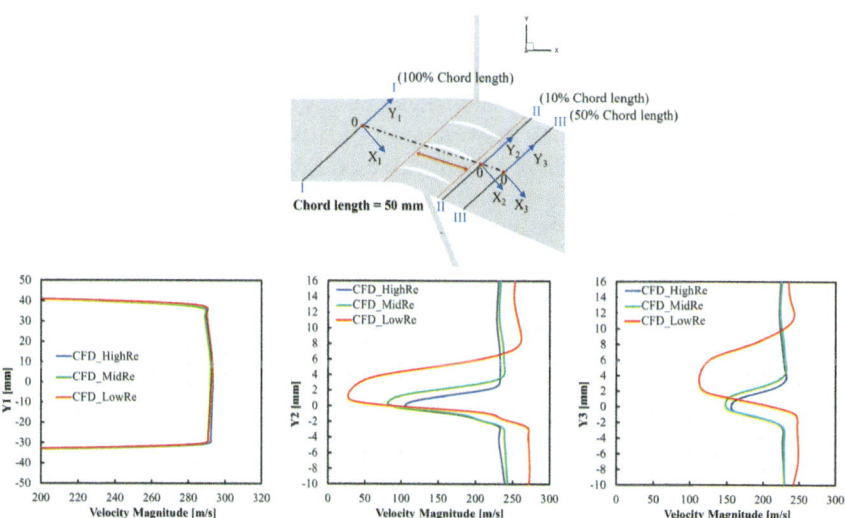

Fig. 12 Comparison of velocity at three different traverse location for chosen Reynolds cases, (Traverse I) 100% chord upstream, (Traverse II) 10% chord downstream and (Traverse III) 50% chord downstream the middle profile

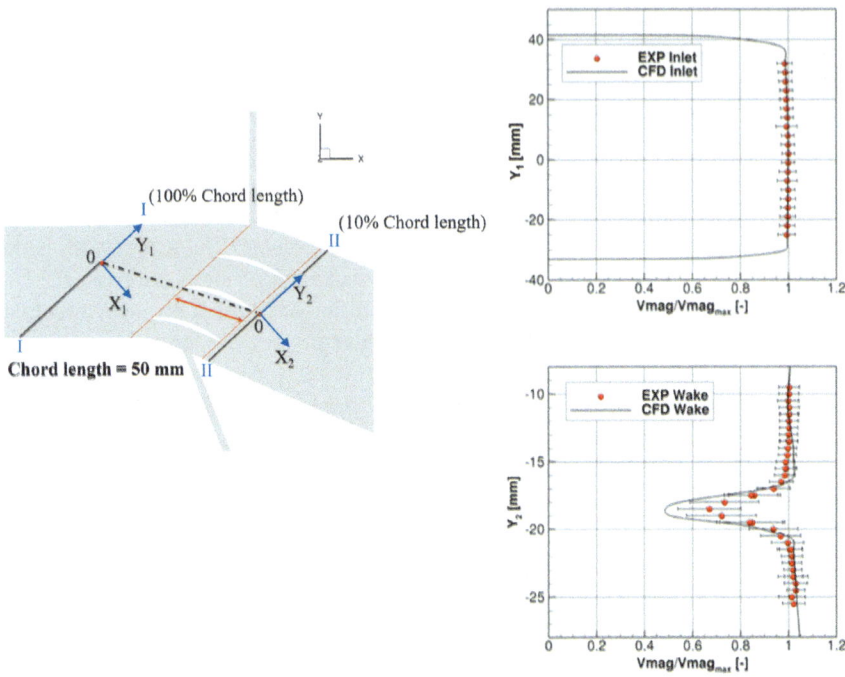

Fig. 13 Comparison of LDA measurement (Normalised velocity distribution with standard deviation) with numerical prediction for *MidRe* case at (Traverse I) one chord upstream and (Traverse II) 10% chord downstream of middle profile

wake at traverse (II) have been accurately predicted by the chosen turbulence model. Therefore further investigation would be carried out at traverse (III) as marked in Fig. 12. The depth of the wake has been over-predicted by the chosen turbulence model, which indicates the chosen turbulence model predicts high anisotropy in the wake. To compare the losses due to boundary layer separation, it is necessary to investigate the total pressure losses downstream of the trailing edge of the middle profile for different Reynolds conditions based on Eq. (6).

$$\text{Pressure loss coefficient } (L_P) = \frac{Pt_1 - Pt_2}{Pt_1 - Ps_1} \quad (6)$$

where Pt_1 is the inlet total pressure and Ps_1 is the inlet static pressure estimated at the location one chord upstream of the blade passage, Pt_2 is outlet total pressure estimated with mass flow averaged along the traverse locations as shown in Fig. 14.

A higher total pressure loss coefficient indicates greater energy loss in the cascade, implying lower efficiency. It is obvious that from the Fig. 14, that the flow separation is larger and losses are higher when the Reynolds number is decreased from *High* Re to *Low* Re numbers. The comparison of the averaged total pressure loss coefficient along the traverse marked in Fig. 14 has been described in Table 5. From the Table 5, the

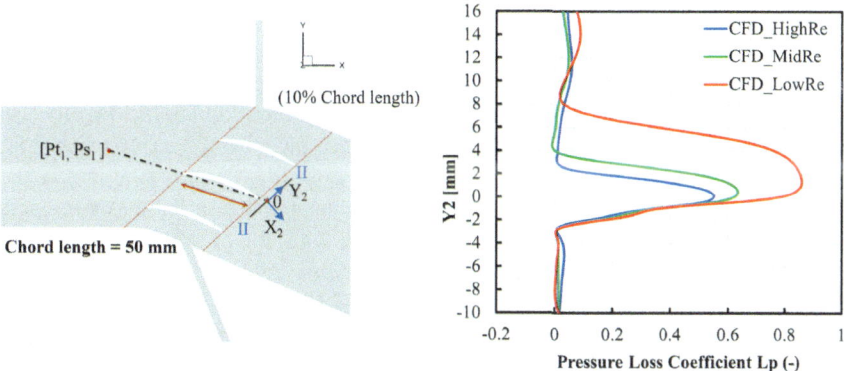

Fig. 14 Comparison of total pressure losses estimated at traverse (II) locations at 10% chord length downstream the trailing edge of middle profile

Table 5 Comparison of averaged total pressure loss coefficient for chosen Re number cases

Configuration	Averaged total pressure loss coefficient
High Reynolds	0.166
Mid Reynolds	0.181
Low Reynolds	0.276

averaged total pressure loss coefficient increases with decrease in Reynolds number indicating higher losses. When shock waves interact with thicker boundary layers at *Low* Re condition, the separation increases, which is reflected in the wake as higher losses. Therefore laminar separation is more pronounced when Reynolds number is decreased and laminar SBLI is not preferred at *Low* Re condition.

6 Conclusion

The investigations presented in this chapter have been performed on a stator cascade optimized by Rolls Royce Deutschland during the TEAMAero project. The main design objective of this research was accomplished by designing a new rectilinear three-blade test section. Effect of end-wall suction and blade holders for upper and lower profiles on shock structure in the test section was investigated. A novel approach of inlet valve has been positioned upstream of the test section to control inflow conditions to achieve the required Re number in blade passage for blow-down type wind tunnel configuration. The Re number is reduced by the reduction of total pressure downstream of the inlet valve located upstream of the test section. The inlet Mach number is adjusted by the control valve downstream of the test section and the outlet static pressure modification. However, it should be noted that this is not possible to set the Re number precisely, but the Reynolds number is an effect of reduced total pressure (reduced density) upstream of the test section.

The numerical model has been validated with the experimental results from the test section mounted at the IMP-PAN transonic blow-down wind tunnel facility. The numerical simulations using the EARSM turbulence model could accurately predict the flow structure on the suction side of the middle profile which is the focus of this research. The isentropic Mach number at the mid-span location on the suction side of the middle profile has been estimated based on the static pressure measurements from the pressure taps. The strength of the shock wave decreases with the reduction of Re number. To visualise the separation due to SBLI, wall shear stresses have been compared at different Re conditions. The separation bubble starts earlier when the Re number is decreased whereas the reattachment happens earlier for higher Re cases. However, for *Low* Re case the boundary layer remains separated till training edge on the suction side of middle profile. The position of the shock wave on the suction side of the middle profile for *High* Re is at (x/C = 0.47), *Mid* Re at (x/C = 0.46) and *Low* Re at (x/C = 0.45), which affirms the shear stress predictions.

The separation bubble and corner separation due to side walls are visualized using the iso-surface of the negative value of the velocity component. To quantify the passage contraction due to separation AVDR values have been compared at chosen traverse location. AVDR increases with the reduction of the Re number, which indicates that the blade passage above the suction side of the middle profile gets more contracted with the reduction of the Re number. To visualize the flow structure above the suction side of the middle profile, an oil flow visualization has been compared with velocity streamlines plotted above the suction side for *Mid* Re case. It is shown that the chosen numerical model could accurately predict the location and size of the separation bubble and corner flow structure. A detailed boundary layer investigation has been carried out at 9 traverses on the suction side of the middle profile to present the evolution of the boundary layer and its interaction with shock waves. The boundary layer is thicker when the Re number is reduced although a significant difference can be seen downstream of the shock wave. The integral parameter comparison of the boundary layer depicts the differences when the Re number is reduced. The shape factor plots affirms the separation and reattachment positions of boundary layer predicted using shear stresses.

To estimate the losses due to boundary layer separation wake downstream the blade passage has been compared for chosen Re numbers at two traverse locations. The difference from boundary layer thickness is translated to wake where the wake thickness is increased when the Re number is decreased. A similar trend is visible at further downstream traverse. The inflow conditions have been estimated at one chord length upstream of the blade passage to have identical inlet conditions for the chosen Re comparison study. The LDA measurements have been carried out for the mid-Re case at upstream traverse and wake traverse. The numerical predictions from the designed test section model has good agreement with LDA traverse measurement at inlet traverse. The shape and width of the wake have been correctly estimated by the chosen turbulence model whereas the depth of the wake has been over-predicted by the EARSM model. Based on the wake estimation at 10% downstream of the blade passage, total pressure losses have been compared for chosen Reynolds cases. It is observable that a larger separation of boundary layer due to SBLI results in higher losses at *Low* Re compared to *Mid* and *High* Re cases. Therefore one could conclude that the laminar SBLI is unfavorable when the Re number is decreased, due to increased losses compared to higher Re number conditions.

Acknowledgements This project has received funding from the European Union Horizon 2020 research and innovation programme under the Marie Sklodowska-Curie grant agreement 860909 TEAMAero (Towards Effective Flow Control and Mitigation of Shock Effects in Aeronautical Applications). Numerical simulations were performed in the Computational Centre of TASK (Trojmiejska Akademicka Siec Komputerowa) and supported by PL-Grid Infrastructure.

References

1. Hilgenfeld, L., Cardamone, P., Fottner, L.: Boundary layer investigations on a highly loaded transonic compressor cascade with shock/laminar boundary layer interactions. Proc. Inst. Mech. Eng., Part A: J. Power Energy **217**(4), 349–356 (2003)
2. Wang, Y., Rao, A.G., Eitelberg, G.: Study of shock wave control by suction & blowing on a highly-loaded transonic compressor cascade. Int. J. Turbo Jet-Eng. **30**(1), 79–90 (2013)
3. Bell, R.M., Fottner, L.: Investigations of Shock/Boundary-Layer Interaction in a Highly Loaded Compressor Cascade, vol. 78781. American Society of Mechanical Engineers (1995)
4. Hergt, A., et al.: The present challenge of transonic compressor blade design. J. Turbomach. **141**(9), 091004 (2019)
5. Heners, J.P., et al.: Evaluating the aerodynamic damping at shock wave boundary layer interacting flow conditions with harmonic balance. J. Eng. Gas Turb. Power **145**(3), 031012 (2023)
6. Davidson, T.S., Babinsky, H.: Influence of transition on the flow downstream of normal shock wave–boundary layer interactions. In: 54th AIAA Aerospace Sciences Meeting (2016)
7. Klinner, J., et al.: Investigation of shock-induced flow separation over a transonic compressor blade by conditionally averaged PIV and high-speed shadowgraphs (2018)
8. Szwaba, R., Kaczynski, P., Doerffer, P.: Roughness effect on shock wave boundary layer interaction area in compressor fan blades passage. Aerosp. Sci. Technol. **85**, 171–179 (2019)
9. Becker, B., Reyer, M., Swoboda, M.: Steady and unsteady numerical investigation of transitional shock-boundary-layer-interactions on a fan blade. Aerosp. Sci. Technol. **11**(7–8), 507–517 (2007)
10. Flaszynski, P., et al.: Laminar-turbulent transition tripped by step on transonic compressor profile. J. Therm. Sci. **27**(1), 1–7 (2018)
11. Li, L., et al.: Unsteady effects of wake on downstream rotor at low Reynolds numbers. Energies **15**(18), 6692 (2022)
12. Sieverding, C.: Aerodynamic development of axial turbomachinery blading. Thermodyn. Fluid Mech. Turbomach. **1**, 513–565 (1985)
13. Baines, N.C., et al.: A short-duration blowdown tunnel for aerodynamic studies on gas turbine blading. In: Turbo Expo: Power for Land, Sea, and Air, vol. 79597. American Society of Mechanical Engineers (1982)
14. Hirsch, C.: Advanced methods for cascade testing. NASA STI/Recon Tech. Rep. N **94**, 15119 (1993)
15. Sonoda, T., et al.: Advanced high turning compressor airfoils for low Reynolds number condition—Part I: Design and optimization. J. Turbomach. **126**(3), 350–359 (2004)

16. Hergt, A., Steinert, W., Grund, S.: Design and experimental investigation of a compressor cascade for low Reynolds number conditions. In: 21th International Symposium on Air Breathing Engines, Busan (South Korea), ISABE-2013-1104 (2013)
17. Schreiber, H.-A., et al.: Advanced high-turning compressor airfoils for low Reynolds number condition—Part II: experimental and numerical analysis. J. Turbomach. **126**(4), 482—492 (2004)
18. Sonoda, T., Schreiber, H.-A.: Aerodynamic characteristics of supercritical outlet guide vanes at low Reynolds number conditions. J. Turbomach. **129**(4), 694–704 (2007)
19. Joseph, A., Pawel, F., Piotr, D., Michal, P.: Low Reynolds number effect on highly loaded compressor stator cascade. In: 15th European Conference on Turbomachinery Fluid dynamics & Thermodynamics. European Turbomachinery Society (2023)
20. Flaszynski, P., et al.: Shock wave boundary layer interaction on suction side of compressor profile in single passage test section. J. Therm. Sci. **24**(6), 510–515 (2015)
21. Doerffer, P., et al., (eds.): Transition Location Effect on Shock Wave Boundary Layer Interaction: Experimental and Numerical Findings from the TFAST Project, vol. 144. Springer Nature (2020)
22. Kahraman, S., et al.: Numerical investigations of transitional sbli on a highly loaded transonic compressor passage. In: Turbo Expo: Power for Land, Sea, and Air, vol. 87080. American Society of Mechanical Engineers (2023)
23. Jameson, A., Schmidt, W., Turkel, E.: Numerical solution of the Euler equations by finite volume methods using Runge Kutta time stepping schemes. In: 14th Fluid and Plasma Dynamics Conference (1981)
24. Menter, F.R.: Two-equation eddy-viscosity turbulence models for engineering applications. AIAA J. **32**(8), 1598–1605 (1994)
25. Menter, F.R., et al.: A correlation-based transition model using local variables—Part I: Model formulation. J. Turbomach. **128**(3), 413–422 (2006)
26. Celik, I.B., et al.: Procedure for estimation and reporting of uncertainty due to discretization in CFD applications. J. Fluids Eng.—Trans. ASME **130**(7) (2008)
27. Blasius, H. The Boundary Layers in Fluids with Little Friction, No. 1256. National Advisory Committee for Aeronautics (1950)
28. Barenblatt, G.I.: Scaling laws for fully developed turbulent shear flows. Part 1. Basic hypotheses and analysis. J. Fluid Mech. **248**, 513–520 (1993)

Open Access This chapter is licensed under the terms of the Creative Commons Attribution 4.0 International License (http://creativecommons.org/licenses/by/4.0/), which permits use, sharing, adaptation, distribution and reproduction in any medium or format, as long as you give appropriate credit to the original author(s) and the source, provide a link to the Creative Commons license and indicate if changes were made.

The images or other third party material in this chapter are included in the chapter's Creative Commons license, unless indicated otherwise in a credit line to the material. If material is not included in the chapter's Creative Commons license and your intended use is not permitted by statutory regulation or exceeds the permitted use, you will need to obtain permission directly from the copyright holder.

Experimental and Numerical Investigations of SBLI and Flow Control on a Transonic Compressor Cascade

Edwin J. Munoz Lopez and Alexander Hergt

Abstract The flow through a transonic compressor cascade is inherently unsteady due to the shock-boundary layer interactions (SBLI) on the blade. Despite decades of research, few details are known about the mechanisms that cause such behavior. This chapter presents a multidisciplinary study aiming to elucidate these mechanisms and optimize flow control methods to mitigate their effects. For this purpose, the Transonic Cascade TEAMAero was first optimized and its performance was validated experimentally. The cycle of shock oscillation was then compared using advanced experimental measurement techniques and high-fidelity LES. This comparison revealed a continuous propagation of pressure waves from a point upstream of the trailing edge. The interaction of these waves with the main bow shock at different points of the cycle was then linked to the frequencies observed in the oscillation spectrum. A configuration of this cascade with two roughness patches was finally optimized using a novel procedure developed. The optimal configurations obtained show how the targeted design of these devices can simultaneously mitigate shock oscillations and improve performance. This chapter demonstrates how the combined application of advanced numerical and experimental techniques needs to be intensified as researchers search for a global theory of SBLI in compressor blades.

Keywords Transonic flow · Shock induced separation · Advanced experimental methods · Flow control optimization

---

E. J. Munoz Lopez (✉) · A. Hergt
German Aerospace Center (DLR), Cologne, Germany
e-mail: edwin.munozlopez@dlr.de

A. Hergt
e-mail: alexander.hergt@dlr.de

1 Introduction

Unsteady shock-boundary layer interactions (SBLI) are ubiquitous to sonic flows. The inevitable presence of these two complex features allows many hypotheses why their interaction is highly unsteady. These interactions were first observed from as early as the 1940s [1, 2], though the measurement techniques at the time did not allow for detailed analyses on the sources of unsteadiness. Only in recent decades and with the help of advanced numerical and experimental methods, comprehensive hypotheses have been formulated and tested to explain this phenomenon [3, 4]. These hypotheses typically involve mechanisms of propagation of disturbances in the downstream direction from the incoming flow, or in the upstream direction from the turbulent flow downstream of the shock, or in both directions from the interaction of the shock with its own separation bubble [5–9].

Within the context of compressor blades, the unsteady SBLIs are known to reduce the performance and robustness of typical designs, among other issues [10, 11]. In recent decades, the prevalent theory of shock oscillation for applications of aerodynamic profiles seems to be Lee's trailing edge feedback mechanism [5, 12, 13]. According to Lee, the vortical structures created at the shock foot interact with the trailing edge to create pressure waves that influence the position of the shock continuously. Previous numerical studies of different fidelity levels have found evidence to this theory on different compressor blade designs [14, 15]. However, experimental studies in the past have also shown evidence of disturbances originating and propagating from a region downstream of the shock, but upstream of the trailing edge [16]. Due to the complexity posed by such configurations, further research is required to achieve a more complete understanding of all the mechanisms at play.

A high level of interest is also present in the study and application of flow control methods (FCM) to mitigate these oscillations. Early studies of these devices were focused on vortex generators (VG) with configurations of co-rotating or counter-rotating vortices [17, 18]. More recently, shapes have emerged that are specifically conceptualized to mitigate unsteady SBLI effects, such as the shock control bump (SCB) [19, 20]. In principle, these shapes aim to fix the shock in place and also generate vortices to aid flow reattachment. Regardless of the FCM to be studied, it is clear that their design requires a robust optimization procedure. This is necessary to methodically explore the design space for a set of optimal designs that could mitigate shock oscillations and improve the performance of the given design.

The Institute of Propulsion at the DLR has fortunately accumulated an extensive amount of experience in the fields pertaining to the study of SBLIs and FCMs on compressor cascades. This is observed in the experimental studies with advanced measurement techniques in Hergt et al. [11, 22], Klinner et al. [16, 21], but also in the numerical studies with advanced numerical methods for blade design in Voß et al. [23], Aulich et al. [24], Schnoes et al. [25] and for high-fidelity simulations in Klose et al. [26], Morsbach et al. [27], Bergmann et al. [28]. This expertise was fully leveraged within the context of the doctoral work performed for the TEAMAero research consortium. The content of this work is only briefly summarized in this chapter, with readers being referred to a future publication of the whole monograph for further detail. This chapter then begins by presenting a brief overview of the baseline compressor cascade designed for the consortium, following a detailed analysis of the unsteady SBLIs on this cascade with advanced experimental and numerical techniques, and finally presenting the novel numerical optimization procedure developed to mitigate them.

2 The Transonic Cascade TEAMAero

The Transonic Cascade TEAMAero (TCTA) is a new compressor cascade optimized at the DLR with the typical design tools developed and available at the institute of propulsion. The main purpose of this new design was to serve as a baseline geometry for the different objectives previously outlined. This section then first provides a brief overview of the design process before presenting the experimental campaign carried out at the DLR's Transonic Cascade Wind Tunnel (TGK) to validate its performance.

2.1 Optimization Process Overview

For a detailed breakdown of the optimization performed, the reader is referred to Munoz Lopez et al. [29, 30]. The process chain built for this optimization consisted of three main procedures: generating a new cascade geometry, meshing the cascade, and estimating its performance through RANS simulations. The simulations were performed with the DLR's in-house CFD solver, TRACE, and applying the k-ω SST model coupled with the $\gamma - Re_\Theta$ transition model [31, 32]. The result from this process chain was managed by the DLR's optimization suite, AutoOpti, which was set to minimize two objectives: the flow losses at the aerodynamic design point (ADP), and over the working range (WR) at a constant inlet Mach number (M_1^{ADP}). The flow losses were quantified at each operating point by the total pressure loss coefficient (ω).

The ADP for the optimization was set to a Mach number of 1.20, an inflow angle of 145.8°, and an axial velocity density ratio (AVDR) of 1.20. The latter being a parameter that quantifies the amount of flow area contraction at the outlet compared to the inlet due to the growth of the sidewall's boundary layer (BL) [33]. This effect was simulated in the numerical domain with contracting inviscid sidewalls. The operating point was further constrained by the aerodynamic loading, as measured with the de Haller number (DH). After finding the ADP by adjusting the outlet pressure in the simulations, the WR was determined by methodically increasing and decreasing the inflow angle until the cascade stalled or choked, respectively.

Fig. 1 Transonic cascade TEAMAero—design definition

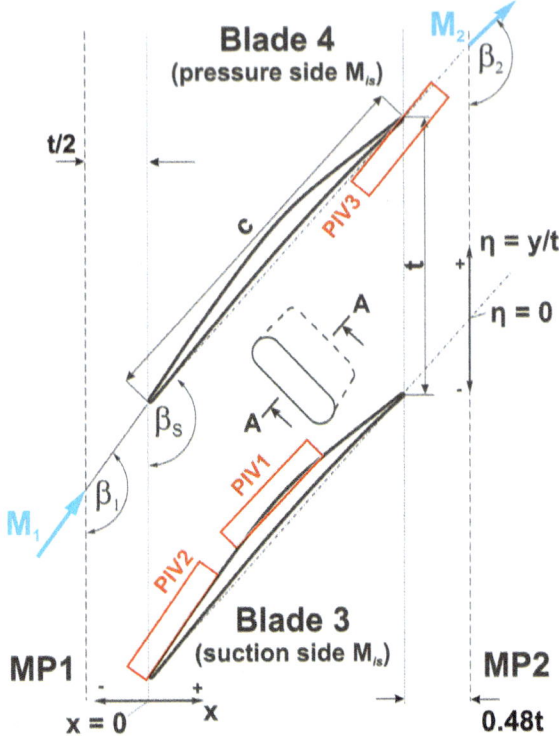

The optimization results yielded several candidate designs from the Pareto front. A number of post-optimization analyses at different off-design points (ODP) were performed to pick the candidate with the most robust performance for the final TCTA design. This design is shown in Fig. 1, as manufactured for the experimental campaign. The cascade's main geometrical properties are summarized in Table 1, along with its validated performance for an AVDR of 1.05. These conditions were found to be the most stable for measurements in the wind tunnel, as discussed in the following sections. The performance properties also correspond to test number (TN) 125, as described in Munoz Lopez et al. [34].

2.2 Experimental Validation

This work package was finalized with an experimental campaign at the DLR's Transonic Cascade Wind Tunnel (TGK) in order to validate the performance of the TCTA cascade [34]. For this purpose, the final cascade design from Fig. 1 was manufactured with a total of six blades and installed in the TGK in Cologne. The experimental configuration of the cascade in the wind tunnel is shown in Fig. 2. This figure shows

Table 1 Transonic cascade TEAMAero—properties of interest

General properties	
Blade chord, c (mm)	100
Pitch, t (mm)	65
Stagger angle, β_{st} (°)	135.8
Blade span, s (mm)	168
Aerodynamic design point properties	
Inflow Mach, M_1	1.20
Reynolds number, Re (10^6)	1.35
Inflow angle, β_1 (°)	145.7
Axial velocity density ratio, AVDR	1.05
de Haller number, DH	0.582
Flow turning, FT (°)	10.1
Total pressure loss coefficient, ω	0.1132

the different features of the wind tunnel used to adjust the operating conditions in the test section, as well as the motorized wake probe assembly used to measure the conditions at the outlet.

The focus of this experimental campaign was to validate the steady performance of the cascade, as optimized in Sect. 2.1. In order to perform these measurements, a number of steady measurement techniques were employed at the inlet measurement plane (MP) 1, outlet MP2, and over the blade suction and pressure surfaces. These included pressure tap measurements at these locations of interest, but also three-hole probe measurements at the MP2, and Laser-2-Focus (L2F) measurements at the MP1. The latter is a type of laser anemometry technique developed at the DLR in

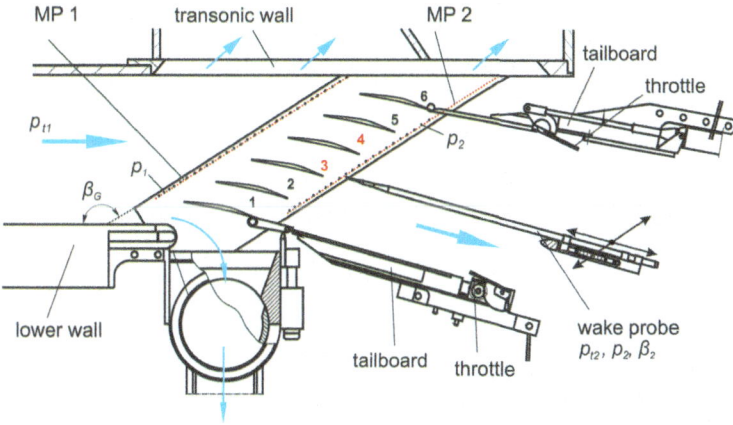

Fig. 2 Experimental configuration of the TCTA installed in the TGK

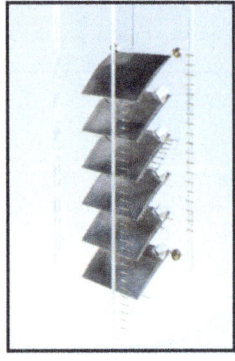
(a) TCTA cascade

(b) TGK test section with the TCTA cascade installed

Fig. 3 TCTA assembly and wind tunnel configuration for the experiments

Schodl [35] that allows the measurement of the inflow angle without obstructing the supersonic flow. For more details on the application and challenges of using these techniques to measure this inherently unsteady SBLI flow, the reader is referred to the extensive discussion published in Munoz Lopez et al. [34]. The standalone cascade assembly and the fully instrumented configuration for the tests are shown in Fig. 3.

The main results validated the expected losses of the cascade within a margin of error of 3–6% throughout the working range at an AVDR of 1.05, as shown in Fig. 4a. In this figure, the measurement points are labeled with their test numbers (TN), where TN125 corresponds to the ADP of the cascade and TN78 indicates the onset of choke. The cascade also shows a high aerodynamic loading in Fig. 4b, with more than 10° flow turning at the ADP. This high loading is achieved over a wide working range of at least 2°. The TN27 taken from choked conditions at an AVDR of 1.00 is also shown. This measurement demonstrates the difficulty of capturing the correct onset of choke numerically, which depends strongly on the AVDR of the cascade. The reader is referred to Munoz Lopez et al. [34] for further comparisons of flow visualizations, the isentropic Mach number (M_{is}) distributions, the wake loss profiles, and others.

3 The Unsteady SBLI on the Transonic Cascade TEAMAero

The flow through a transonic compressor cascade is highly unsteady, and the TCTA is no exception. The study of these unsteady SBLIs is a difficult undertaking both experimentally and numerically due to the small temporal and spatial scales involved. This means that advanced configurations are required in order to observe, quantify, and simulate such interactions. In this section, a short overview of the detailed study of

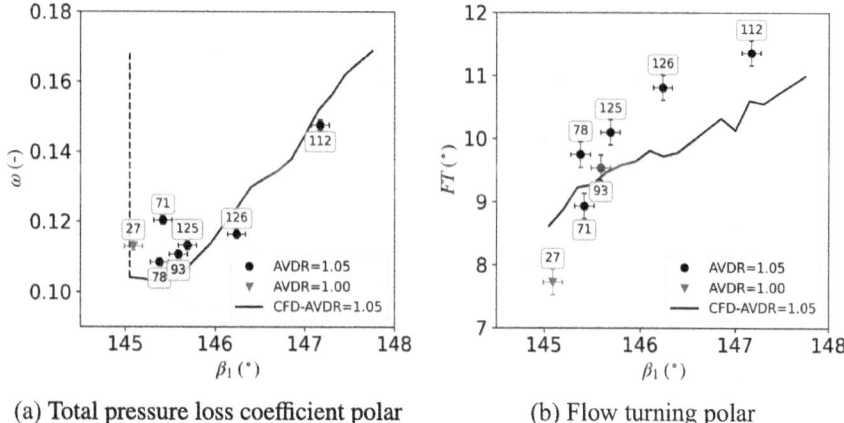

(a) Total pressure loss coefficient polar (b) Flow turning polar

Fig. 4 Working range performance of the TCTA: CFD and experimental results comparison

the unsteady SBLI on the TCTA will be presented [36]. This is Part 3 of a three-part publication detailing the oscillation of the shock via advanced experimental techniques in Part 1 [37] and advanced numerical methods in Part 2 [38]. In Part 3, the results from high-fidelity LES simulations, high-speed shadowgraph (HSS), and particle image velocimetry (HSPIV) measurements are analyzed together. This section briefly discusses the preliminary studies performed, before presenting the advanced numerical and experimental methods, followed by a description of the shock oscillation cycle, the validation of the LES results with the experimental measurements, and finally a summary of the mechanisms of unsteadiness identified.

The preliminary studies consisted of two experimental campaigns performed to quantify the interactions to be studied. The first one consisted of capturing standard 2D-2 component and high-resolution snapshot PIV images, as shown in Klinner et al. [39]. These images provided a detailed look at the averaged flow field through the cascade, revealing a mean shock position of 52.6% with respect to the blade's chord. proper orthogonal decomposition (POD) analyses also revealed some of the main spatial modes of oscillation of the shock, although these could not be linked to any frequencies yet. The flow through the cascade was then measured via HSS at 20 kHz, as described in Munoz Lopez et al. [40]. This was done over three adjacent cascade passages to study the joint movement of their shock structures. These studies revealed that the shock in the passage of the TCTA oscillates at a broad frequency band between 500 and 550 Hz, which is modulated by a high-frequency tone at 1140 Hz and its harmonics. The information gathered helped prepare the advanced experimental and numerical investigations presented in the following sections.

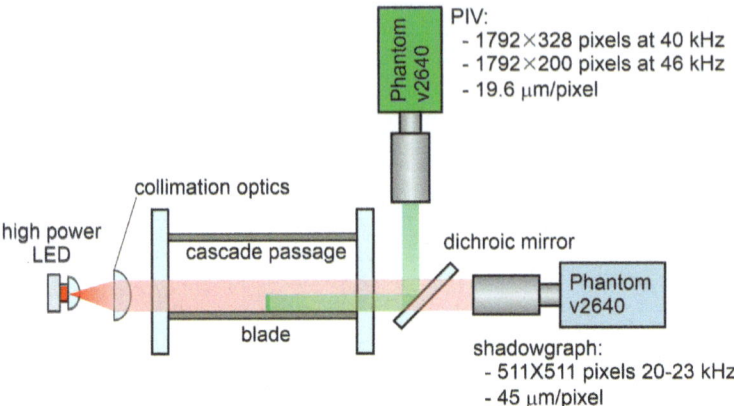

Fig. 5 Experimental configuration for simultaneous capturing of time-resolved shadowgraphy and PIV

3.1 Advanced Experimental and Numerical Methods for SBLI Flows

In order to gather the most amount of information possible on the unsteady SBLI over the TCTA, advanced numerical and experimental techniques were applied on the ADP of the cascade. By leaning on both set of results, the limitations of the experimental measurements could be overcome, and the high-fidelity simulations could be thoroughly validated. This section first provides a brief overview of the experimental configuration used to capture HSS and HSPIV images of the shock at the TCTA's ADP, before describing the numerical configuration for the main LES simulation performed on the ADP of the cascade.

3.1.1 Advanced Experimental Methods for SBLI Flows

The main objective was to obtain detailed footage in different regions of interest that could be compared directly to the LES simulations. This was done with HSPIV recordings at the regions marked in Fig. 1. These are relatively small due to the limitations of current hardware, allowing fields of view of $31 \times 6.4 \, mm^2$ with a pixel density of 19.6 μm/pixel. In order to complement these recordings, simultaneous HSS recordings are performed with a bigger field of view of $33 \times 33 \, mm^2$ and a magnification of 45 μm/pixel on which the shock position can be tracked. Even though hardware issues did not allow the perfect synchronization of these recordings, the two sets of high-speed measurements allow for a unique insight into the flow in the passage. The configuration required for such measurements with two Phantom v2640 HS cameras capturing sampling rates of 20 or 23 kHz is shown in Fig. 5.

A typical measurement run with this setup would consist of 8 different bursts of up to 10 thousand shadowgraph images and 20 thousand PIV images per burst. The PIV images are evaluated in window sizes of either 64 × 32 or 64 × 16 pixels to obtain validation rates higher than 95% of the cross-correlated image pairs. The shadowgraph images on the other hand are typically processed directly from the tiff raw data files with different regions of interest depending on the desired evaluation. For instance, to apply either shock-tracking algorithms over the entire shock region, or smaller windows for the evaluation of PSD contours or wave propagation analyses. For more details on the configuration and the processing of the results, the reader is referred to the cited publications.

3.1.2 Advanced Numerical Methods for SBLI Flows

The simulation results used for this study were calculated with TRACE's discontinuous Galerkin spectral method (DGSEM) solver. The implementation and validation of this method for has been described in detail in Klose et al. [26], Morsbach et al. [27], Bergmann et al. [28]. For the numerical part of this work package, an extensive comparison was performed between URANS and LES simulations of different resolutions [38]. For the analyses presented in this section, only the results from the fully resolved LES mesh are used. Its main feature is the long simulation time computed, which was required to capture converged statistics on the main low frequency band of shock oscillations around 500 Hz.

In this simulation, the operating conditions are set to match those of the experiments from TN125, as presented in Sect. 2.2. The only deviation is the periodic boundary conditions in the spanwise direction, which sets an AVDR of 1.00 instead of 1.05. This implies a higher loading of the cascade and a more upstream mean position of the shock over the blade. Nevertheless, the conditions are similar enough to provide a good comparison with the experiments. The rest of the domain consists of an inlet 1.0 chord length upstream of the leading edge, and an outlet 1.5 chord lengths downstream of the trailing edge of the blade. These boundaries apply 1D non-reflecting boundary conditions via the method of flow characteristics in Schlüß et al. [41]. The operating conditions are set by specifying the total pressure, total temperature, and inflow angle at the inlet, as well as the static pressure at the outlet.

The mesh used for the simulation is generated as a 2D slice with the Gmsh package and contains 108,564 rectangular elements with structured refinements around the shock locations. The mesh is then extruded by 10% of the chord length in the spanwise direction, with elements set to a polynomial order of 3 to provide a 4th order accurate spatial discretization. The simulation is advanced in time explicitly via a 3rd order Runge-Kutta scheme. The maximum non-dimensional cell sizes in the mesh are within recommended levels in the streamwise, wall-normal, and spanwise directions: $\Delta \xi^+_{max} = 35$, $\Delta \eta^+_{max} = 2$, and $\Delta \zeta^+_{max} = 25$. These values are normalized by the polynomial order given to compare with finite volume simulations. Finally, the domain is sampled with 1D probe lines and a coarser 2D regular grid for analysis. A

Fig. 6 Coarse TCTA mesh for simulations with TRACE's DGSEM LES solver

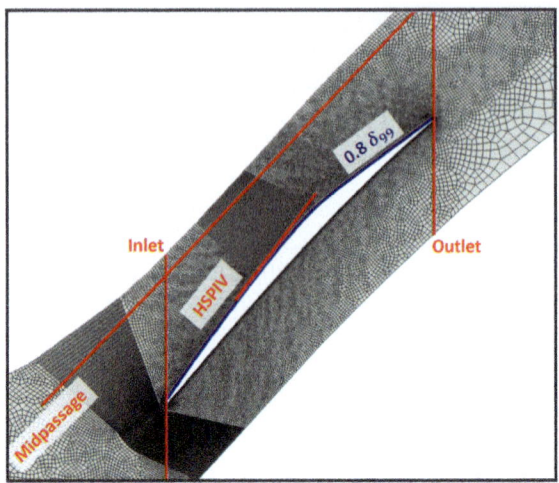

coarse version of the domain and some of the most relevant probe lines sampled are shown in Fig. 6.

3.2 The Cycle of Shock Oscillation

The experimental results amount to more than 56 s of footage from the 14 measurement runs with 8 bursts each. The LES results on the other hand correspond to more than 100 convective time units (CTU) of simulation time, but still amount to 'only' 30 ms of footage. The results must then be analyzed with care in order to draw meaningful conclusions from them. The first evaluation is performed by tracking the shock position over time for both datasets. For the HSS results, this is done via an image gradient sensor over pixel rows of the images. Outliers are searched along the pixel rows and then again along image sequences to obtain the most accurate measurement of the shock position. For the LES results, the shock can be tracked with high precision along the midpassage probe line shown in Fig. 6 by applying a modified ducros sensor as in Pirozzoli et al. [42].

The resulting signals can then be evaluated with power spectral densities (PSD) by applying the Welch method, as shown in Fig. 7. The figure shows that the oscillation of the shock is much more tonal in the LES than the experiments. The peak of the main oscillation is also strongly marked in the LES at 614 Hz, while the experiments show a broad peak around 513 Hz. Even if these peaks differ by about 100 Hz, they still lie remarkably close to each other with respect to any other such comparison found in the literature. This figure also shows a second PSD for the LES results obtained from an imperfect shock tracking from a probe line closer to the blade surface and applying a density gradient sensor. These results more closely resemble those of the

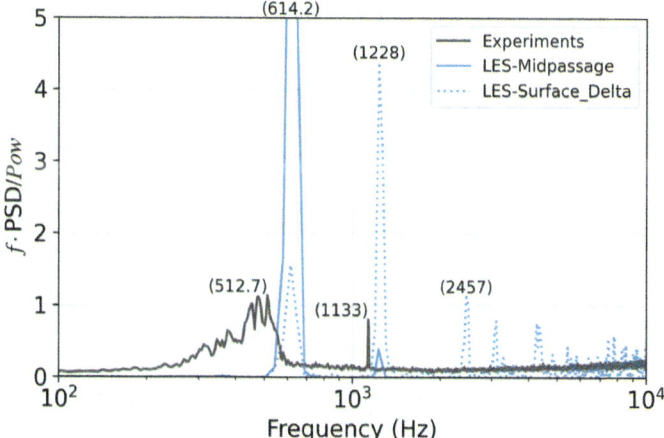

Fig. 7 Power spectral densities of the shock position in the experiments and the LES results

experiments, with stronger high-frequency tones and a broader low-frequency peak of oscillation.

This simple comparison informs the rest of the results in this section, whereby the authors demonstrate that these two datasets show the same mechanisms of shock oscillation and not necessarily that they are replicas of each other. This would be impossible due to the different factors affecting the flow in the experiments: the lack of perfect periodicity, the sidewall suction in the passage, the limited stiffness of the blade, and others. For a more comprehensive description, the readers are referred to the discussions in Munoz Lopez et al. [34, 36].

The shock oscillation in the passage of the TCTA is now described by leaning on the detail offered by the LES results. This is done in Fig. 8 with roughly equi-temporal Mach contours of one entire cycle of shock oscillation. In these snapshots, some of the main features of the mechanisms to be investigated can already be observed. The shock starts at its most upstream position, and a weak oblique shock is observed to be propagating upstream in the sonic region of the flow over the blade. An acoustic wave just upstream of the trailing edge has also formed from the apparent interaction of the separated flow convecting downstream from the previous cycle. These waves take the form of shocklets that start interacting with the main shock as it moves into the passage. The strong interaction between these features seems to produce a momentary choking that culminates in a strong oblique shock propagating upstream. The resulting breakdown of the flow under the shock foot produces a column of separated flow that convects downstream to complete the cycle.

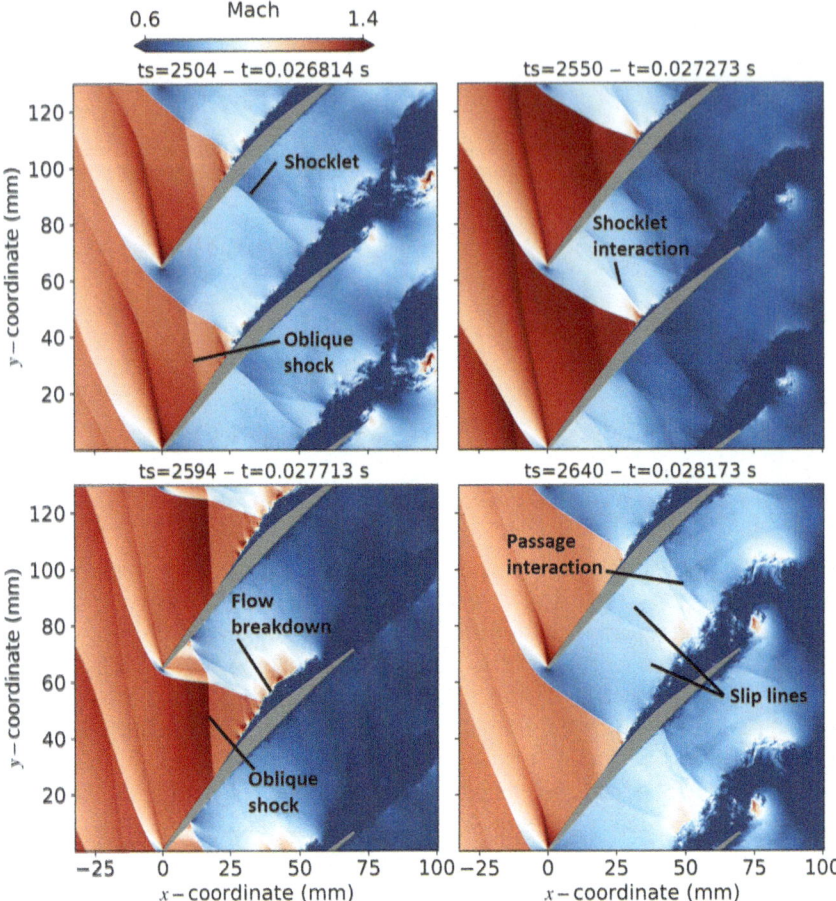

Fig. 8 Mach contours of one shock oscillation cycle over the TCTA

3.3 Validating LES Simulations with Experimental Measurements

The cycle of oscillation from the LES results observed in the previous section is now validated by comparing it directly with the experimental results via qualitative and quantitative methods. For this purpose, Fig. 9 shows a cycle of shock oscillation as observed from two similar cross-correlated sequences of HSS and HSPIV images. In this sequence, the same shocklet features previously noted can be observed to propagate upstream in the subsonic region of the flow. The movement of the shock into the passage also coincides with the interaction of the shocklets and the increase of the flow separation under the shock foot. This interaction is also observed to generate weak and strong oblique shocks that propagate upstream in the sonic region

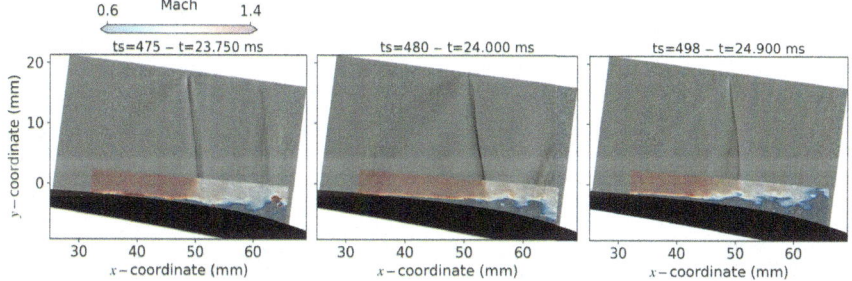

Fig. 9 Sequence of similar shock oscillation events from correlated HSS and HSPIV footage

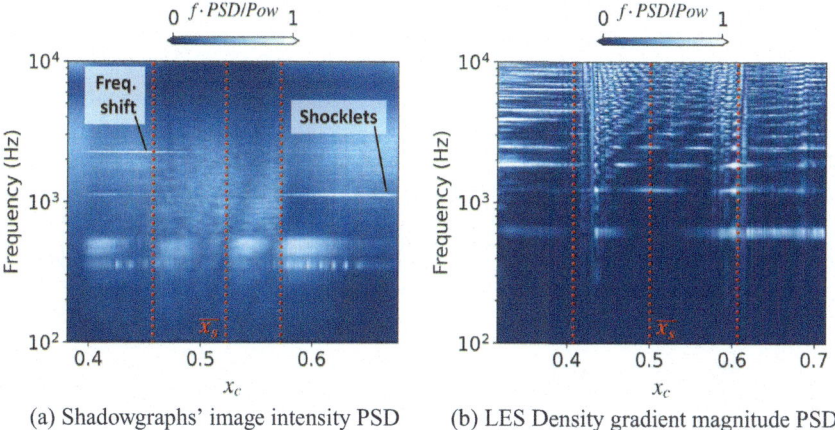

(a) Shadowgraphs' image intensity PSD (b) LES Density gradient magnitude PSD

Fig. 10 PSD contours comparison of the shock oscillation for experiments and LES results

of the flow. The different features shown in the figure and the sequence of the events then mirror very closely the ones identified from the LES results in Fig. 8.

One quantitative comparison is shown in this section with Fig. 7, which shows the PSD contours of the image gradient for the experiments and the density gradient magnitude for the LES. These contours were evaluated over a few pixel rows for the shadowgraphs and a similar interpolated probe line from the LES. The max., min. and mean positions of the shock are marked by red dotted lines in both contours. The contour from the experiments clearly shows that the region after the shock is dominated by the high-frequency tone at 1133 Hz. This frequency must then belong to the upstream movement of the shocklets, which occurs continuously throughout the cycle and at much higher rates than the movement of the main shock itself. The region affected by the shock is difficult to interpret given that it acts as a source and sink to the other features, although its main lower frequency band can still be identified. More interestingly, the region upstream of the shock shows a shift of energy towards the harmonic of the high-frequency tone at 2260 Hz.

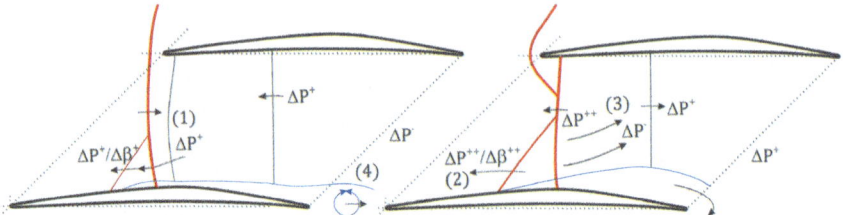

Fig. 11 Observed mechanisms of shock oscillation at the TCTA's ADP

Although the PSD of the LES results again shows itself to be much more tonal, it also shares a lot of the key traits previously noted. For instance, the high-frequency tones are again present downstream of the shock, and there is again a shift towards the higher tones of the most energetic frequencies. The region inside the shock is also difficult to interpret due to the appearance and disappearance of different frequency tones. However, wave propagation analyses via contours of the cross-correlation coefficient along the same pixel rows and probe lines reveal similar speeds of propagation of these shocklets. Further comparisons between both sets of data are shown and discussed in more detail in Munoz Lopez et al. [36]. These results demonstrate that the LES and the experiments are showing similar mechanisms of oscillation and prepares them for further analysis in the following section.

3.4 Mechanisms of SBLI Unsteadiness

By validating the LES results and leaning on the detailed data available of the whole flow field, the propagation of the pressure and flow incidence disturbances in the passage can be studied in detail. This analysis derived in the identification of at least two mechanisms of shock oscillation that are responsible for the different frequencies observed in the spectra from Fig. 7.

The mechanisms are summarized in Fig. 11. This schematic shows how the propagation of pressure waves in the subsonic region of the flow is continuous and occurs at high rates of propagation. When the shock is not inside the passage, the propagation of these waves continues in the sonic region in the form of weak oblique shocks (1). These shocks are only possible due to the easy propagation of these disturbances in the weak and uniform laminar boundary layer upstream of the shock. As the oblique shocks propagate upstream, they transmit the increase in pressure in the sonic region, but they also inevitably carry an increase in the flow angle upstream of the shock and therefore its position. This is how the movement of the shock is modulated by small oscillations at the high-frequency tones observed in the spectra, extending the mechanism of oscillation proposed in Hergt et al. [43].

When the propagation of the shocklets coincides with the main shock in the passage, strong interactions between these features lead to a strong pressure build-up that culminates in a stronger oblique shock propagating upstream (2). Just like before, as this oblique shock propagates, it inevitably carries both an increase in static pressure and the flow angle ahead of the shock. This oblique shock eventually reaches the leading edge of the adjacent upper blade and causes a strong oscillation of the operating point on the entire periodic domain. As the main shock moves upstream, the blocks of separated flow convecting downstream interact strongly with the flow in the passage (3). This generates regions of uneven flow properties that culminate in the pressure waves propagating upstream. The point at which these disturbances started propagating upstream is not unique, but was generally just upstream of the trailing edge in the region of 80% chord length.

In order to close the cycle, it was observed how the separated flow interacts strongly with the trailing edge of the profile, leading to a temporary vortex shedding from the trailing edge of the cascade (4). This vortex shedding seems to be responsible for the decreased static pressure at the outlet of the cascade that forces the shock to start moving back into the passage. Therefore, the manner and lag at which this interaction at the trailing edge occurs may be directly responsible for the actual main frequency of oscillation of the shock. Although the mechanisms described are inherently different from Lee's trailing edge feedback theory, this feature of the profile is still shown to have considerable significance in the way the whole cycle unfolds. For a lengthier discussion of the results and the mechanisms proposed, the reader is again referred to Munoz Lopez et al. [36].

4 Mitigating SBLI Unsteadiness with Flow Control Methods

In this last work package, the experience gathered from the previous sections is applied for the optimization of flow control methods (FCM) on the TCTA to mitigate its shock oscillations. For this purpose, a novel procedure has been developed to generate, mesh, and simulate the flow through any given blade with FCMs. This section then begins with a brief description of this optimization procedure developed, followed by the results obtained for a new TCTA configuration with roughness patches and reduced SBLI unsteadiness. For more details on the subject matter, the readers are referred to a later publication of the doctoral thesis by the authors.

4.1 A Novel Process Chain to Optimize Flow Control Methods

The design of compressor blades via numerical optimization typically requires an extensive amount of work and development before any shape can start being optimized. Most of this relates to the fully automated process chain necessary to parametrize, mesh, simulate, and evaluate any given optimization member. In this section, the process again employs the DLR's optimization suite, AutoOpti, along with the in-house CFD solver, TRACE. The latter was again used for (U)RANS simulations applying Menter's 2003 k-ω-SST model coupled with the 2009 version of his $\gamma - Re_\Theta$ transition model. For more details on these well-established packages, the reader is referred to the respective literature [23, 25, 44–46]. This section briefly presents the two main new software packages developed for the optimization of FCMs on compressor blades.

4.1.1 VortexGen: Parametrization and Generation of Blades with Flow Control Methods

The parametrization of the blade is performed with the in-house software BladeGen, which builds the blade with cubic B-spline surfaces defined by their control points and outputs the result in CAD files [23]. The surface is defined along the chord by the normalized u-coordinates, where values equal to 0, 1, 2, and 3 indicate the start of the pressure surface, leading edge, suction surface, and trailing edge, respectively. The normalized v-coordinate defines the hub and tip of the blade from 0 to 1. In order to add the FCMs, the newly developed software, VortexGen, receives the CAD output from BladeGen and applies the following routine for each FCM: a reference point is searched, the B-spline mesh in the region is refined, and the points within the region are modified based on the desired shape. With this framework, the FCMs shown in Fig. 12 and described below were implemented:

- **Roughness patch (RP)**: Defined by 5 design parameters, including a reference location (u, v), two lengths in spanwise and chordwise directions normalized by the chord (l_u, l_v), and the equivalent sand-grain roughness height (k_s).
- **Dome or dimple**: Defined by 5 parameters, including the reference location for the dome's center $(u, v)^{\text{center}}$, the maximum height (h), the radius from the center (r), and a stretch factor in the spanwise v-direction $(a_{uv} = r_v/r)$. The height of the dome can be negative to create a dimple instead of a bump.
- **Ramp vortex generator (RVG) or plow**: Defined by a total of 9 design parameters. These include a reference location (u, v), a ramp angle (θ), its maximum height (h), total length and width (l, w), a width ratio parameter (p_w), and two lengths $(l_{\text{tip}}, l_{\text{edge}})$. The ratio p_w defines the angle of the ramp with respect to the chordwise direction.
- **Shock control bump (SCB) or plow**: Defined similarly as in Ogawa et al. [19] with a total of 12 parameters. These include a reference location (u, v); a length

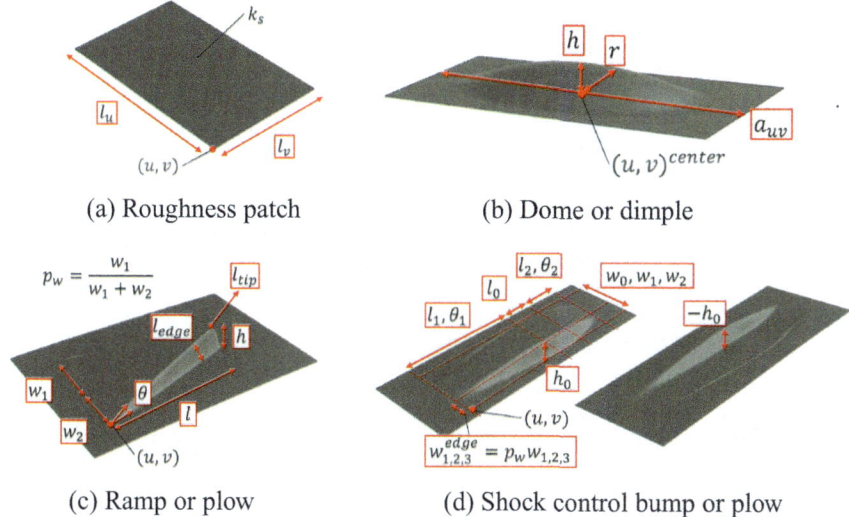

Fig. 12 Flow control methods parametrized and implemented in VortexGen for optimization procedures

and an angle for the front (l_1, θ_1) and rear (l_2, θ_2) ramps; a length and a height for the flat middle section (l_0, h_0); a total of three widths for each of these sections (w_0, w_1, w_2); and finally, a ratio parameter (p_w).

4.1.2 Centaur: Unstructured Meshing for Flow Control Methods

Just like the generation of the geometry itself, the meshing of a blade with FCMs can be an intricate procedure to automate. For this process chain, the unstructured mesher was used from the commercial software with scripting capabilities, Centaur. The domain is meshed with triangular prism layers for the blade's suction and pressure surfaces, two structured prism layer regions of hexahedral elements for the leading and trailing edges, and tetrahedral elements elsewhere. A number of refinement regions cover the bow shock area and the wake of the blade. A mesh convergence study was performed with the baseline TCTA configuration, from which appropriate mesh size and refinement factors were defined.

In order to automatically import and mesh any given TCTA configuration with FCMs, the python program, CentWrap, was developed. This program reads the output from VortexGen and creates an adapted set of scripts with refinement regions for the given FCMs and their wakes. The refinement sizes imposed are based on the most limiting features of the FCM, such as its height, width, rounding edge, or others depending on the FCM type. A series of mesh convergence studies were performed

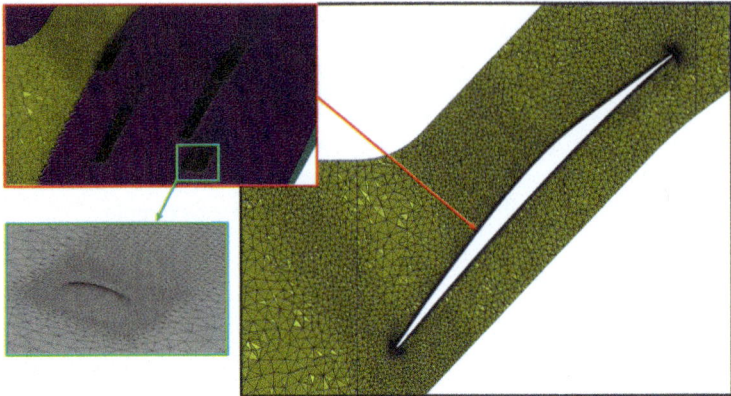

Fig. 13 Sample TCTA configuration with multiple flow control methods meshed automatically with Centaur and CentWrap

to ensure that the appropriate features and scaling factors were selected. A sample mesh of the TCTA with FCMs is shown in Fig. 13.

4.2 Optimization Strategy and Design Space

The remaining sections will focus on the optimization of a TCTA configuration with two RPs. The optimization of the remaining FCMs shown in Sect. 4.1.1 is left for the final publication of this doctoral work. In order to ensure that the SBLI effects were captured the best way possible within the scope of an optimization, a lot of effort was spent into testing and calibrating the numerical methods used. Afterward, a general optimization strategy was devised that could be adapted based on the type and number of FCMs to be optimized. This section presents an overview of said optimization strategy, along with a detailed rundown of the definition of the objective functions and the design space for the optimizations.

4.2.1 Simulating Unsteady SBLI Effects for Numerical Optimization

As shown in Sect. 3, capturing the unsteady effects from the SBLI in transonic cascade flows is not a trivial task. Recent experience comparing URANS with LES simulations revealed that excessively small timesteps ($\Delta t \approx 7.5$ ns) are required in URANS to capture similar oscillations [38]. This is a prohibitive timescale for the scope of the optimizations due to the long simulation times required to obtain converged statistics of the shock oscillation. The best option available was then to force the shock to oscillate with URANS simulations at more affordable timesteps

($\Delta t \approx 20\,\mu s$). With such timesteps, the simulation converges to the desired operating conditions, which are searched thanks to a new PID boundary conditions controller (BCC) developed. This controller can modify the input values at any boundary based on the rolling output of the simulation.

After the simulation converges to the desired operating conditions, the inflow angle at the inlet was modified by a delta based on a given input signal with zero average value. This signal was synthesized from a superposition of sine functions, where phase shifts are randomized and the amplitudes were taken from the spectrum of the shock oscillation in the experiments. This way, the shock inside the passage could be forced to oscillate with similar properties as in the experiments. These simulations were run for approximately 2000 timesteps, or about 20 cycles of the main target frequency from the experiments at nearly 500 Hz.

4.2.2 A Framework to Optimize Flow Control Methods

A general strategy was created to allow the optimization of different FCMs for the given objectives. This was achieved with a new Python program, OptiFCM, that is called by the optimizer and is able to set up and manage different optimization configurations. This program builds the process chain required for the imported configuration with the help of a number of custom classes that extend the functionalities offered by the job management software, FireWorks [47]. The latter allows the management of the process chain, so that if any issues arise or one of the constraints is not met, the process chain is defused. The results available are then returned to the optimizer to train the metamodel accordingly. The optimizations themselves follow a simple strategy, as visualized in Fig. 14 and described below:

1. A new member is provided by the optimizer to OptiFCM with a set of design parameters in order to create and launch the process chain.

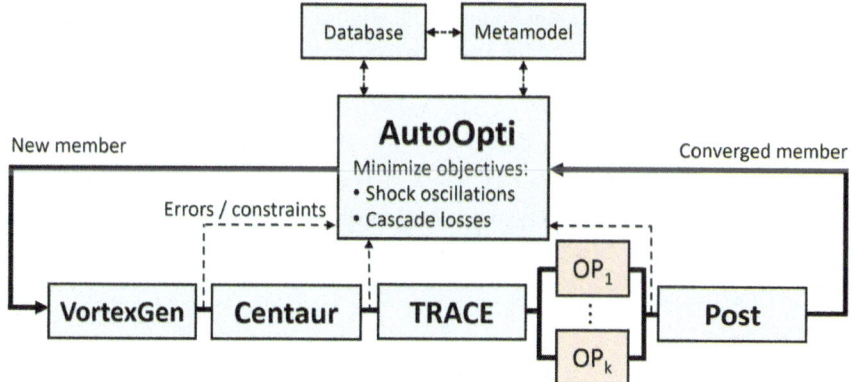

Fig. 14 Optimization strategy flowchart to mitigate SBLI unsteadiness with flow control methods

2. The VortexGen input files are created and the different geometrical constraints for the FCM types requested are calculated. The process chain is defused if one of the constraints is not met.
3. The output files from VortexGen are read by CentWrap and the scripts are generated to mesh the configuration with Centaur. The output files are checked to make sure the geometry was meshed without issues.
4. The output is compiled into a baseline TRACE simulation configuration, which is used to calculate each operating point requested. For the optimization results shown in this chapter, two operating points were simulated: the ADP, and an operating point near stall and at equal M_1 labeled as ODP1. However, the process chain supports any number of operating points desired. The simulations are checked for convergence and compliance to the constraints after every calculation.
5. The output of the converged URANS simulations is then post-processed to calculate the objective functions, which are discussed in the following sections.
6. Lastly, the results are passed to the optimizer, which updates the members database and the metamodel for the creation of the new members.

4.2.3 Optimization Objective Functions and Design Space

The goal of the optimization is to mitigate the unsteadiness of the shock in the passage, while maintaining or improving the performance of the cascade throughout its working range. In order to achieve this goal, the optimization was set to minimize two objective functions with the following formulations:

$$f_1 = \frac{\overline{\sigma}_{x_s}^{ADP} + \overline{\sigma}_{x_s}^{ODP1}}{2}, \qquad f_2 = \frac{\overline{\omega}^{ADP} + \overline{\omega}^{ODP1}}{2}, \qquad (1)$$

where the first objective focuses on the standard deviation of the shock position signal captured inside the passage, σ_{x_s}. The standard deviations of the entire shock front tracked at each span (k) and pitch (j) locations of the 3D rectilinear probe grid sampled are averaged based on the variances of the signal as follows:

$$\overline{\sigma} = \sqrt{\frac{\sum^{n=j \times k} \sigma_{j,k}^2}{n}} \qquad (2)$$

This same value is then averaged across the different operating points calculated. The objective f_2 on the other hand focuses on the total pressure loss coefficient of the cascade, as averaged over the course of the URANS with oscillating inlet conditions:

$$\overline{\omega} = \frac{\overline{P}_{01} - \overline{P}_{02}}{\overline{P}_{01} - \overline{P}_1} \qquad (3)$$

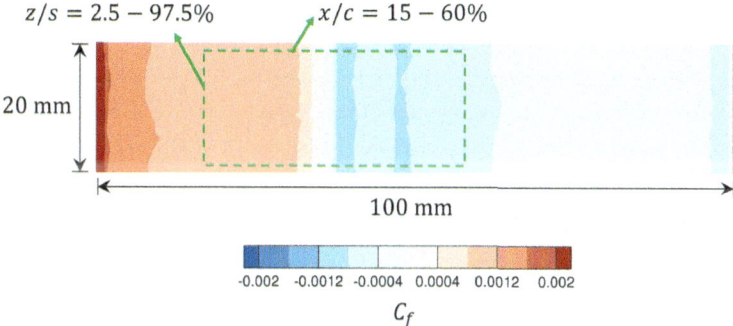

Fig. 15 Design space on the blade surface of the TCTA at the ADP for the optimization of flow control methods

The design space and constraints for the optimization are then defined for the two roughness patches to be optimized. The computational domain used for the simulations consists of periodic boundaries in the span and pitch directions of the cascade. The blade span was 20% of the blade chord, while the inlet and outlet boundaries are placed at one axial chord length upstream and downstream of the leading and trailing edges, respectively. The reference position is then freely set in u between 15 and 60% of the chord length, as shown in Fig. 15. The patches were centered in the v direction based on the value of its span length, l_v, which was freely set between 5 and 95%. The last free parameter, l_u, then determines the maximum u-position of each patch, u_{max}. To ensure that the patches lie within the desired region and don't overlap, two constraints are added on u_{max} and the evaluated overlapping area between them. The de Haller number and mass flow are also constrained to have deltas no higher than $\pm 2.5\%$ with respect to the baseline to ensure that the operating conditions haven't changed considerably.

4.3 Results of an Optimized Roughness Patch Configuration

For this optimization, the design parameters for the RPs to optimize were set according to Fig. 15, with the length l_u limited to between 0.005–0.45, l_v and 0.025–0.950, and k_s to 0.0–70 (μm). With these settings, the configuration was optimized until a total of 310 members were calculated providing a well-populated Pareto front and satisfactory convergence criteria for the Kriging metamodel. The results will be presented in this section starting with the Pareto front itself, followed by a categorization of the best configurations obtained, and finally some of the perceived mechanisms of unsteady SBLI mitigation being applied.

4.3.1 Pareto Front Overview

The Pareto front obtained from the optimization is shown in green markers in Fig. 16. The members have been plotted by their fitness values in terms of the shock position standard deviation and the averaged cascade losses in the x- and y-axes, respectively. The "failed" members plotted are those that have reached the end of the process chain successfully, but have in the end failed the constraint on the averaged inlet Mach number. Most of these have highly unstable configurations and Mach numbers that are lower than the 1.19 limit set, and therefore show lower cascade losses than the rest. The reference member shown corresponds to the baseline configuration without any roughness patches applied on the suction surface.

Figure 16 quickly offers two main conclusions: the turbulent boundary layer is able to support and mitigate shock oscillations considerably, and the reduced severity of said oscillation can decrease the losses downstream of the cascade. That is, the unsteady flow generated from a severe oscillation of the shock naturally leads to higher losses from the increased turbulence and additional mixing that may occur in the wake. This despite the inevitable increase in viscous losses on the blade surface and the size of the boundary layer prior to the shock. Finally, depending on the intensity of the shock control action, the configuration can be tuned to equally optimal solutions from the Pareto front that either control the movement of the shock more strongly with higher losses or the other way around.

4.3.2 Categorizing the Best Members Calculated

The design parameters of the two RPs of the 16 Pareto front members with respect to the two fitness functions is shown in Fig. 17. From these graphs, it can be noted

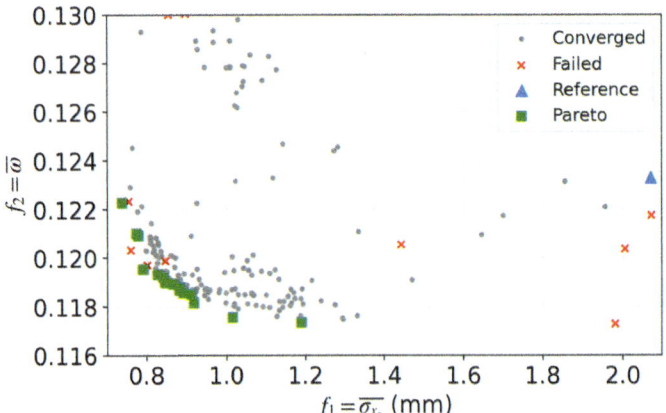

Fig. 16 Roughness patches optimization's Pareto front: avg. shock position standard deviation (x-axis) and avg. cascade losses (y-axis)

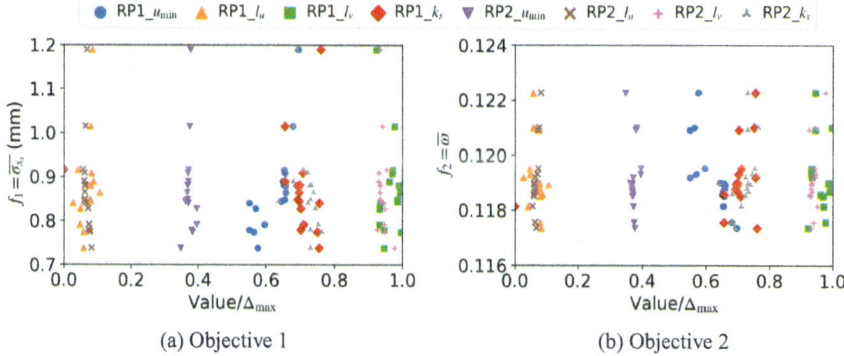

Fig. 17 Objective results of Pareto front members with respect to their scaled design parameters

Table 2 Optimization results for main members of the pareto front

	u_1	l_{u1}	l_{v1}	k_{s1} (μm)	u_2	l_{u2}	l_{v2}	k_{s2} (μm)	f_1 (mm)	f_2
Member 308	2.41	0.0311	0.900	52.9	2.30	0.0416	0.929	51.0	0.74	0.1223
Member 200	2.46	0.0428	0.879	53.3	2.32	0.0353	0.889	48.3	1.19	0.1174

that the optimizer converged to very similar values for most parameters. The main differences refer to the parameters of the first patch, RP1, placed further downstream than the second, RP2. Some configurations show that RP1 was not active, either with a k_s of nearly 0, or placed downstream of the shock ($\Delta u_1 > 0.65$), where it would have little effect on the separated flow. The figure also shows that a higher k_{s2} reduces the shock oscillation at the cost of higher losses. It can also be noted that the average non-scaled value of k_s and l_u for the RP2 of all members was 50.2 μm and 3.45 mm, respectively. Both shapes were also determined to be more effective when they extended over most of the blade span.

Two configurations are now looked at in more detail from the Pareto front: Member 308, with one of the lowest averaged standard deviations of the shock position and highest averaged losses, and Member 200, with the opposite characteristics. The design parameters and main optimization results for these two members are summarized in Table 2. The averaged solution of the URANS ADP simulation with oscillating inlet conditions is also shown for both members in Fig. 18. The figure shows the contours of the Mach number at the periodic boundary, as well as the friction coefficient over the blade surface. The areas where the roughness patch treatments were applied are also marked with black edges on the blade surface.

From both figures, it is clear that the most upstream roughness patch, RP2, trips the laminar boundary layer over the surface. In addition, the configuration with the stronger shock oscillations has RP1 placed far downstream of the shock. When RP1

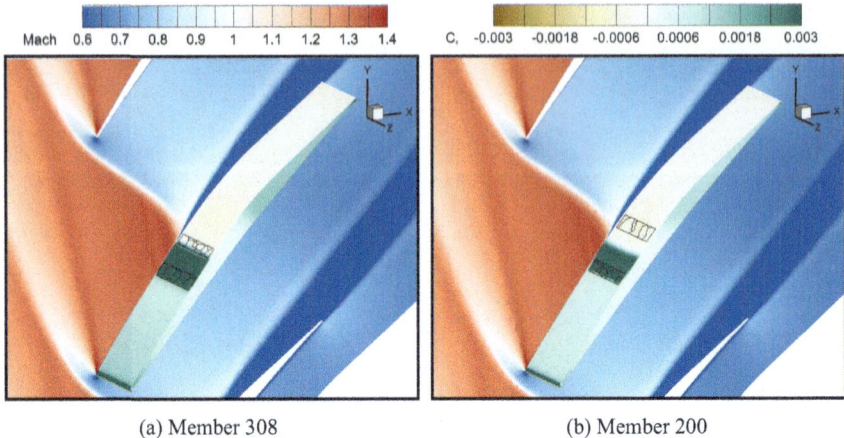

Fig. 18 Mach number and friction coefficient contours of averaged flow solutions of main Pareto front members at the ADP

is placed further upstream (Member 308), the oscillation of the shock is reduced by more than 64% when compared to the baseline configuration. The losses on the other hand have only marginally decreased by 0.8% of their original value. For Member 200 on the other hand, the oscillations of the shock are only reduced by 42%, but the averaged cascade losses have been reduced by 4.8% with respect to the baseline. Lastly, for both configurations, the ODP shows very similar contours (not shown), with the exception that the shock is further upstream and therefore the friction coefficient over the entirety of RP1 is negative. The shock position is still downstream of RP2, so that the incoming boundary layer is still turbulent.

4.3.3 On the Mitigation of Unsteady SBLI Effects

It has been determined that a well-placed RP can mitigate the oscillation of the shock and reduce the losses in the wake of the cascade. It can then be concluded that a turbulent boundary layer is able to withstand the oscillation of the shock better than a laminar one. To investigate this further, Fig. 19a is shown with the BL profiles just prior to the shock at 0.45 chord length from the URANS ADP simulations. In this figure, it can be observed that the laminar BL of the baseline configuration is already separated. This is due to the increased amplitude of oscillation of the shock, but also due to the earlier separation from the pressure gradient imposed by the shock. The BL for the two optimized members are shown to be more resilient, although they are difficult to compare with each other due to the different placement of the two RPs.

To take a closer look at the differences between the designs of both optimized members, their boundary layer profiles right at the beginning and end of RP2 are shown in Fig. 19b. In this graph, it is clear that how the boundary layer height, δ_{99},

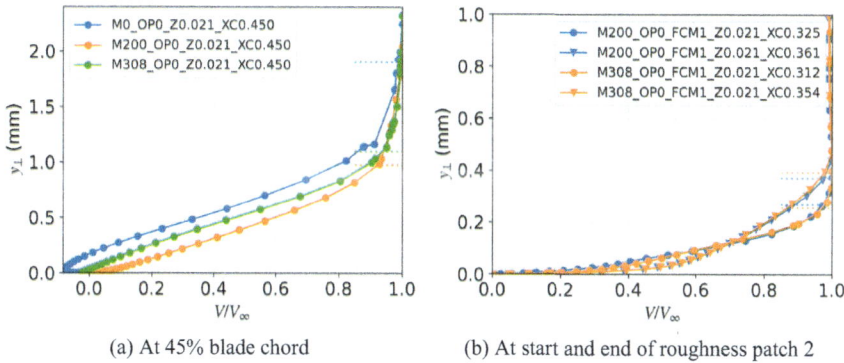

Fig. 19 Boundary layer profiles comparison for the TCTA's baseline, Member 200, and Member 308 configurations

increases considerably over the length of the patch. However, the shorter and slightly lower roughness height of Member 200 seems to contribute to this member having a marginally smaller boundary layer and higher momentum near the BL edge. These slight differences may be playing a factor in the way and the location where the boundary layer separates near the shock. Therefore, it is not just the tripping of the boundary layer, but also the location and strength of the roughness applied to the flow that has contributed to the lower shock oscillations measured for this member.

5 Conclusions

In this chapter, the authors have given a brief overview of the comprehensive work package performed by the DLR for the TEAMAero consortium. This contribution aimed to improve the understanding of the unsteady SBLIs in transonic compressor cascades and propose novel methodologies to mitigate its effects. This was achieved by designing the Transonic Cascade TEAMAero, studying its cycle of shock oscillation, and finally by optimizing different configurations of flow control methods on its surface. The application throughout these work packages of both advanced numerical and experimental techniques allowed for detailed analysis of the results. This approach enhanced the conclusions obtained with respect to those that would have been possible if performed as isolated work packages.

The most relevant results to highlight showed that the oscillation of the shock on the TCTA is linked to mechanisms of interaction of the separated flow with the cascade's passage. Afterward, it was shown how a well-crafted optimization procedure can be developed to both mitigate the unsteady effects from these interactions and improve the performance of the cascade. These results exemplify the need to further integrate experimental and numerical methods in search of a deeper understanding of SBLI in transonic compressor cascades. Only the strong collaboration of numerous stakeholders and disciplines will allow the consolidation of a global theory of SBLI in transonic compressor cascades.

Acknowledgements This work has received funding from the European Union's Horizon 2020 research and innovation program under grant agreement no. 860909. The LES simulations presented were performed on the national supercomputer HPE Apollo Hawk at the High Performance Computing Center Stuttgart (HLRS) under the grant number TCTA-SBLI/44238.

Competing Interests The authors have no conflicts of interest to declare that are relevant to the content of this chapter.

References

1. Ferri, A.: Investigations and experiments in the guidonia wind tunnel. Tech Memo NACA **901**, 1–33 (1939)
2. Fage, A., Sargent, R.F.: Shock-wave and boundary-layer phenomena near a flat surface. Math. Phys. Sci. (1947). https://doi.org/10.1098/rspa.1947.0058
3. Dussauge, J.P., Piponniau, S.: Shock/boundary-layer interactions: possible sources of unsteadiness. J. Fluids Struct. (2008). https://doi.org/10.1016/j.jfluidstructs.2008.06.003
4. Dolling, D.S.: Fifty years of shock-wave/boundary-layer interaction research: What next? AIAA J. (2001). https://doi.org/10.2514/2.1476
5. Lee, B.H.: Oscillatory shock motion caused by transonic shock boundary-layer interaction. AIAA J. (1990). https://doi.org/10.2514/3.25144
6. Pirozzoli, S., Grasso, F., Gatski, T.B.: DNS analysis of shock wave/turbulent boundary layer interaction at M = 2.25. In: 4th International Symposium on Turbulence and Shear Flow Phenomena (2005). https://doi.org/10.1615/tsfp4.2010
7. Dupont, P., Piponniau, S., Sidorenko, A., Debiève, J.F.: Investigation by particle image velocimetry measurements of oblique shock reflection with separation. AIAA J. (2008). https://doi.org/10.2514/1.30154
8. Crouch, J.D., Garbaruk, A., Magidov, D., Travin, A.: Origin of transonic buffet on aerofoils. J. Fluid Mech. (2009). https://doi.org/10.1017/S0022112009006673
9. Touber, E., Sandham, N.D.: Large-Eddy simulation of low-frequency unsteadiness in a turbulent shock-induced separation bubble. Theor. Comput. Fluid Dyn. (2009). https://doi.org/10.1007/s00162-009-0103-z
10. Epstein, A.H., Kerrebrock, J.L., Thompkins, W.T.: Shock structure in transonic compressor rotors. AIAA J. (1979). https://doi.org/10.2514/3.61134
11. Hergt, A., Klinner, J., Wellner, J., Willert, C., Grund, S., Steinert, W., Beversdorff, M.: The present challenge of transonic compressor blade design. J. Turbomach. (2019). https://doi.org/10.1115/1.4043329
12. Lee, B.H.: Self-sustained shock oscillations on airfoils at transonic speeds. Prog. Aerosp. (2001). https://doi.org/10.1016/S0376-0421(01)00003-3
13. Hartmann, A., Feldhusen, A., Schröder, W.: On the interaction of shock waves and sound waves in transonic buffet flow. Phys. Fluids (2013) https://doi.org/10.1063/1.4791603
14. Priebe, S., Wilkin, D., II, Breeze-Stringfellow, A., Mousavi, A., Bhaskaran, R., d'Aquila, L.: Large Eddy simulations of a transonic airfoil cascade. In: Proceedings of ASME Turbo Expo 2022 (2022). https://doi.org/10.1115/GT2022-80683
15. Ranjan Majhi, J., Venkatraman, K.: On the nature of transonic shock buffet in an axial-flow fan. AIAA J. (2023). https://doi.org/10.2514/1.j063318
16. Klinner, J., Hergt, A., Grund, S., Willert, C.E.: High-speed PIV of shock boundary layer interactions in the transonic buffet flow of a compressor cascade. Exp. Fluids (2021). https://doi.org/10.1007/s00348-021-03145-3

17. Pearcey, H.: Introduction to shock-induced separation and its prevention by design and boundary layer control. In: Boundary Layer and Flow Control (1961)
18. Mccullough, G., Nitzberg, G., Kelly, J.: Preliminary investigation of the delay of turbulent flow separation by means of wedge-shaped bodies. Res. Memo. NACA **A50L12**, 1–28 (1951)
19. Ogawa, H., Babinsky, H., Pätzold, M., Lutz, T.: Shock-wave/boundary-layer interaction control using three-dimensional bumps for transonic wings. AIAA J. (2008). https://doi.org/10.2514/1.32049
20. John, A., Qin, N., Shahpar, S.: Using shock control bumps to improve transonic fan/compressor blade performance. J. Turbomach. (2019). https://doi.org/10.1115/1.4042891
21. Klinner, J., Hergt, A., Grund, S., Willert, C.E.: Experimental investigation of shock-induced separation and flow control in a transonic compressor cascade. Exp. Fluids (2019)
22. Hergt, A., Klinner, J., Willert, C., Grund, S., Steinert, W.: Insights into the unsteady shock boundary layer interaction. In: Proceedings of ASME Turbo Expo 2022 (2022). https://doi.org/10.1115/GT2022-82720
23. Voß, C., Aulich, M., Kaplan, B., Nicke, E.: Automated multiobjective optimisation in axial compressor blade design. In: Proceedings of ASME Turbo Expo 2006 (2006). https://doi.org/10.1115/GT2006-90420
24. Aulich, M., Siller, U.: High-dimensional constrained multiobjective optimization of a fan stage. In: Proceedings of ASME Turbo Expo (2011). https://doi.org/10.1115/GT2011-45618
25. Schnoes, M., Voß, C., Nicke, E.: Design optimization of a multi-stage axial compressor using throughflow and a database of optimal airfoils. J. Glob. Power Propul. (2018). https://doi.org/10.22261/jgpps.w5n91i
26. Klose, B.F., Morsbach, C., Bergmann, M., Hergt, A., Klinner, J., Grund, S., Kügeler, E.: A numerical test rig for turbomachinery flows based on large Eddy simulations with a high-order discontinuous Galerkin scheme—Part II: Shock capturing and transonic flows. J. Turbomach. (2023). https://doi.org/10.1115/1.4063827
27. Morsbach, C., Bergmann, M., Tosun, A., Klose, B.F., Bechlars, P., Kügeler, E.: A numerical test rig for turbomachinery flows based on large Eddy simulations with a high-order discontinuous Galerkin scheme—Part 3: Secondary flow effects. In: Proceedings of ASME Turbo Expo 2023 (2023). https://doi.org/10.1115/GT2023-101374
28. Bergmann, M., Morsbach, C., Klose, B.F., Ashcroft, G., Kügeler, E.: A numerical test rig for turbomachinery flows based on large Eddy simulations with a high-order discontinuous Galerkin scheme—Part I: Sliding interfaces and unsteady row interactions. J. Turbomach. (2023). https://doi.org/10.1115/1.4063734
29. Munoz Lopez, E.J., Hergt, A., Grund, S.: The new chapter of transonic compressor cascade design at the DLR. In: Proceedings of ASME Turbo Expo 2022 (2022). https://doi.org/10.1115/GT2022-80189
30. Munoz Lopez, E.J., Hergt, A., Grund, S., Gümmer, V.: The new chapter of transonic compressor cascade design at the DLR. J. Turbomach. (2023). https://doi.org/10.1115/1.4056982
31. Menter, F.R., Kuntz, M., Langtry, R.: Ten years of industrial experience with the SST turbulence model turbulence heat and mass transfer. Turbul. Heat Mass Transf. **4**(625–632), 2003 (2023)
32. Langtry, R.B., Menter, F.R.: Correlation-based transition modeling for unstructured parallelized computational fluid dynamics codes. AIAA J. (2009). https://doi.org/10.2514/1.42362
33. Starken, H., Schimming, P., Breugelmans, F.A.: Investigation of the axial velocity density ratio in a high turning cascade. In: Proceedings of ASME Turbo Expo 1975 (1975). https://doi.org/10.1115/75-GT-25
34. Munoz Lopez, E.J., Hergt, A., Ockenfels, T., Grund, S., Gümmer, V.: The current gap between design optimization and experiments for transonic compressor blades. Int. J. Turbomach. Propul. Power (2023). https://doi.org/10.3390/ijtpp8040047
35. Schodl, R.: Laser dual-beam method for flow measurements in turbomachines. In: Proceedings of ASME Turbo Expo 1974 (1974). https://doi.org/10.1115/74-GT-157
36. Munoz Lopez, E.J., Hergt, A., Klinner, J., Klose, B., Willert, C., Guemmer, V.: The unsteady shock-boundary layer interaction in a compressor cascade—Part 3: Mechanisms of shock oscillation. In: Proceedings of ASME Turbo Expo 2024 (2024)

37. Klinner, J., Munoz Lopez, E.J., Hergt, A., Willert, C.: The unsteady shock-boundary layer interaction in a compressor cascade—Part 1: Measurements with time-resolved PIV. In: Proceedings of ASME Turbo Expo 2024 (2024)
38. Klose, B.F., Morsbach, C., Bergmann, M., Munoz Lopez, E.J., Hergt, A., Kügeler, E.: The unsteady shock-boundary layer interaction in a compressor cascade—Part 2: High-fidelity simulation. In: Proceedings of ASME Turbo Expo 2024 (2024)
39. Klinner, J., Munoz Lopez, E.J., Hergt, A., Willert, C.: High-resolution PIV measurements of the shock boundary layer interaction within a highly loaded transonic compressor cascade. In: 15th International Symposium on Particle Image Velocimetry (2023). Available via eLib. https://elib.dlr.de/197278/. Cited 11 June 2024
40. Munoz Lopez, E.J., Hergt, A., Klinner, J., Grund, S., Flamm, J., Gümmer, V.: Investigations of the unsteady shock-boundary layer interaction in a transonic compressor cascade. In: Proceedings of ASME Turbo Expo 2023 (2023). https://doi.org/10.1115/GT2023-102622
41. Schlüß, D., Frey, C., Ashcroft, G.: Consistent non-reflecting boundary conditions for both steady and unsteady flow simulations in turbomachinery applications. In: ECCOMAS Congress 2016—Proceedings 7th European Congress on Computational Methods in Applied Sciences and Engineering (2016). https://doi.org/10.7712/100016.2342.5411
42. Pirozzoli, S.: Numerical methods for high-speed flows. Ann. Rev. Fluid Mech. (2011). https://doi.org/10.1146/annurev-fluid-122109-160718
43. Hergt, A., Klose, B., Klinner, J., Bergmann, M., Munoz Lopez, E.J., Grund, S., Morsbach, C.: On the shock boundary layer interaction in transonic compressor blading. In: Proceedings of ASME Turbo Expo 2023 (2023). https://doi.org/10.1115/GT2023-103218
44. Siller, U., Voß, C., Nicke, E.: Automated multidisciplinary optimization of a transonic axial compressor. In: 47th AIAA Aerospace Sciences Meeting (2009). https://doi.org/10.2514/6.2009-863
45. Becker, K., Heitkamp, K., Kuegeler, E.: Recent progress in a hybrid-grid CFD solver for turbomachinery flows. In: Procedings of 5th European Conference on Computational Fluid Dynamics ECCOMAS (2010)
46. Ashcroft, G., Heitkamp, K., Kügeler, E.: High-order accurate implicit Runge-Kutta schemes for the simulation of unsteady flow phenomena in turbomachinery. In: Proceedings of 5th European Conference on Computational Fluid Dynamics ECCOMAS (2010)
47. Jain, A., Ong, S.P., Chen, W., Medasani, B., Qu, X., Kocher, M., Brafman, M., Petretto, G., Rignanese, G.M., Hautier, G., Gunter, D., Persson, K.A.: Fireworks: a dynamic workflow system designed for high throughput applications. Concurr. Comp.-Pract. E (2015). https://doi.org/10.1002/cpe.3505

Open Access This chapter is licensed under the terms of the Creative Commons Attribution 4.0 International License (http://creativecommons.org/licenses/by/4.0/), which permits use, sharing, adaptation, distribution and reproduction in any medium or format, as long as you give appropriate credit to the original author(s) and the source, provide a link to the Creative Commons license and indicate if changes were made.

The images or other third party material in this chapter are included in the chapter's Creative Commons license, unless indicated otherwise in a credit line to the material. If material is not included in the chapter's Creative Commons license and your intended use is not permitted by statutory regulation or exceeds the permitted use, you will need to obtain permission directly from the copyright holder.

Surface Roughness Effect on Shock Boundary Layer Interaction on Compressor Rotor Profile

Ahmed H. Hanfy, Pawel Flaszyński, Piotr Kaczyński, and Piotr Doerffer

Abstract High-pressure ratios in transonic compressor rotors and fan blades pose challenges such as supersonic speeds and shock waves, which can result in boundary layer separation and potential performance issues. This study presents experimental research exploring the effect of surface roughness on shockwave-boundary layer interaction (SBLI). By employing various measurement techniques, the study provides a comprehensive overview of how surface roughness affects SBLI, enriching the dataset for internal flows across four specimens with different roughness parameters. The findings indicate that lower roughness surfaces lead to larger separation bubbles and more typical shock structures, whereas increased roughness results in smaller separation bubbles but higher flow instability.

Keywords Transonic flow · Shock induced separation · Boundary layer transition · Measurements · Flow visualization

15.1 Introduction

Advancements in propulsion turbojet engines prioritize efficiency, driving continual refinement of fan blade designs to mitigate pressure losses from viscous effects and shocks. Recent innovations focus on transonic laminar profiles, offering potential drag reductions of up to 15%. These profiles sustain laminar boundary layers along a significant portion of the chord length. However, the transition from laminar to turbu-

---

A. H. Hanfy (✉) · P. Flaszyński · P. Kaczyński · P. Doerffer
Institute of Fluid-Flow Machinery, Polish Academy of Sciences (IMP PAN), Gdańsk, Poland
e-mail: ahmed.hanfy@imp.gda.pl

P. Flaszyński
e-mail: pflaszyn@imp.gda.pl

P. Kaczyński
e-mail: pkaczynski@imp.gda.pl

P. Doerffer
e-mail: doerffer@imp.gda.pl

lent boundary layers significantly impacts profile drag, especially at lower Reynolds numbers encountered at high altitudes, where transition locations are pushed downstream. In transonic compressor rotor or fan blades operating under higher pressure ratios with fewer stages, challenges escalate due to increased flow velocities inducing supersonic speeds and shock waves at the blade tips, leading to boundary layer separation and risk engine performance [1].

Controlling the impact of the laminar shock boundary layer interaction has been a focal point of extensive research for many years. Techniques such as tripping the laminar boundary layer into the turbulent flow upstream of the shock wave using steps or roughness patches have been investigated extensively [2–5].

Understanding the influence of surface roughness on laminar-turbulent transition is of paramount importance, especially in turbomachinery where blades rapidly accumulate roughness due to deposition, mechanical damage, chemical effects, or flow control device employment. Surface roughness introduces flow instabilities that can significantly alter laminar-turbulent transition locations [6, 7].

Despite the negative impact of and roughness on the performance and lifespan of gas turbines, engineered roughness has been explored as a form of passive flow control in many turbomachinery applications. This approach is beneficial in overcoming laminar boundary layer separation at low Reynolds numbers by inducing early transition [8]. For instance, in studies such as those by Boese [9] and Leipold [10], have shown that distributed roughness applied to the entire suction side of highly loaded compressor profiles can suppress laminar separation and reduce the total pressure loss coefficient by up to 5% in some cases, within the range of $1.5e+5 < Re < 11e+5$ using v-groove type riblets.

Although significant progress has been made in research, a comprehensive understanding of the role of roughness and its scales is still necessary. This gap exists mainly due to a lack of systematic studies on the structure of the boundary layer and the variety of roughness types that influence flow dynamics within the roughness sublayer. The current roughness scale, typically defined by sand-grain roughness height, which often fails to fully characterize roughness in many cases. Therefore, there is a need for a universal roughness scale that can describe all types of roughness and be applicable across any flow regime [11].

In unperturbed boundary layers with minimal flow disturbances, the natural laminar-turbulent transition occurs through the initiation of exponentially growing 2D waves, known as Tollmien-Schlichting (TS) waves, followed by a 3D instability stage leading to turbulent boundary layer formation. However, sources of high flow disturbance, such as freestream turbulence exceeding 1% [12] or surface roughness [13, 14], accelerate laminar-turbulent transition, bypassing the primary instability stage. Even smaller roughness levels may amplify primary instabilities linearly [15], while larger roughness bypasses primary instabilities by distorting the flow locally [13, 15, 16].

Given the complex interplay between surface roughness and shock wave boundary layer interaction, this study aims to provide an overview of the current understanding of roughness effects on SBLI. Leveraging insights from experimental studies

Table 15.1 Parameters of the TFAST cascade

Parameter	Unit	Value
Profile Chord	mm	100
Stagger angle	°	44.39
Pitch to Chord ratio	–	0.6
Thickness to Chord ratio	–	0.033
Blade inlet angle	°	50.91
Blade exit angle	°	33.22
Deflection angle	°	14.86
AVDR	–	1.22
Reynolds number	–	1.48e6
Turbulence intensity	%	2.4–2.7

conducted at IMP PAN transonic wind tunnel, this work seeks to elucidate key aerodynamic characteristics and enrich the SBLI dataset for internal flows.

A single passage test section is designed to reproduce the SBLI on the suction side of a transonic fan profile. The test section design based on the specifications outlined by Piotrowicz [17], derived from data for the compressor cascade geometry provided by Rolls-Royce Deutschland (RRD) during the EU FP7 TFAST predecessor project. The parameters are provided in Table 15.1.

This chapter is divided into two main sections. Section 15.2 provides a detailed overview of the test section setup, including the roughness parameters, implementation, experimental methods and data post-processing techniques. Section 15.3 concerns the aerodynamic behavior and the effects of roughness on the shock structure, induced separation, and boundary layer characteristics downstream of the shock on the suction surface of the fan profile.

15.2 Experimental Setup and Post-processing Procedures

The test section, as illustrated in Fig. 15.1, consists of two transonic fan profiles. The upper profile acts as a shock and wake generator, creating a normal shock wave that interacts with the boundary layer on the suction side of the lower profile, typically occurring at approximately $0.48x/c$. Additionally, two controllable suction slots are positioned on either side of the lower profile to minimize secondary flow effects and regulate the blockage of the test section passages to achieve the design Axial Velocity Density Ratio (AVDR).

The design of the test section incorporates two screw mechanisms. The first mechanism adjusts the channel above the upper profile, controlling the mass flow through the passages in conjunction with the suction slots. The second mechanism controls

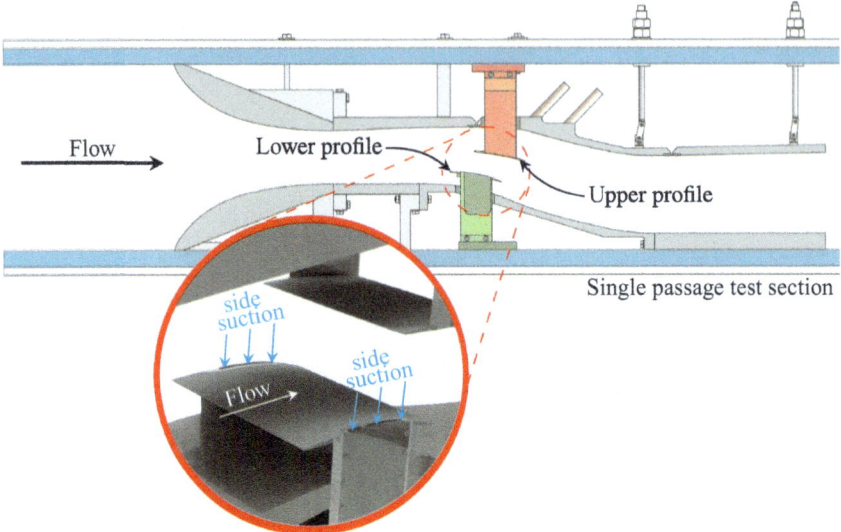

Fig. 15.1 TFAST test section

the outlet pressure of the test section and is used to maintain the required flow conditions.

The inflow Mach number (M = 1.22) is determined by averaging the isentropic Mach values calculated from five pressure taps placed sequentially in the flow direction on the lower wall of the test section between the nozzle throat and the lower profile. The inlet turbulence intensity is averaged from vertical traverse measurement 20 mm upstream of the lower profile using Laser Doppler Anemometry (LDA). The measured turbulence intensity is in the range of 2.4–2.7%, indicating the onset of a bypass transition on the lower profile regardless of the shock effects.

15.2.1 Planned Measurements and Challenges

Figure 15.2 illustrates the typical shock structure in the test section. The lower profile generates a detached shock wave (Leading-edge shock) that interacts with the test section's upper wall and is reflected (Reflected shock). An expansion wave is formed at the leading-edge of the lower profile, creating a Mach line. Simultaneously, the upper profile generates a normal shock (Passage shock).

The upper part of the leading-edge shock and the corresponding reflected shock exhibit an unsteady nature due to their interaction with the boundary layer of the test section's upper wall. Conversely, the Mach wave remains relatively steady. However, the passage shock, which is the primary focus of the study, is highly unsteady due to its interaction with the boundary layer on the lower profile.

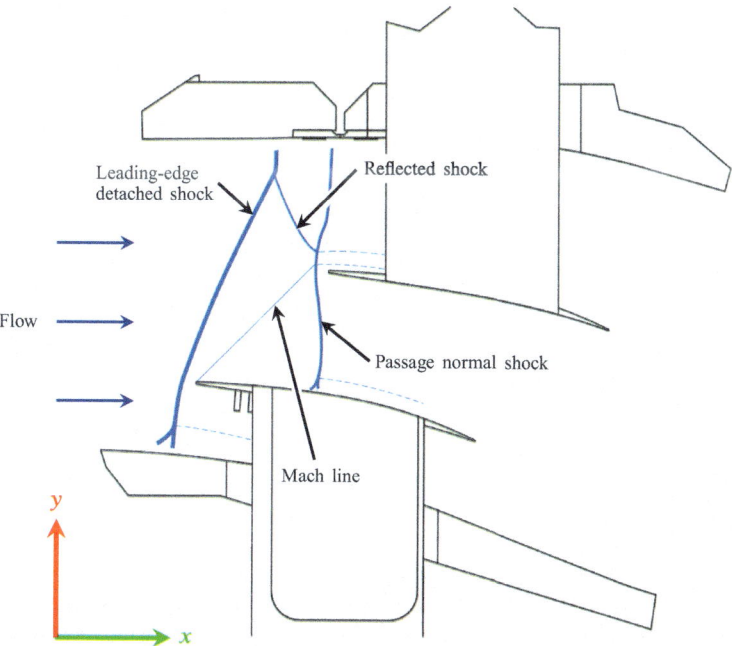

Fig. 15.2 Shock structure

To gain different perspectives and a better understanding of flow behavior, multiple measurement techniques are employed. Static pressure on the lower profile is measured using twelve pressure taps positioned along the mid-span of the suction side, beneath the shock interaction. These pressure taps cover 46% of the chord length to determine the shock location.

To visualize the flow structure on the suction side of the lower profile, a 10–35% concentration of titanium dioxide oil paint was used. Recorded images were then post-processed using matrix transformation and scaling techniques to enhance measurability, as described in Subsect. 15.2.1.2.

Schlieren system in Z-Configuration is used to capture the flow structure. This system utilized a Canon EOS M50 for snapshots and a Fastec Imaging HiSpec 2G mono high-speed camera for unsteady analysis detailed in Subsect. 15.2.1.2.

LDA is employed to measure the inflow conditions as mentioned in the above section. Additionally, LDA is used to measure the boundary layer on the lower profile. For boundary layer measurements, only a single component is operated to enable near-wall measurements. *Coherent Innova 70C* 5W Argon-Ion laser as laser source is coupled with *FiberFlow* system from *Dantec Dynamice* to split the laser beam. The seeding particles are generated by *Flow Tracker 700 CE* using *DEHS* (Di-Ethyl-Hexyl- Sebacic) oil with an average particle size of 2 μm.

Due to the curvature of a fan profile, especially downstream of the shock, defining the boundary layer measurement position posed a challenge. A Python script [18] is

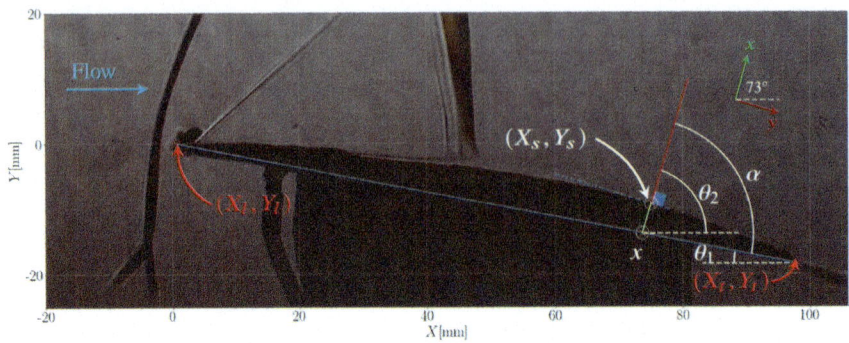

Fig. 15.3 Defining surface point based on laser coordinates

developed and uses the profile CAD model information, facilitating precise determination of measurement locations.

Figure 15.3 provides a visual demonstration of point localization on the lower profile using a schlieren image of the actual test section and shock configuration.

The coordinates of the leading-edge (X_l, Y_l) and trailing edge (X_t, Y_t) in the new system are defined, from the laser traverse system. Subsequently, the Python script manages the calculation of the surface point coordinates in this new system $Ps = (X_s, Y_s)$ and the LDA traverses, according to information such as $(||Ps - x||, \theta_2)$ from profile data points or CAD.

Considering the size of the seeding particles and based on CFD modeling of the test section by Piotrowicz [19], the evaluated boundary layer thickness at $0.242x/c$ (close to $0.25x/c$) is 0.25 mm. This is associated with a measured free-stream velocity of $U_\infty = 404$ m/s) the Stokes number upstream the shock can be estimated as $S_{tk} \approx 22.36 \gg 1$. This indicates that particles will diverge from the stream, especially where the flow decelerates abruptly, resulting in high tracing errors and the accumulation of oil particles on the surface of the profile. Consequently, measurements near the wall upstream of the shock are significantly difficult. Conversely, downstream of the shock, the flow is expected to become turbulent due to the interaction. With an estimated Stokes number of $S_{tk(0.75x/c)} \approx 0.52 < 1$, indicating that particles closely follow fluid streamlines and are affected by most turbulent structures. However, there remains a notable error exceeding 1%.

Furthermore, during measurements very close to the wall, reflections from the profile's surface can significantly lower the signal-to-noise ratio. Additionally, the low seeding levels in this region result in reduced particle counts, leading to lower confidence in the measured mean velocity and significant velocity variation across the measurement volume, thereby increasing scatter. Moreover, employing the back-scatter configuration for boundary layer measurements poses a challenge due to the minimal amount of light scattered back by the seeding particles [20].

To address the issues, the measurements are conducted 0.2 mm above the wall, with an augmented amount of seeding injected solely for near-boundary measure-

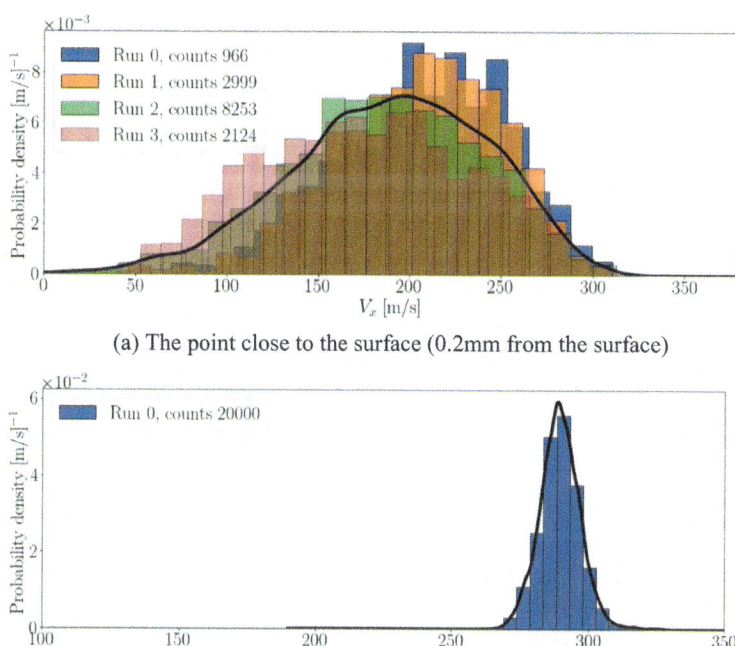

Fig. 15.4 Velocity range of detected seeding elements at $0.75x/c$

ments. To mitigate uncertainty, the measurement is repeated multiple times at the designated locations near the profile surface. Here the results of the measured location at $0.75x/c$ will be presented reflecting the boundary layer jump and mixing loss due to SBLI. Figure 15.4a illustrates the data obtained from near-wall measurements conducted with multiple trials to enhance the statistical evaluation of the flow's average speed. Figure 15.4b shows the free stream velocity range where the seeding is rich and velocity variation is narrow. In both cases, the mean velocity value was estimated according to the thick black line, which represents the weighted average of all measurements taken at the measuring point.

15.2.1.1 Profiles Manufacturing/Machining and Roughness Measurement

With a primary interest in the SBLI on the suction side of the lower profile, where the shock interaction occurs, four copies of the lower profile were manufactured with Titanium alloy to ensure a similar response to the machining process as the actual fan blade. The profiles are manufactured through wire electrical discharge machining. These replicas were crafted with an original shape error of less than ±0.02 mm (±20 μm).

Various surface textures were defined and applied by TEAMAero partner RRD. Three of the profiles are entirely textured (P1, P2, and P4), while one was left not machined as a reference (P3), as depicted in Fig. 15.5. The evaluation of roughness is conducted using 2D Stylus diamond equipment.

The texturing of the profiles led to an increase in shape deviation, with values reaching approximately ±0.1 mm (±100 μm) from the original model shape. The evaluation of roughness encompassed both chord-wise (or stream-wise) and spanwise (or crossflow) directions at four locations.

A summary of the machining process and the resulting roughness parameters is provided in Table 15.2. Where Ra_{avg} represents the arithmetic average roughness values average of both spanwise and chordwise.

15.2.1.2 Image Analysis Techniques for Better Understanding of Flow Structure

Recent advancements in computer vision techniques significantly enhance image processing capabilities, surpassing the limitations of traditional measurement methods. These improvements allow for direct observation of aerodynamic parameters without intruding the flow, offering a more comprehensive understanding of the studied phenomena.

15.2.2 The Development of Line-Scanning Technique

Inspired by the line-scanning technique for processing schlieren images, an advanced method for shock tracking has been developed. This boosts shock localization accuracy and evaluates detection precision. Additionally, the method characterizes intricate shock structures and estimates oblique shock wave angles to ensure the flow conditions. The assessment of detected shock locations can also be used for supervised learning purposes.

The schlieren data was processed using a Python script [21]. This script identifies the maximum density gradient area variation (the largest valley as shown in Fig. 15.6) within a user-defined region, automatically and accurately, while ignoring other flow disturbances and considering the shock location history. Additionally, the script can pinpoint uncertain shock locations for complex structures.

To define the shock angle, the tracking region and the number of slices to be tracked are manually specified. The script then divides the shock region into slices where the shock is tracked. For a series of points, the line is interpolated using the vertical least squares method, and the shock angle is defined for each snapshot.

Moreover, the script uncertainty estimation accounts for potential inaccuracies in pinpointing the shock wave's position, considering factors like valleys of nearly

Surface Roughness Effect on Shock Boundary...

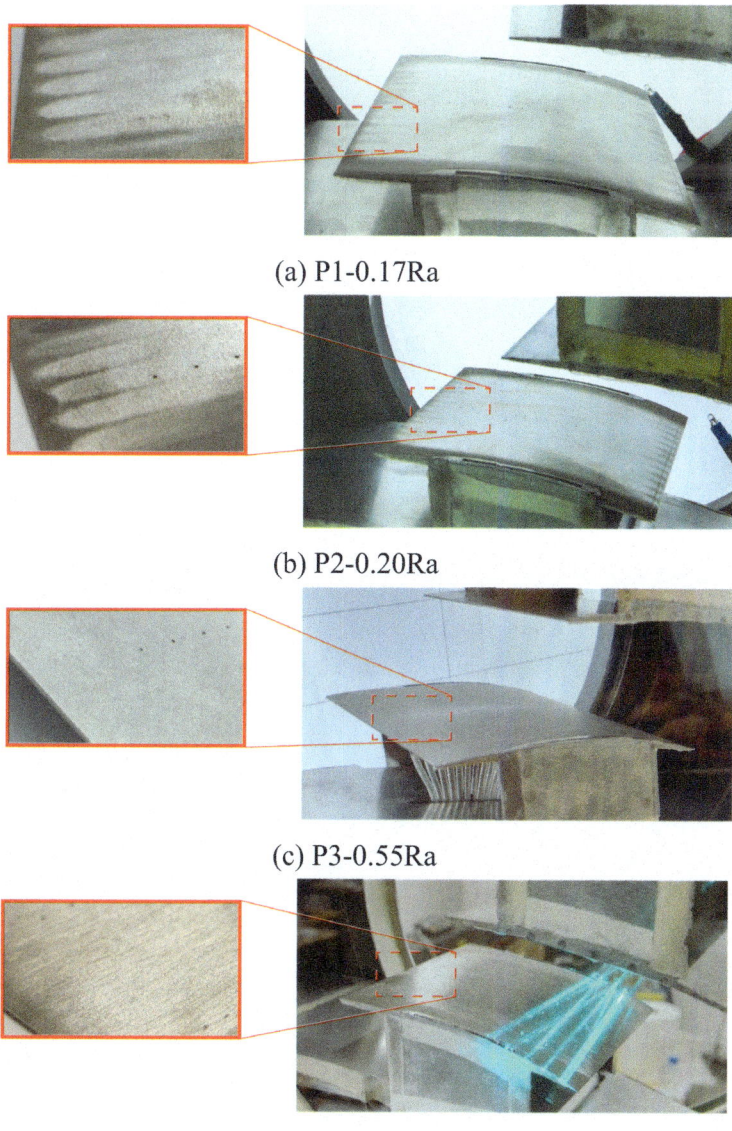

(a) P1-0.17Ra

(b) P2-0.20Ra

(c) P3-0.55Ra

(d) P4-1.48Ra

Fig. 15.5 Lower profile models

Table 15.2 Profiles roughness parameters

Profile	Machining process	Ra_{avg} (µm)
P1-0.17Ra	Polishing	0.17
P2-0.20Ra	Polishing	0.20
P3-0.55Ra	Not applicable	0.55
P4-1.48Ra	Polishing	1.48

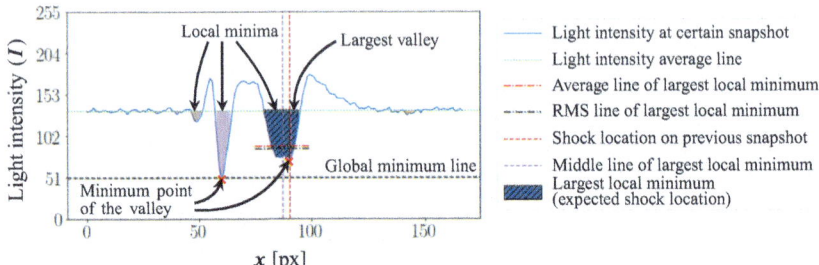

Fig. 15.6 Snapshot slice features and shock detection

equal size, variations in sub-valley selection compared to the largest one, and the presence of nearly identical sub-valleys.

15.2.3 Digital Image Processing for Oil Visualization

Obtaining suitable optical access to the test section is crucial for capturing three-dimensional flow behavior, such as depicting shock structures and surface-flow responses to shock boundary layer interactions. This requires specialized facility design and arrangement, which can be particularly challenging for test sections with complex geometries, such as blade cascades and non-flat nozzle walls.

Side wind tunnel windows provide insight into the test section however, the camera perspective distorts the captured image. Image distortion can lead to unclear flow distribution (i.e., the lengths and angles are distorted according to the distance from the camera, the focal lens and camera angle) and inaccurate estimation of angles or areas of regions of interest.

The proposed method in Hanfy et al. [22] aims to flatten the suction side of a constant chord airfoil using a single camera view, as shown in Fig. 15.7. The initial image captured by the camera is depicted in Fig. 15.7b, and the true scale-mapped surface of the airfoil, obtained after processing, is shown in Fig. 15.7c.

The process begins by defining the two vanishing points and the two chord lines (the near chord and far chord lines to the camera). The script then uses the pre-defined

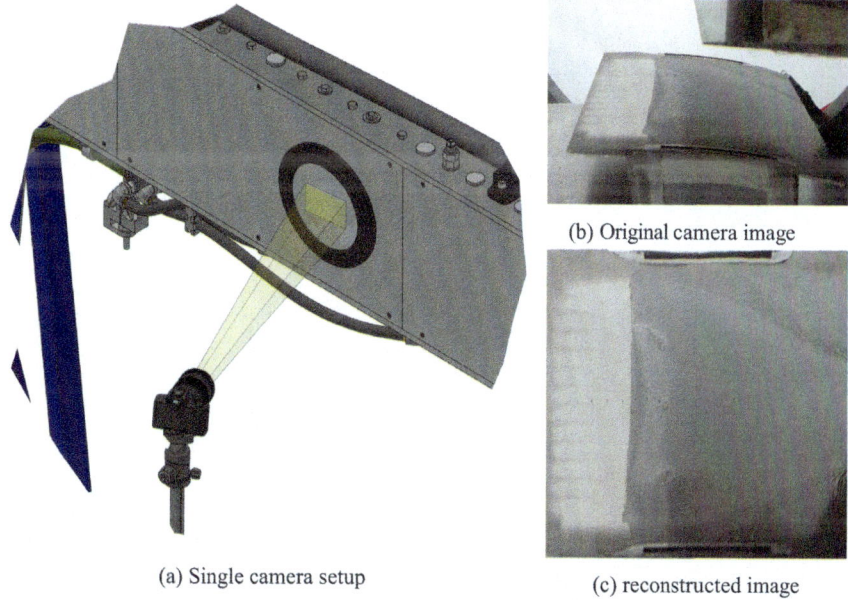

Fig. 15.7 **a** Single camera setup. **b** Original camera image. **c** Reconstructed image

profile surface points as Cartesian coordinates,[1] with the chord line as the reference plane for the airfoil, to be mapped in homogeneous coordinates of the perspective view using the cross-ratio method [23]. Afterward, the curved surface is segmented, and each plane segment is transformed and scaled. Finally, the neighboring segments are stitched together to create the flat image as in Fig. 15.7c.

15.2.3.1 Parameters Considered for Flow Adjustment

Achieving the desired design conditions while ensuring that the flow is influenced solely by profile roughness requires careful attention to the system's sensitivity to the test section control parameters, such as side wall suction, profiles relative location to the nozzle, and inflow angle.

One of the key parameters in this study is the sidewall suction, illustrated in Fig. 15.1. This approach is commonly employed to manage the blockage resulting from the unavoidable growth of the boundary layer on the cascade sidewalls. The degree of this blockage is typically assessed using the AVDR parameter, which quantifies the effective flow area contraction between the inlet and outlet of the cascade.

---

[1] These points can be generated either from the CAD model or from profile coordinates, where the number of points represents the divisions of the curved surface.

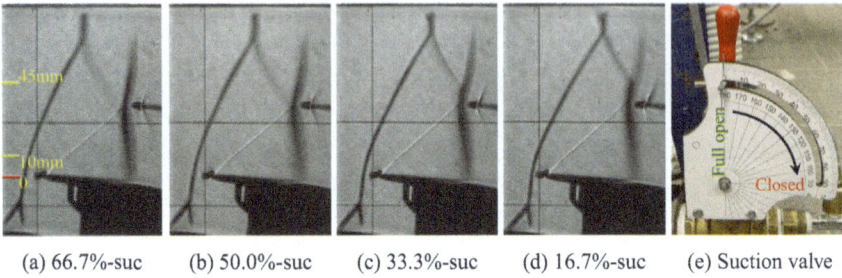

(a) 66.7%-suc   (b) 50.0%-suc   (c) 33.3%-suc   (d) 16.7%-suc   (e) Suction valve

Fig. 15.8 Suction effect on shock structure in the test section

Table 15.3 Leading-edge shock inclination to the horizontal

Case	Valve opening	Inclination difference (°)	Angle variation (σ) (°)
66.7%-suc	30°	+2.19	0.42
50.0%-suc	45°	+1.63	0.42
33.3%-suc	60°	+0.00	0.45
16.7%-suc	75°	−1.18	0.41

The amount of sucked mass flow via the slots is regulated with a ball flow control valve, as shown in Fig. 15.8e. The effect of side wall suction on shock structure and behavior is investigated.

Increased suction shifts the shock system downstream due to the rise in mass flow. To establish comparable operating conditions across all configurations, the flow conditions were carefully adjusted to closely match a specific isentropic Mach number distribution, aligned with the aerodynamic design point of the blading, where the passage shock is positioned near $0.48x/c$. Consequently, the back pressure of the test section was fine-tuned for each control valve setting. While this adjustment did not significantly impact the pressure distribution over the lower profile, it did cause the leading-edge shock to move upstream.

Figure 15.8 showcases the average image of time-resolved schlieren images captured at a frame rate of 2 kHz for 4 s, across four control valve positions expressed as a percentage of valve opening.

From the Figure, the change in shock structure due to suction is evident. For suction levels higher than 33.3% of full valve opening, the leading-edge shock is pushed upstream and exhibits a less inclined angle compared to lower suction cases. The reflected shock attaches to the passage shock below the upper profile, contrary to the modeled conditions. Regardless of the leading-edge shock angle, no significant variations in structure are observed for valve openings above 50%.

The average inclination angle of the leading-edge shock (relative to the horizontal) was evaluated within 10–45 mm above the leading-edge using the script mentioned in Subsect. 15.2.1.2. Table 15.3 shows the difference in inclination angle from the model angle. It can be observed that the inclination decreases by almost 2 °C with

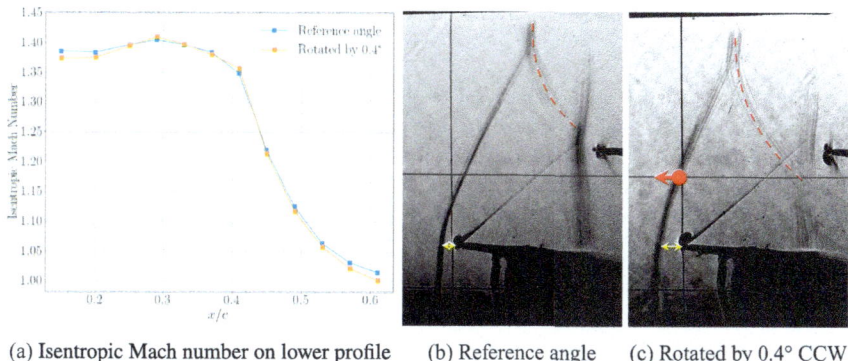

(a) Isentropic Mach number on lower profile   (b) Reference angle   (c) Rotated by 0.4° CCW

Fig. 15.9 Sensitivity of the system to rotation, (- -) is the mean line of reflected shock

higher suction. The leading-edge shock oscillation exhibits a narrow root mean square (RMS) variation in angle (σ), averaging 0.43°.

Further study on the system's sensitivity to small changes in the position of the profiles relative to the nozzle ($\approx \pm 1\%$ x/c in both horizontal and vertical directions) was conducted. The results indicated that the profile position does not significantly alter the overall flow structure or the inflow Mach number. However, it does affect the Mach number distribution over the lower profile just upstream of the passage shock within the range of $0.3-0.4 x/c$.

On the other hand, the inflow Mach number (at $\approx 0.15-0.2$ x/c) is highly sensitive to the profile inflow angle. Even a small change of 0.4 °CCW can alter the inflow Mach number by 0.03, as shown in Fig. 15.9a. This angular adjustment also impacts the entire flow structure, as illustrated in Fig. 15.9b and c. When rotated, the blockage in the lower passage pushes the leading-edge shock upstream and as a consequence, the reflected shock attaches to the passage shock below the upper profile. In this case, to balance the mass flow between passages, the suction slots should be regulated.

Moreover, a second passage shock starts to appear downstream and gets stronger by narrowing the upper profile channel due to the blockage of the channel.

15.3 Results and Discussion

The roughness study was carried out by replacing the lower profile with one of the profiles detailed in Sect. 15.2.1.1. Key operating conditions, including the profile location and inflow angle, were carefully verified to ensure consistency across all tests. For each profile, the wall suction was uniformly set at 33% of the suction valve opening. The findings from the study are summarized as follows.

15.3.1 Main Flow Structure

Figure 15.10 depicts the distribution of isentropic Mach numbers derived from pressure measurements on the suction side of the lower profile. The graph confirms a coinciding shock position on the suction side for all configurations with similar pressure distribution, despite that, P2-0.20Ra exhibits a slight difference in isentropic Mach number for x/c in range from 0.2 to 0.4 where the pressure distribution is slightly different in comparison with the other cases which may be referred to the surface quality.

Figure 15.11 presents snapshot images of schlieren visualization, averaged from over 50 randomly captured snapshots for each mounted profile. These snapshots

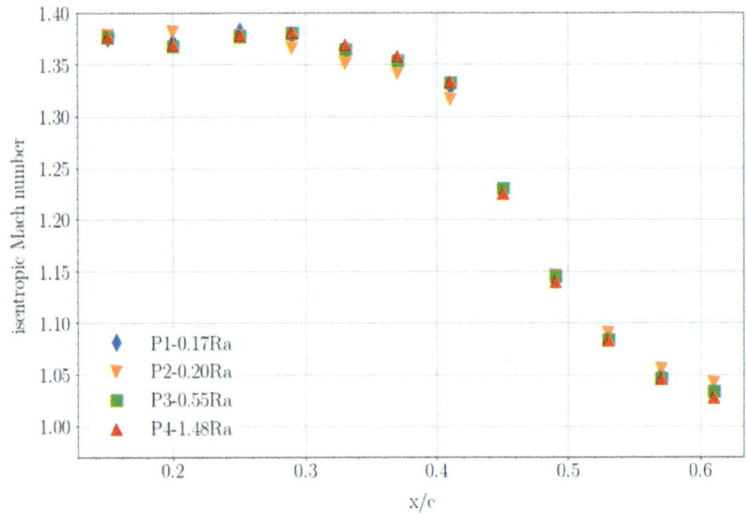

Fig. 15.10 Isentropic Mach number on the suction side of lower profile

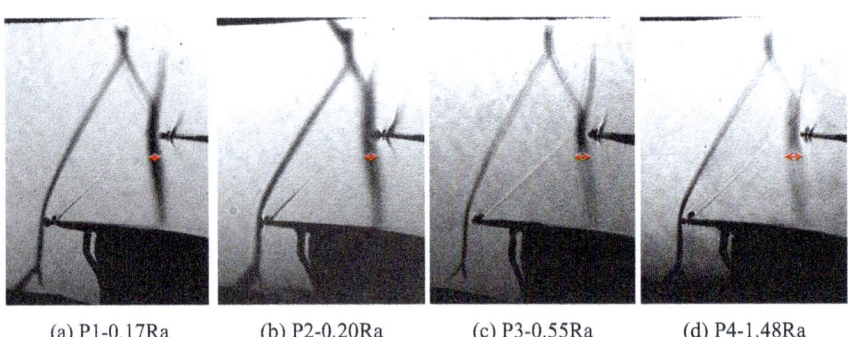

(a) P1-0.17Ra (b) P2-0.20Ra (c) P3-0.55Ra (d) P4-1.48Ra

Fig. 15.11 Schlieren images for the four profiles

exhibit a similar structure across all profiles. However, a slight variation is observed in P2-0.20Ra (shown in Fig. 15.11b), although the leading-edge shock location from the lower profile is similar to the other cases. The shock displays a greater inclination, resulting in a reflected shock attachment to the passage shock, notably higher than in other cases. A detailed analysis presented in Sect. 15.2.3.1 indicates that such behavior of the leading-edge shock can be influenced by either side wall suction or profile shape because it is directly connected to the mass flow in the passage, not the profile roughness.

Furthermore, in the cases of P3-0.55Ra and P4-1.48Ra, as shown in Fig. 15.11c and d, particularly in the latter, the passage shock exhibits a slightly greater displacement (indicated by the red arrow) compared to the lower roughness cases depicted in Fig. 15.11a and b. This could be attributed to a shift in the transition location, as the roughness may induce an earlier transition, thereby affecting shock stability.

15.3.2 Shock Induced Separation

Figure 15.12 illustrates the oil visualization on the four profiles. In cases P1-0.17Ra, P2-0.20Ra, and P4-1.48Ra, the oil was applied downstream of the separation line up to the trailing edge to preserve the roughness effect upstream of the shock. Conversely, for P3-0.55Ra, the oil visualization covers the entire profile.

The stagnation zone, marking the onset of separation, occurs earlier in P1-0.17Ra and P3-0.55Ra (illustrated in Fig. 15.12a and c) in comparison with other cases, appearing around $0.33 x/c$. This suggests lower shear stress upstream leading to a larger separation bubble.

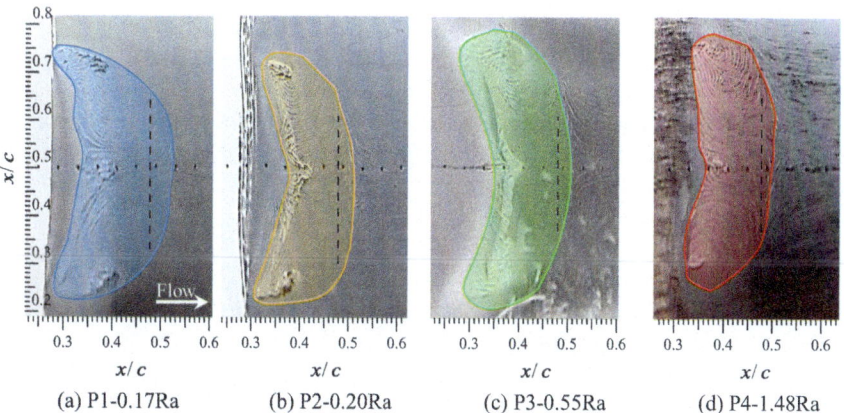

Fig. 15.12 Oil visualization focus on the separation bubble generated due to boundary layer interaction on the lower profile suction side, (- -) is the mean line of shock location

Table 15.4 Separation size

	P1-0.17Ra	P2-0.20Ra	P3-0.55Ra	P4-1.48Ra
As/Ap (%)	8.8	6.9	8.9	8.1

In the case of P2-0.20Ra (Fig. 15.12b), the separation line forms a V-shape, starting from the span periphery at approximately $0.31x/c$ and the middle at roughly $0.39x/c$. The separation bubble bit spans on one side although symmetrical pressure distribution in suction.

P3-0.55Ra displays a much more uniform shape of the separation bubble and smoother streak lines, which are more symmetric and evenly distributed. In the case of P4-1.48Ra (Fig. 15.12d), the stagnation zone location is less clearly visible but it is estimated to be between 0.33 and $0.37x/c$. Higher levels of disturbance within the separation bubble. This may be attributed to surface roughness, which disrupts the oil streak patterns.

Table 15.4 compares the estimated size of the separation bubble from oil visualization, where A_s is the separation area and A_p is the total area of the suction side. For P1-0.17Ra, the separation size confirms the expectation that the smoother profile delays the transition, suggesting laminar interaction and a larger separation bubble. In contrast, P2-0.20Ra shows a significantly smaller separation bubble. Despite the roughness sizes being close and the machining processes being similar between P1-0.17Ra and P2-0.20Ra, the underlying cause of the observed differences remains under investigation for a conclusive explanation.

The unmachined surface of P3-0.55Ra supports flow uniformity on the surface, resulting in a separation bubble size similar to that of P1-0.17Ra. The high uncertainty in defining the boundaries of the separation bubble makes it difficult to confirm whether the smaller size of the separation bubble in P4-1.48Ra, as indicated in the table, is due to the transition location. The relatively rough profile surface is expected to cause more disturbances and induce an earlier transition, leading to a more favorable interaction and a smaller separation bubble.

Indeed, disturbances are present and can be observed on the left side of the profile surface (as indicated by the node at the upper part of Fig. 15.12d) and in the middle of the profile, where the saddle point is slightly shifted from the center.

15.3.3 Boundary Layer State

The boundary layer was carried out at $0.75x/c$ up to 20 mm above the profile surface to make sure both the boundary layer and free stream profile are well captured by traverse as detailed in Sect. 15.2.1.

Based on the turbulent boundary layer power low approximation, an optimization function was used to estimate the best-fit power value to the boundary layer [24].

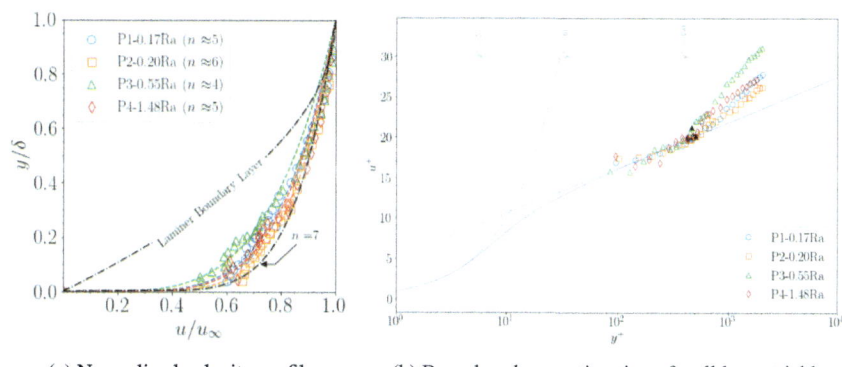

(a) Normalized velocity profiles (b) Boundary layer estimation of wall law variables

Fig. 15.13 Boundary layer parameters at $0.75x/c$

Table 15.5 Boundary layer parameters at $0.75x/c$

Profile	δ_{99} (mm)	δ_{99}^* (mm)	δ_{99}^{**} (mm)	H	H_{est}
P1-0.17Ra	4.94	0.813	0.532	1.53	1.40
P2-0.20Ra	4.58	0.663	0.439	1.51	1.34
P3-0.55Ra	5.31	0.956	0.595	1.61	1.48
P4-1.48Ra	4.39	0.672	0.430	1.56	1.38

Figure 15.13a shows the normalized values of the boundary layer thickness and the estimated values for each profile. All values confirm the development of a fully turbulent boundary layer at that location.

The integral parameters presented in Table 15.5 reveal that P3-0.55Ra exhibits the highest boundary layer thickness, significantly surpassing P1-0.17Ra, which has the closest highest boundary layer thickness by almost 0.4mm. The reason behind this disparity is currently under investigation.

It's worth noting that the shape factor values ranging from 1.51 to 1.61 may not accurately reflect the true shape, as they are based on the linearity assumption between the points and the no-slip condition at the wall. However, using an optimization function based on the power law for turbulent boundary layers, the boundary layer shape and corresponding power were estimated. The results, ranging from 1.3 to 1.4, indicate a consistent estimation, except for the overestimation in P3-0.55Ra. This overestimation suggests an ongoing transition in that case, which may explain the size of the boundary layer.

In addressing the overestimation of the boundary layer in P3-0.55Ra, the wall parameters are illustrated in Fig. 15.13b. Notably, the minimum y^+ measured within the boundary layer falls within the range of 85–100. The log layer boundary is denoted by the black marker in the figure, which closely aligns with the y^+ values with $B = 5.5$. Interestingly, P3-0.55Ra exhibits the lowest measured y^+, indicating a strong adverse pressure gradient in the outer layer. This suggests that despite the presence of the reattachment zone, the flow is still influenced by separation. The behavior of the other profiles shows a similar trend, particularly in the outer layer closer to the log line.

15.4 Conclusions

This chapter summarizes an experimental study of the surface roughness effect on SBLI conducted in the IMP PAN transonic wind tunnel. A dedicated single passage test section replicates the SBLI on the suction side of a representative fan profile, utilizing four manufactured specimens with varying surface textures: two with refined smooth surfaces, one with high roughness, and one left unmachined. Various measurement techniques and analysis methods were employed to assess the aerodynamic perspectives of laminar-turbulent transition induced by roughness and its effects on SBLI.

The first smooth profile demonstrated typical shock flow structure and unsteadiness but exhibited a large separation bubble, indicating lower shear stress, a delayed transition and laminar interaction. This was corroborated by boundary layer measurements showing a significant increase in thickness downstream of the shock. The second smooth profile presented slight deviations in pressure distribution and higher upstream shock wave inclination.

The unmachined profile displayed the typical shock structure, with a separation bubble similar to the first smooth profile, indicating transition delay. A significant increase in the boundary layer thickness at the reattachment zone indicates an ongoing transitional effect. The rough profile exhibited a smaller separation bubble. Disturbances and a thickened boundary layer were observed upstream of the shock, but there was a smaller jump in the boundary layer thickness downstream.

Overall, the findings highlight the complex influence of surface roughness on SBLI, demonstrating variations in flow behavior, separation bubbles, and boundary layer characteristics across different surface textures. Despite the minor aerodynamic effects due to the low roughness length scales relative to the boundary layer scales, these results raise additional questions, highlighting the need for further investigation and increased research efforts.

Acknowledgements We thank the Institute of Manufacturing and Materials Technology at Gdańsk University of Technology for their support in evaluating the profiles' shape and roughness.

References

1. Becker, B., Reyer, M., Swoboda, M.: Steady and unsteady numerical investigation of transitional shock-boundary-layer-interactions on a fan blade. Aerosp. Sci. Technol. **11**(7–8), 507–517 (2007). https://doi.org/10.1016/j.ast.2007.05.002
2. Grothe, P., Flaszynski, P., Szwaba, R., Piotrowicz, M., Kaczynski, P., Tartinville, B., Hirsch, C., & Hergt, A.: WP-3 internal flows-compressors. In: Doerffer, P., Flaszynski, P., Dussauge, J.-P., Babinsky, H., Grothe, P., Petersen, A., Billard, F. (eds.) Transition Location Effect on Shock Wave Boundary Layer Interaction: Experimental and Numerical Findings from the TFAST Project, pp. 229–296. Springer International Publishing (2021). https://doi.org/10.1007/978-3-030-47461-4_4
3. Flaszynski, P., Doerffer, P., Szwaba, R., Piotrowicz, M., Kaczynski, P.: Laminar-turbulent transition tripped by step on transonic compressor profile. J. Therm. Sci. **27**(1), 1–7 (2018). https://doi.org/10.1007/s11630-018-0977-4
4. Giepman, R., Louman, R., Schrijer, F., van Oudheusden, B.: Experimental investigation of boundary layer tripping devices for shock wave—boundary layer control. In: 45th AIAA Fluid Dynamics Conference (2015). https://doi.org/10.2514/6.2015-2780
5. Klinner, J., Hergt, A., Grund, S., Willert, C.E.: Investigation of shock-induced flow separation over a transonic compressor blade by conditionally averaged PIV and high-speed shadowgraphs. In: 19th International Symposium on Applications of Laser Techniques to Fluid Mechanics (2018). https://api.semanticscholar.org/CorpusID:127457320
6. Schneider, S.P.: Effects of roughness on hypersonic boundary-layer transition. J. Spacecraft Rockets **45**(2) (2008). https://doi.org/10.2514/1.29713
7. Corke, T.C., Bar-Sever, A., Morkovin, M.V.: Experiments on transition enhancement by distributed roughness. Phys. Fluids **29**(10) (1986). https://doi.org/10.1063/1.865838
8. Bons, J.P.: A review of surface roughness effects in gas turbines. J. Turbomach. **132**(2), 21004 (2010). https://doi.org/10.1115/1.3066315
9. Boese, M., Fottner, L.:. Effects of riblets on the loss behavior of a highly loaded compressor cascade. In: American Society of Mechanical Engineers, International Gas Turbine Institute, Turbo Expo (Publication) IGTI, 5 B (2002). https://doi.org/10.1115/GT2002-30438
10. Leipold, R., Boese, M., Fottner, L.: The influence of technical surface roughness caused by precision forging on the flow around a highly loaded compressor cascade. J. Turbomach. **122**(3) (2000). https://doi.org/10.1115/1.1302286
11. Kadivar, M., Tormey, D., McGranaghan, G.: A review on turbulent flow over rough surfaces: fundamentals and theories. Int. J. Thermofluids **10** (2021). https://doi.org/10.1016/j.ijft.2021.100077
12. Jacobs, R.G., Durbin, P.A.: Simulations of bypass transition. J. Fluid Mech. **428** (2001). https://doi.org/10.1017/S0022112000002469
13. Anika, N.N., Djenidi, L., Tardu, S.: Bypass transition mechanism in a rough wall channel flow. Phys. Rev. Fluids **3**(8) (2018). https://doi.org/10.1103/PhysRevFluids.3.084604
14. Vadlamani, N.R., Tucker, P.G., Durbin, P.: Distributed roughness effects on transitional and turbulent boundary layers. Flow Turbul. Combust. **100**(3) (2018). https://doi.org/10.1007/s10494-017-9864-4
15. Reshotko, E.: Transient growth: a factor in bypass transition. Phys. Fluids **13**(5) (2001). https://doi.org/10.1063/1.1358308
16. Rizzetta, D.P., Visbal, M.R.: Direct numerical simulations of flow past an array of distributed roughness elements. AIAA J. **45**(8) (2007). https://doi.org/10.2514/1.25916
17. Piotrowicz, M., Flaszyński, P., Doerffer, P.: Investigations of shock wave boundary layer interaction on suction side of compressor profile. J. Phys.: Conf. Ser. **530**(1) (2014). https://doi.org/10.1088/1742-6596/530/1/012068
18. Hanfy, A., Flaszynski, P., Doerffer, P., Kaczynski, P.: CoGenerator Liberary (1.3.0). Zenodo (2024). https://doi.org/10.5281/zenodo.12538002

19. Piotrowicz, M., Flaszynski, P.: Numerical investigations of shock wave interaction with laminar boundary layer on compressor profile. J. Phys: Conf. Ser. **760**(1), 012023 (2016). https://doi.org/10.1088/1742-6596/760/1/012023
20. Dantec Dynamics: BSA Flow Software User's Guide (Version 5.11.00.11) (2013)
21. Hanfy, A., Flaszynski, P., Kaczynski, P., Doerffer, P.: Shock Oscillation Analysis Library (v2.0.0) (2024). https://doi.org/10.5281/zenodo.11197727
22. Hanfy, A., Flaszynski, P., Doerffer, P., Kaczynski, P.: Curved Surface Reconstraction Liberary (0.0.5). Zenodo (2024). https://doi.org/10.5281/zenodo.12207363
23. Courant, R., Robbins, H., Stewart, I.: What is mathematics?: an elementary approach to ideas and methods. Am. Math. Month. **4**(1), 172—179. Oxford University Press (1996)
24. Virtanen, P., Gommers, R., Oliphant, T.E., Haberland, M., Reddy, T., Cournapeau, D., SciPy 1.0 Contributors: SciPy 1.0: Fundamental algorithms for scientific computing in Python. Nat. Methods **17**, 261–272 (2020). https://doi.org/10.1038/s41592-019-0686-2

Open Access This chapter is licensed under the terms of the Creative Commons Attribution 4.0 International License (http://creativecommons.org/licenses/by/4.0/), which permits use, sharing, adaptation, distribution and reproduction in any medium or format, as long as you give appropriate credit to the original author(s) and the source, provide a link to the Creative Commons license and indicate if changes were made.

The images or other third party material in this chapter are included in the chapter's Creative Commons license, unless indicated otherwise in a credit line to the material. If material is not included in the chapter's Creative Commons license and your intended use is not permitted by statutory regulation or exceeds the permitted use, you will need to obtain permission directly from the copyright holder.

Shock Oscillation Mechanisms of Highly Separated Transitional Shock-Wave/Boundary-Layer Interactions

Philipp Nel, Anne-Marie Schreyer, and Marius Swoboda

Abstract At cruise altitude, low Reynolds numbers result in a laminar boundary layer on the suction side of a transonic fan blade, extending to the shockwave/boundary-layer interaction. For transitional SBLIs with significant shock-induced separation, a shock oscillation mechanism occurs, characterized by the growth and natural suppression of the upstream laminar section of the separation bubble. The authors utilize a combination of numerical and experimental techniques across various cases, including a canonical case, cascades, and a 3D fan, to investigate the phenomenon. To validate the dynamic mechanism observed in large eddy simulations, experiments using high-speed Schlieren, spark light sh dowgraphy and PIV were conducted. The characteristic length scale for the oscillation mechanism, based on the travel distance of the laminar separation shock, is a key finding. The mechanism existence strongly depends on free stream turbulence and the boundary layer state. Oscillation frequencies are much lower for the turbulent oncoming boundary layer compared to the laminar case, which shows a strong link between the large scale movement of the laminar separation shock, the separation bubble, and reflected shock movement. In contrast, the turbulent interaction shows significantly less reflected shock travel distance. Preliminary full span LES simulations corroborate the link of findings to the application.

Keywords Transonic flow · Shock induced separation · Large eddy simulations · Compressor cascade

---

P. Nel (✉) · M. Swoboda
Rolls-Royce Deutschland Ltd. & Co. KG., Blankenfelde-Mahlow, Germany
e-mail: Nel@IST.RWTH-Aachen.de

M. Swoboda
e-mail: Marius.Swoboda@Rolls-Royce.com

A.-M. Schreyer
Mechanical, Automotive and Aeronautical Engineering, Munich University of Applied Sciences, Munich, Germany
e-mail: anne-marie.schreyer@hm.edu

© The Author(s) 2025
P. Flaszynski et al. (eds.), *Towards Effective Flow Control and Mitigation of Shock Effects in Aeronautical Applications*, Notes on Numerical Fluid Mechanics and Multidisciplinary Design 201, https://doi.org/10.1007/978-3-031-86605-0_16

1 Introduction

The focus of this chapter is to investigate the shock oscillation mechanism caused by a shock-wave/boundary-layer interaction (SBLI), which is suppressed by tripping the boundary layer, along with its numerical and experimental validation. This mechanism involves a highly separated transitional SBLI, with transition occurring on the shear layer of the upstream section of the separation bubble. The motivation for this work arises from problematic SBLIs in transonic fans at cruise altitude. In smaller transonic fans (e.g., corporate jets), which fly higher than commercial aircraft, a laminar boundary layer persists on the suction side of the fan, leading to an SBLI with a laminar oncoming boundary layer. Trends in compressor design aim for higher loading, resulting in stronger shocks and increased shock-induced separation [11]. These low-Reynolds conditions can cause shock oscillations and excite structural blade modes, leading to fatigue. Engine tests show that promoting transition on the suction side upstream of the shock mitigates structural excitation at high blade loading. Transition strips are used to reduce vibration caused by shock oscillations, though the exact mechanism of oscillation remains unclear, as does the reason for the strips' effectiveness. Across the working range of a transonic fan, shock structures interact with the boundary layers of neihbouring blades. The interaction is sensitive to Mach number and boundary layer state, inducing pressure fluctuations and vibrational stresses on the blades, reducing component lifespan [7]. Different shock structures result in varying behaviour due to differences in separation severity and pressure distribution. Common shock structures are shown in Fig. 1. The relationship between shock structure and unsteadiness is unclear and will be investigated. Shock oscillations can fluctuate within 10% of the blade chord length and have been linked to changes in inflow angle and operating point in cascade experiments [11]. These oscillations are common at sea level, whereas transonic fans at cruise altitude experience quieter inflow conditions. Flow control devices like vortex generators and roughness elements have been explored to mitigate oscillations by promoting early turbulence transition, but their effect on reducing oscillation amplitude and dominant frequencies is limited [4, 13].

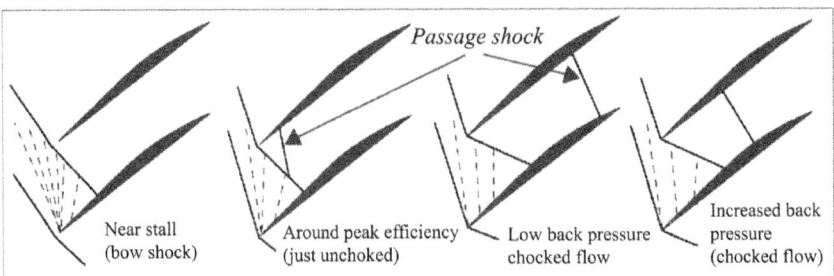

Fig. 1 Shock structures across the working range of transonic compressors (adapted from Denton and Xu [3])

Although LES simulations and experiments have also extensively been performed on transonic cascades, the focus is on a canonical research configuration derived to validate the mechanism observed in LES. The sea-level cascades used proved ineffective of replicating the shock oscillation mechanism at altitude, and a canonical experiment was required. This configuration featured a high Mach number, low turbulence levels, a stable incident shock, and a strong enough oblique SBLI to create separation with transition on the shear layer of the upstream laminar separation bubble. This work also presents the physical realization of the canonical case experiment, with LES used for validation. The experiments and simulations on the canonical configuration demonstrate the silencing of this shock oscillation mechanism across various cases, and that LES can resolve the mechanism, forming a basis for industrial applications aimed at optimizing flow control strategies. The chapter concludes with a preliminary application to a highly loaded 3D fan demonstration case.

2 Preliminary Investigations and Definition of Research Configuration

In this section, we start by highlighting the need for the use of LES as opposed to Reynolds-averaged Navier-Stokes (RANS) or unsteady RANS (URANS). After this, the canonical configuration is extracted using LES, starting from a multi-passage transonic cascade. The study shows that the same shock oscillation mechanism as in the cascade simulations can be observed in the final extracted canonical research configuration, which is a highly separated (strong) oblique SBLI on a flat plate, with transition occurring on the shear layer of the upstream part of the shock-induced separation.

An evaluation of Reynolds averaged methods U/RANS for capturing shock oscillations in highly loaded transonic fans was conducted through a parameter study [15] considering a range of settings affecting the numerics and physics at the highly loaded conditions where shock oscillations are expected. The impact of multi-grid cycles in the simulations was significant. Running RANS without multi-grid cycles resulted in a steady solution across all conditions. The transition model, coupled with multi-grid cycles, also had a significant impact, creating chaotic oscillations. The URANS simulations damped out these non-physical oscillations within the inner iteration of the implicit time-stepping scheme, which could be seen in the fact that the solution was only different over the inner iterations but that the final solution after each time step was the same when having multigrid cycles in URANS. The effect of multi-grid settings on the RANS solution's sensitivity was also evident in the significant changes in oscillation pattern and amplitude. Consequently, Large Eddy Simulations were employed to resolve the shock oscillation mechanism.

2.1 Cascade LES Investigations

To find the appropriate boundary conditions for LES, a full-span fan setup was simulated in RANS at the condition where the altitude strain gauge vibration tests showed the largest excitation.

The resulting flow structure shows a pre-shock Mach number around 1.6 and a passage-shock flow structure with large separation (Fig. 2). A quasi-2D LES domain with inviscid endwalls is then created to match the observed flow features of the full-span fan under these highly loaded conditions. This domain follows an inviscid end wall contraction [15] in order to be consistent with calculations in previous work [13] and the existing experimental setups at IMP PAN and DLR [4]. In the preliminary LES calculations, Δx^+ and Δz^+ values are around 25 for the case at Re =350,000, and 50 for the case at Re = 1.4 million, with Δy^+ ranging between 1 at the wall and the aforementioned values away from the interaction. Due to the large separation, a length of 20% of the chord is selected as the span-wise dimension, which showed independence of the span-wise extent [14]. In order to investigate the oscillation mechanism present at altitude conditions, the LES simulations are based on a clean inflow.

The LES simulations did show a significant shock oscillation. However, the question remained whether the shock oscillation mechanism resolved by LES was physical in nature.

Therefore, the final validation of LES should focus specifically on the mechanism of shock oscillation. However, since no experimental data on this mechanism was available, the first step in validation was to evaluate the isentropic Mach number distribution predicted by LES and different subgrid-scale (SGS) models. Using the Lufthansa Technik Cascade case, LES were conducted at Mach 1.12, and the results of different SGS models were compared against both experimental data and RANS results [13]. The results revealed that the upstream influence of the SBLI played an important role in the observed shock oscillation mechanism. The isentropic Mach number distribution showed that the Smagorinsky subgrid-scale model overpredicted the upstream influence and resulted in excessive smearing at the shock, which did not align with the experimental data. It was shown in Nel et al. [14] that the Smagorinsky model caused a larger shock oscillation than the implicit LES because of this, with

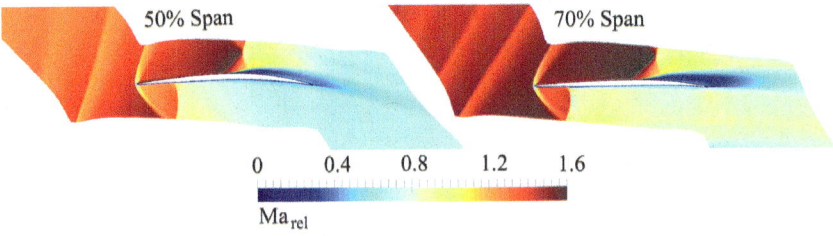

Fig. 2 Flow in transonic fan: full span RANS simulation, cuts at 50 and 70% span showing contours of Mach number

the amplitude increasing as the Smagorinsky constant was increased. Conversely, the upstream influence was adequately captured in the implicit LES case as well as with the WALE and SIGMA subgrid-scale models, where a reduction in isentropic Mach number distribution between $x/c = 0.2$ and 0.36 due to the upstream influence, similar to the experimental results, was observed. When comparing RANS results with LES and the experimental data, RANS failed to predict this dip in the upstream isentropic Mach number distribution.

The reason for this discrepancy became evident when analyzing the density gradient contours, which revealed the upstream separation shock, directly linked to the upstream extent of the separation bubble. The upstream extent predicted by RANS was significantly smaller than that predicted by LES, confirming the limitations of RANS in this scenario [14].

Testing with laminar and turbulent oncoming boundary layers is performed to understand whether the nature of the observed shock oscillation is of the expected type. This means that the shock oscillation mechanism should be one which is present for a laminar boundary layer, but suppressed for a turbulent boundary layer. For the implicit LES case with a laminar suction side boundary layer, which produces similar results as the SIGMA and WALE SGS-models, the shock oscillation takes the baseline form of Fig. 3.

2.1.1 Turbulent Boundary Layer Effect

Turbulence was promoted by equispaced step elements on the blade surface. The elements applied in the Re = 350,000 case were 1.2, 2, and 3 times the boundary layer height of the laminar case at the same location. The resulting shock structure and its effect on the separation were observed through averaged volumetric numerical data. Each consecutive increase in element height achieved a greater reduction in the separation bubble size.

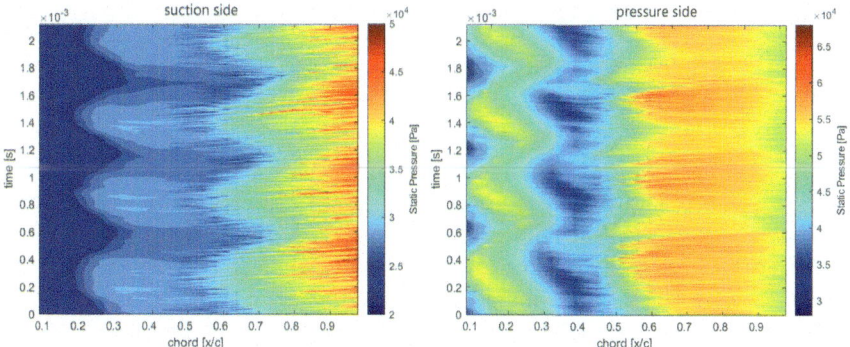

Fig. 3 TFAST cascade at Re 350,000: pressure distribution for implicit LES and laminar suction side boundary layer [13]

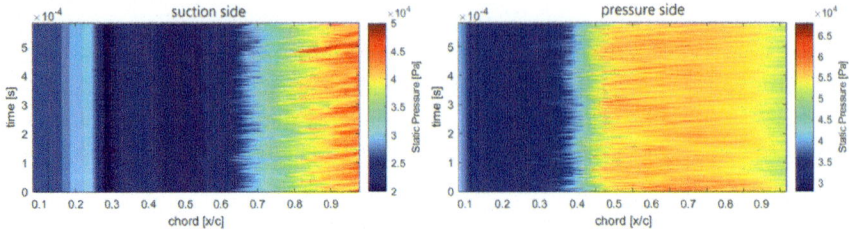

Fig. 4 TFAST cascade at Re 350 k: pressure distribution for case D [13]

In the case where the boundary layer was not fully turbulent by the time it reached the SBLI, the shock oscillation became more irregular, with the dominant frequency increasing from a distinct St = 0.1 (based on chord length) to a less periodic frequency around St = 0.13, with fluctuations inside the upstream section of the separation at around St = 0.7. The shock travel significantly reduced.

In cases where the element height was larger (2 and 3 times the boundary layer height), the shock stopped oscillating (Fig. 4). This also held for a case with a more upstream shock position to test the sensitivity [13].

The Strouhal number of the shock oscillation remains constant when comparing the original Reynolds number case (Re 1.4M) with a reduced Reynolds number case (Re 350,000), based on chord length. The reason becomes clearer when analyzing the upstream propagation velocity of the separation shock. In both cases, the periodic upstream propagation of the separation shock is approximately 50 m/s. This velocity is likely related to the acoustic wave propagation speed in the shear layer.

2.2 Transonic Cascade Experimental Campaign

An experimental campaign on a transonic cascade was conducted to experimentally find the observed shock oscillation mechanism which is of a type which gets suppressed for a turbulent boundary layer. The shock oscillation behaviour, particularly in terms of difference in amplitude between tripped and laminar cases, were compared with LES simulations. For details on the experimental campaign results and LES cascade simulations, the reader is referred to Nel et al. [13].

The experiment was conducted at six operating points: a bow shock, four passage shock conditions with significant separation, and a swallowed passage shock with even larger, chaotic separation. These operating points are investigated for laminar and tripped boundary layer scenarios. High-speed Schlieren imaging (6 kHz) was employed to capture the behaviour of the shock structure.

Given that separation in the bow shock case is minimal, the behaviour of the laminar separation bubble should have much less influence on bow shock oscillation: LES and DNS simulations of single-passage cascade flows consistently show a steady bow shock [6, 9, 10, 12, 14, 18, 19, 24]. It was demonstrated by an

LES simulation [14] on the Lufthansa Technik Cascade, which exhibits significant shock oscillations in the order of 10% in the experiment, that the LES simulation contrasted, showing insignificant shock travel distance of approximately 1 mm for a chord length of 70 mm. In most experimental tests of transonic cascades, significant shock unsteadiness is observed at any condition (including bow shock), likely due to the non-ideal conditions (inlet turbulence, secondary flows, suction slots, noise). Therefore it should be noted that experimental tests from cascades in transonic wind tunnels cannot be compared directly with idealised LES simulations with quiet inflow which tries to mimic altitude conditions.

The real altitude case has a negligible inflow turbulence compared to wind tunnel tests, and the altitude case is therefore simplified as a laminar inflow in the LES. The cascade LES is a separate investigation and does not seek to be validated directly through the experiment (rather, it is investigated whether the experiment shows the same oscillation mechanism or not).

In the experiment, 6 different shock locations and different step locations were tested. The step locations tested are 8, 16, and 26% x/c, with a step height of 0.1 mm and a width of 3 mm. In the bow shock case, the oscillation is not affected by tripping of the boundary layer and oscillates significantly, contrasting with the quasi-2D LES, where the shock is steady at this condition. In the passage shock case, the shock travel distance increases, due to the larger separation and sensitivity to passage area change. In the swallowed passage shock case, shock oscillations in the low frequency range of 200–500 Hz (St 0.043–0.11) were attenuated by around 50% when tripping the boundary layer, due to the turbulence stabilizing the previously chaotic and massive separation at this condition.

No condition in the experiment had a relatively steady shock, which contrasts with what is expected for a quiet inflow. This shows that additional sources of shock oscillation are present in the experimental cascade. Obvious possible sources include corner and endwall separation as well as inlet velocity fluctuations/turbulence, multi-passage interactions (the Mach number in the lower passages may differ from the upper which may cause interactions between the passages), all of which are avoided in the single passage laminar inflow quasi-2D inviscid endwall LES simulations. It is therefore concluded that a simplified canonical case is required to experimentally investigate the mechanisms of shock oscillation which are inherent to an SBLI with transition on the upstream laminar section of the separation bubble, in order to achieve validation of the mechanism observed in LES, and subsequently return to more complex cases such as compressor cascades or full-span 3D fans (Sect. 3.8).

2.3 Definition of Canonical Research Configuration

The current section focuses on defining a research configuration which can show the shock oscillation mechanism of interest in a practical experimental setup, in order to validate the dynamic behaviour of the shock oscillation mechanism observed in LES simulations.

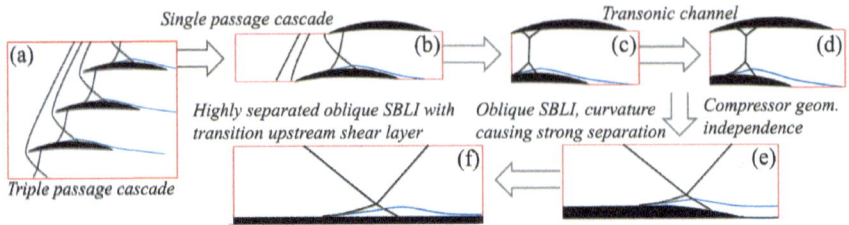

Fig. 5 Simplification of low Reynolds transonic fan shock oscillation problem to most fundamental form [15]

The outcome of the simplification and the intermediate steps are summarised by means of the investigated cases in Fig. 5. The simplification of the problem of shock oscillations on a transonic fan at low Reynolds numbers to a fundamental form that can be studied experimentally, leads to a highly separated (strong) oblique SBLI with transition on the upstream shear layer (configuration f in Fig. 5).

In the Sect. 3, the experiment studying the derived configuration (f) is realised at Delft University of Technology, using high speed Schlieren imaging, Spark Light Shadowgraphy, and Particle Image Velocimetry to study the physics and validate the dynamic behaviour of the mechanism.

In the context of cascade simulations, the domain and target flow structure were defined previously. The first step in simplifying the shock oscillation problem was to reduce the configuration from a multi-passage, where a triple passage was chosen arbitrarily, to a single passage cascade. After 10 convective time steps, results were recorded over two oscillation cycles. The dynamic behavior across the three passages due to shock oscillation was observed. Periodic turbulence was present in the upstream separated shear layer, driven by Kelvin-Helmholtz instabilities that propagated upstream during the oscillation cycle. This was an important recurring feature during the simplification process, as the shock oscillation relied on the presence of such periodic upstream instabilities, which are present in all comparable cases. The behavior and oscillation frequencies observed in the single passage cascade closely matched those of the triple passage case [15]. Consequently, the study of this particular shock oscillation mechanism was simplified to a single passage cascade [14].

Using Dynamic Mode Decomposition to quantify the dominant behaviour of the oscillation in the cascade, it was concluded that the shock oscillation mechanism of interest was present in both the original (Re = 1.4M) and reduced (Re = 350,000) cases, with the Strouhal number of the oscillation being approximately 0.1 based on chord length. The separation shock from the laminar upstream edge of the separation bubble periodically forms, and travels upstream, but carries instabilities with it on the laminar separated shear layer. These instabilities eventually lead to significant turbulence in the upstream part of the separation bubble. At a certain critical upstream position, the turbulence suppresses the upstream section of the separation bubble, causing a partial collapse of the separation bubble and therefore blockage changes and shock movement. The separation shock exists only periodically because the separation bubble collapses rapidly in a smooth (concave) manner. The velocity of the periodically forming upstream-traveling separation shock is approximately 50 m/s for both Reynolds numbers.

When the boundary layer was disturbed by a 2D step in the Re = 1.4M case, a dominant mode existed at the same Strouhal number, and unsteadiness could still be observed in the separation shock foot. However, the resulting shock travel distance was insignificant. This can be explained due to early formation of instabilities on the upstream laminar separated shear layer, which prohibits its growth. In the Re = 350,000 case, where a fully turbulent suction side boundary layer was investigated, the significance of the mode at St = 0.09 was reduced. Although upstream-traveling separation shock waves were still faintly visible, an additional mode at St = 0.023 appeared, likely linked to a low-frequency mechanism inherent to turbulent SBLIs. The shock travel distance was again insignificant.

The shock oscillation problem was further investigated by simplifying it to a more canonical configuration: this was achieved by removing the leading edge and hence its associated shock wave, focusing instead on a single passage shock in a transonic channel. The leading edge shock wave was previously found to be steady, suggesting that neither this shock nor the leading edge geometry was necessary for the shock oscillation mechanism. Consequently, the passage shock became the focus of subsequent investigations. The transonic channel test case revealed the same oscillation mechanism as observed in the cascade, driven by the growth and collapse of a laminar separation bubble due to upstream-propagating shear layer instabilities eventually suppressing the separation. The laminar separation shock periodically formed and moved upstream at a velocity comparable to that in the compressor cascade case. The oscillation mechanism was once again suppressed when the boundary layer was tripped. These findings indicate that the shock oscillation mechanism is not inherent to the compressor-like suction-side blade curvature but rather dependent on the conditions at the interaction, particularly the transition location within a large shock-induced separation. This suggests that further simplification of the problem is necessary to develop a canonical research configuration.

In order to consider a next step of simplification, we regard the findings so far: with regard to compressor cascades, the oscillation mechanism does not occur in cases of small separation, such as in the bow shock condition. Furthermore, the oscillation mechanism is not related to the incident shock movement and is present even though the shock wave from the leading edge is steady. In contrast, an experimental study by the same author [13] found that the leading edge shock wave was unsteady in the experiment, compromising the study due to unexplained additional unsteadiness likely arising from excessive turbulence. For a canonical experiment to be comparable, it would require a reasonably steady impinging shock wave. A steady attached oblique shock wave at a high Mach number is easier to generate and preferred for the cleanest possible experiment. Therefore, an attached oblique shock wave was selected.

For the next case in the simplification, an oblique shock-boundary layer interaction impinging on a curved surface at Mach 2.3 with a deflection angle of 24° from a shock generator was investigated. The curvature induced a larger separation bubble (Fig. 6a) compared to the flat plate case (Fig. 6b).

The larger separation induced by the curvature does not play an integral role in the oscillation mechanism. As shown in Fig. 7, a similar shock oscillation mechanism to the one previously observed exists in both the curved surface and flat plate cases. The reason why the additional separation region caused by the curvature is unnecessary becomes clearer in instantaneous snapshots [15]. In the curved wall case, much of the supplementary part of the separation bubble is not washed away during the downstream/suppression phase of the oscillation. In contrast, the separation bubble is largely suppressed in the flat plate case. Thus, the curvature can be disregarded, allowing for the simplification of the case to a flat plate configuration.

Next, we investigate additional factors that may influence the shock oscillation in the experiment. The thickness of the laminar boundary layer affects both the

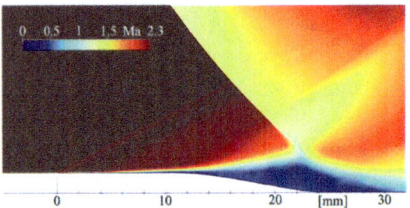

(a) Oblique SBLI impinging on curvature: Ma = 2.3, θ = 24°, time averaged Mach number contours [21].

(b) **Oblique SBLI on flat plate: Ma = 2.3, θ = 24°**, time averaged Mach number contours [21].

Fig. 6 **a** Oblique SBLI impinging on curvature: Ma = 2.3, θ = 24°, time averaged Mach number contours [15] **b** Oblique SBLI on flat plate: Ma = 2.3, θ = 24°, time averaged Mach number contours [15]

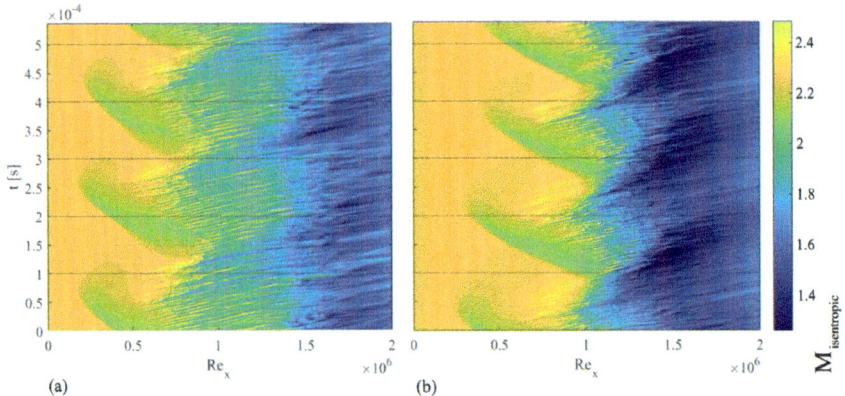

Fig. 7 Isentropic Mach number distribution with time on the wall for the curved surface case (left) and flat plate case (right) at Ma = 2.3, θ = 24° [15]

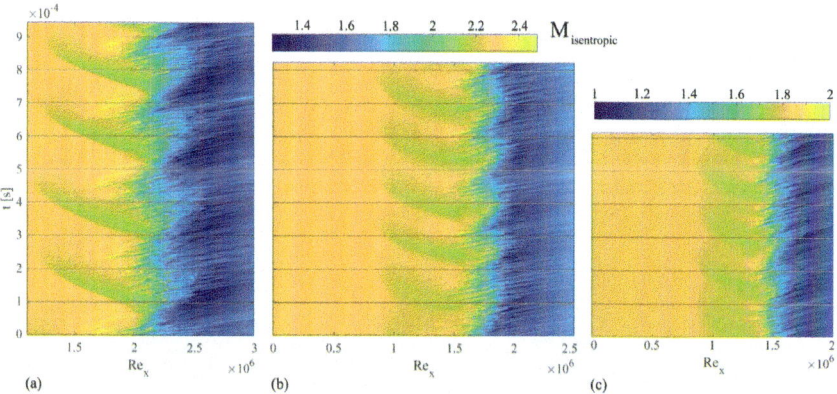

Fig. 8 Isentropic Mach number distribution with time on the wall for **a**: thicker laminar boundary layer (note increased x Reynolds number), Ma = 2.3, θ = 24°, **b**: reduced shock strength case I (Ma = 2.3, θ = 13°) and **c**: reduced shock strength case II (Ma = 1.7, θ = 11°) [15]

oscillation amplitude and frequency. A thicker laminar boundary layer allows for greater upstream influence of the SBLI, leading to an increase in oscillation amplitude and a corresponding decrease in frequency due to the longer growth phase (Fig. 8a). If the shock strength is reduced, the oscillation becomes less pronounced (Fig. 8b), and if the shock is too weak, it becomes more challenging to identify the oscillation mechanism of interest (Fig. 8c).

In the context of an experiment, which will almost certainly involve additional noise (an issue identified in previous experiments by Nel et al. [13]), oscillations from weaker interactions will be difficult to detect, thus reducing the clarity of the results and compromising the comparison with LES. Therefore, a stronger interaction should be the preferred for the purpose of validating the shock oscillation mechanism observed in LES simulations. Furthermore, the distance between the shock generator and the flat plate must be sufficient to ensure a straight impinging shock wave, minimizing the influence of the expansion fan from the shock generator.

2.3.1 Oscillation Mechanism Mesh Sensitivity

To investigate the sensitivity of mesh size on the occurrence of the oscillation mechanism in the newly defined canonical case, a mesh sensitivity study was conducted for the canonical flat plate configuration. Due to the simplicity of the domain, two typical LES grids were compared to a DNS-resolution grid. The grid sizes tested are listed in Table 1. It is important to note that the cell count between the low-resolution LES and DNS grids differs by an order of magnitude. Previous tests on the transonic channel case have shown that the shock oscillation mechanism cannot be resolved when non-dimensional grid size values are excessively large (e.g., larger than 70).

Table 1 Grid resolutions

Mesh	Δx^+	Wall Δy^+	Δz^+	Far Δy^+	Cells (M)
LES (low)	28	1	28	34	16
LES (high)	14	1	14	34	40
DNS	7	1	7	7	220

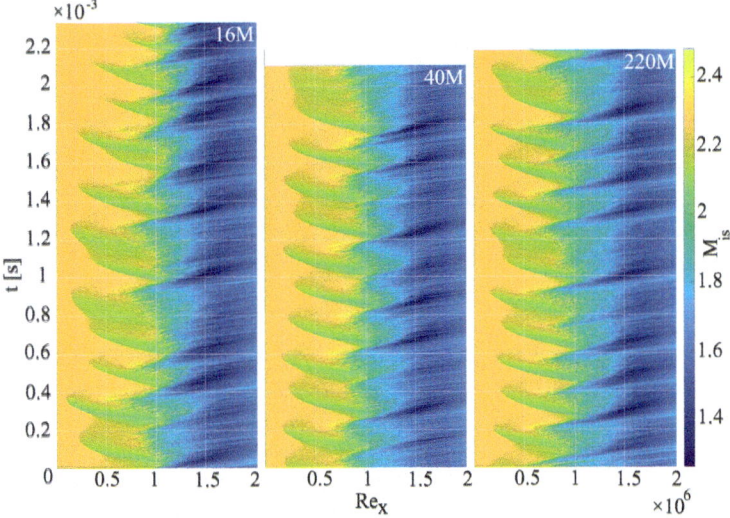

Fig. 9 Grid size comparison (SIGMA SGS): isentropic Mach number distribution with time

Next, the time footprint of the solution along the mid-plane section is examined (Fig. 9). The DNS resolution solution and the 40M mesh show good agreement. The upstream traveling separation shock is clearly visible and almost entirely disappears as the separation bubble collapses. Additionally, the instabilities transported upstream can be observed, retreating during the separation bubble's collapse at approximately twice the speed at which they were transported upstream during the growth phase. Indeed, the collapse phase of the separation bubble is faster than its growth phase. It should also be noted that, although the oscillation is not perfectly periodic, a distinct pattern is identifiable.

No-model LES on the coarse grid exhibited a tendency for the separation bubble to collapse prematurely compared to higher resolution grids. This behavior was also observed in the WALE and SIGMA models on the low-resolution grid.

No early laminar separation bubble collapse was seen for either the DNS-resolution solution, nor for the SIGMA (Fig. 9) or WALE SGS models on the higher resolution grid. The choice between the WALE and SIGMA models appears to have no significant effect on the shock oscillation, at least within the runtime of the current study. However, this investigation suggests that a higher resolution grid is necessary, as it produces results nearly indistinguishable from the DNS solution-at least over the number of oscillation cycles considered-while there is a noticeable difference between the low-resolution grid solutions and those of the DNS or 40M-grid resolution.

The numerical studies conducted thus far have shown that the shock oscillation is primarily a function of the conditions at the shock-wave/boundary-layer interaction, rather than being dependent on compressor-like geometry. The problem of shock oscillation at low Reynolds numbers, arising from a strong transitional SBLI on a transonic fan at altitude, can be simplified to a highly separated transitional oblique shock-wave/boundary-layer interaction with transition occurring on the shear layer of the upstream laminar separation. Moreover, it became evident that the available experimental cascades were too noisy to effectively study the oscillation mechanism observed in the idealized quiet quasi-2D LES simulations, which were conducted to simulate the quiet altitude conditions [13]. Therefore, the next step is to design the physical canonical experiment capable of demonstrating and validating this shock oscillation mechanism.

In this context, we outline the key features that define the shock oscillation mechanism of interest.

The shock oscillation involves a periodically collapsing separation bubble with Kelvin-Helmholtz instabilities on the shear layer upstream of the shock. This separation bubble has thin upstream laminar section which grows in the upstream direction, carrying the instabilities with it upstream of the shock wave impingement. Once a sufficiently upstream position of the laminar separation is reached as it extends upstream, the turbulence generated by these instabilities becomes significant enough to wash away (or suppress) this upstream laminar section and a substantial portion of the bulk separation bubble. At this stage, turbulence is no longer generated upstream of the shock, allowing the separation bubble to grow once again. This upstream growth is clearly marked by the formation of a separation shock that travels at a velocity comparable to the acoustic wave propagation speed in the shear layer. If the boundary layer exhibits significant upstream disturbances, such as those caused by a tripping device, the oscillation amplitude is significantly reduced compared to the laminar case.

The experiment should thus first demonstrate the low-frequency unsteadiness of the reflected shock, which is linked to the unsteadiness of the separation bubble and the separation shock at the same frequency. Secondly, it should show that the transition location is upstream of the shock wave, on the shear layer of the laminar separation bubble, and that this transition location also shifts around, as seen in the numerical simulations. If these conditions can be shown, with a clear qualitative match in behavior, the numerical case simplification approach will be validated, confirming that the resolved mechanism is physical. In the following section, the defined canonical case is realized and studied experimentally.

3 Fundamental Investigation of Shock Oscillation Mechanism

The TST-27 blowdown wind tunnel at the Technical University of Delft (TU Delft) is employed for our experiments on the canonical research configuration. Details on the wind tunnel facility and experimental setup can be found in Nel et al. [16].

A 19° shock generator was selected. The shock impingement point should be located near the leading edge, without the upstream portion of the unsteady laminar separation spilling over it. The final conditions chosen were: $\theta_{SG} = 19°$ (shock generator deflection angle), Ma = 2.3, and P_t = 2.8 bar. The total temperature was assumed to be 293 K. The resulting shock wave impinges 60 mm from the leading edge of a flat plate, with the laminar part of the separation extending as far as 10 mm from the leading edge. The free stream unit Reynolds number was 5.48×10^4 [mm^{-1}]. The free stream velocity is at approximately 550 m/s. In the absence of the incident shock, the laminar boundary layer on the flat plate would have a theoretical thickness of $\delta = 0.162$ mm, a displacement thickness $\delta^* = 0.057$ mm, and a momentum thickness $\theta = 0.022$ mm at 60 mm from the leading edge. The test section model is illustrated in Fig. 10. Both laminar and tripped boundary layer cases were studied.

High-speed Schlieren imaging was used to capture the unsteadiness of the separation bubble (and hence the reflected shock), as well as the periodically visible separation shock and its velocity. Moreover, it allows the comparison of the frequency of dominant oscillation modes in the laminar versus tripped (turbulent boundary layer) cases.

Additionally, spark light shadowgraphy (20 ns light bursts) was utilized to capture the transition location.

To gain an overview of the flow topology, surface oil-flow visualizations were conducted. As is unavoidable in wind tunnel tests, the interaction displayed a 3D nature with significant corner effects. These factors should be considered when comparing the experimental results to the simulations.

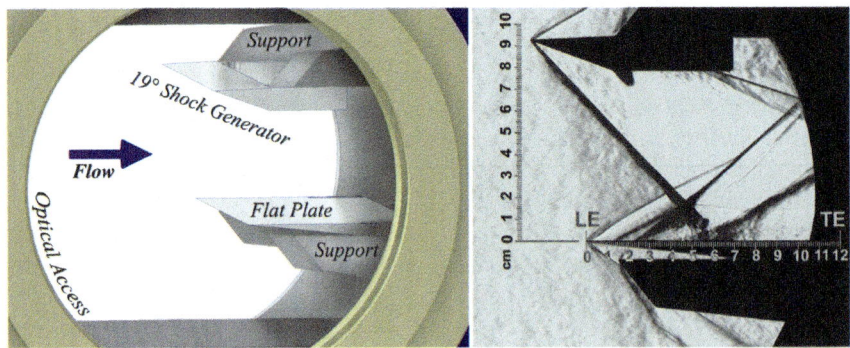

Fig. 10 Left: Model in experimental setup of TU Delft TST-27 Transonic/Supersonic wind tunnel. Right: Schlieren photograph [16]

3.1 High-Speed Schlieren/Shadowgraphy

The laminar separation shock was most clearly visible using a vertical knife edge, which bends light in the horizontal plane, emphasizing the horizontal gradient component and preventing oversaturation of the separation or near-wall shear effects. Therefore, the Schlieren configuration with a vertical knife edge was selected.

With the help of Particle Image Velocimetry [17] to complement the Schlieren data, the key flow features at the mid-plane section during the upstream growth phase of the laminar portion of the separation bubble are annotated in Fig. 11. The forward and backward states of the shock oscillation with a laminar oncoming boundary layer are also depicted. The periodic appearance and upstream propagation of the laminar separation shock influences the separation bubble by entraining generated instabilities into the bubble, which subsequently affects the reflected shock wave. The frequency of this process, which dominates the flow field, is analyzed using Dynamic Mode Decomposition in Subsect. 3.4.

In addition to capturing the oscillation mode associated with the laminar SBLI, another objective of the high-speed Schlieren imaging was to compare the laminar and tripped boundary layer cases. Figure 12a and b present the time-averaged Schlieren sequences for the laminar and tripped cases, respectively. The corresponding standard deviations are shown in Fig. 13a and b.

When comparing the two cases, the time-averaged Schlieren for the tripped case (Fig. 12b) reveals a relatively stable shear layer originating from the turbulent separation bubble, which detaches at approximately $x = 27$ mm. In contrast, the laminar case exhibits significant movement in the laminar separation shock region, emanating between $x = 10$ and 22 mm, as indicated by the standard deviation. For the tripped case, the separation shock exhibits relatively minor deviations, emanating between $x = 20$ and 22 mm.

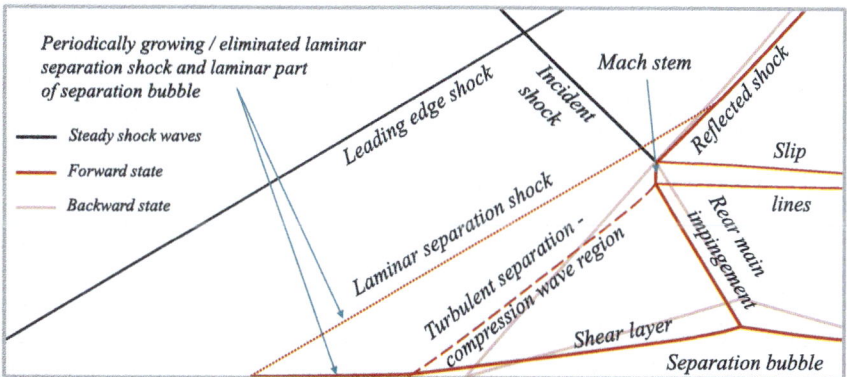

Fig. 11 Flow structure at mid-plane with laminar oncoming boundary layer, showing oscillation forward and backward states

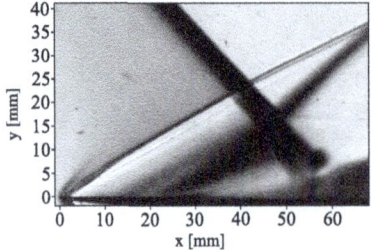
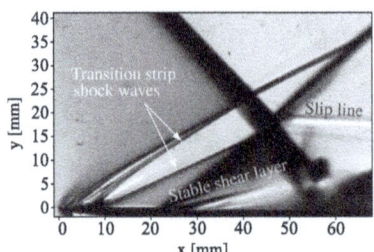

(a) Laminar case: Time averaged high speed Schlieren with vertical knife edge [22].

(b) Tripped case: Time averaged high speed Schlieren with vertical knife edge [22].

Fig. 12 a Laminar case: Time averaged high speed Schlieren with vertical knife edge [16]. **b** Tripped case: Time averaged high speed Schlieren with vertical knife edge [16]

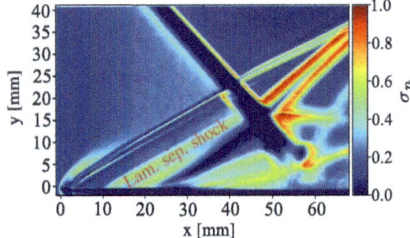
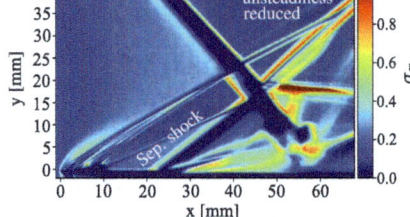

(a) Laminar case: Normalized standard deviation of high speed Schlieren with vertical knife edge [22].

(b) Tripped case: Normalized standard deviation of high speed Schlieren with vertical knife edge [22].

Fig. 13 a Laminar case: Normalized standard deviation of high speed Schlieren with vertical knife edge [16]. **b** Tripped case: Normalized standard deviation of high speed Schlieren with vertical knife edge [16]

It is clear from the standard deviation data that the reflected shock travels a shorter distance in the tripped case compared to the laminar case.

A slip line is clearly visible in the time-averaged Schlieren for the tripped case, due to its relative stability. This slip line emanates from a Mach stem, although the Mach stem is not distinctly visible here due to the spanwise integration of the Schlieren setup. However, the Mach stem is visible in instantaneous high-speed shadowgraphs [16], spark-light shadowgraphs (Fig. 20a), PIV snapshots (Fig. 14), and CFD calculations [16] presented later on.

The averaged and standard deviation results already demonstrate the stabilizing effect of tripping the boundary layer on the separation shock and the reduced travel distance of the reflected shock.

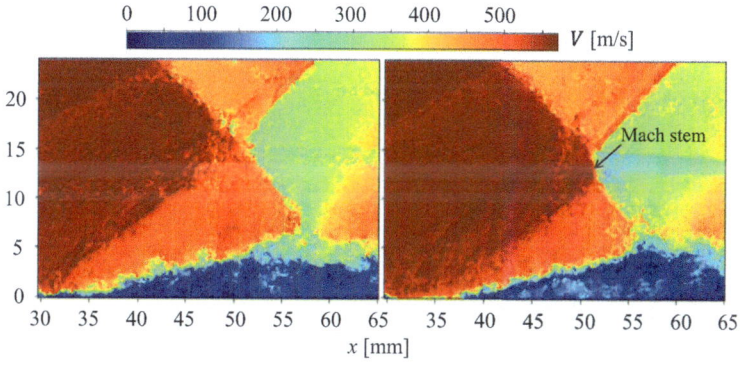

Fig. 14 Example PIV snapshots showing Mach and regular reflection states [17]

3.2 Statistical and Spatial Modal Analysis Using PIV

A Mach stem forms as a result of the strong SBLI. From instantaneous snapshots of the separation bubble region (see e.g. Fig. 14), several key features can be observed: the movement of the reflected shock, variations in the separation bubble size, the formation and disappearance of the Mach stem, and differences in the upstream behavior of the separation bubble. Some snapshots show a distinct separation shock, while others display a smeared compression wave region or the signature of a laminar separation shock.

Throughout the shock oscillation cycle, the Mach stem temporarily vanishes when the reflected shock and separation shock align. The interaction alternates between regular reflection and Mach reflection types during the oscillation.

3.3 Phase Averaged Analysis

Six phase averages were obtained [17], 2 of which are presented in Fig. 15. Phase 1 shows an almost concave upstream part of the separation bubble and a steep separation angle away from the wall, with the flow field exhibiting more closely spaced compression waves at the base of the separation bubble compared to Phase 6. Additionally, the reflected shock is positioned further upstream in Phase 1, whereas it moves downstream in Phase 6. The slip line region is also thicker in Phase 6, indicating a larger Mach stem. Shear layer thickness differences between the different phases were also noted. The variations in shear layer vorticity thickness for different phases, as well as for the time-averaged case, are shown in Fig. 16.

From this figure, it can be noted that the time-averaged result follows a shear layer vorticity thickness growth rate between 0.19 and 0.165, which is consistent with

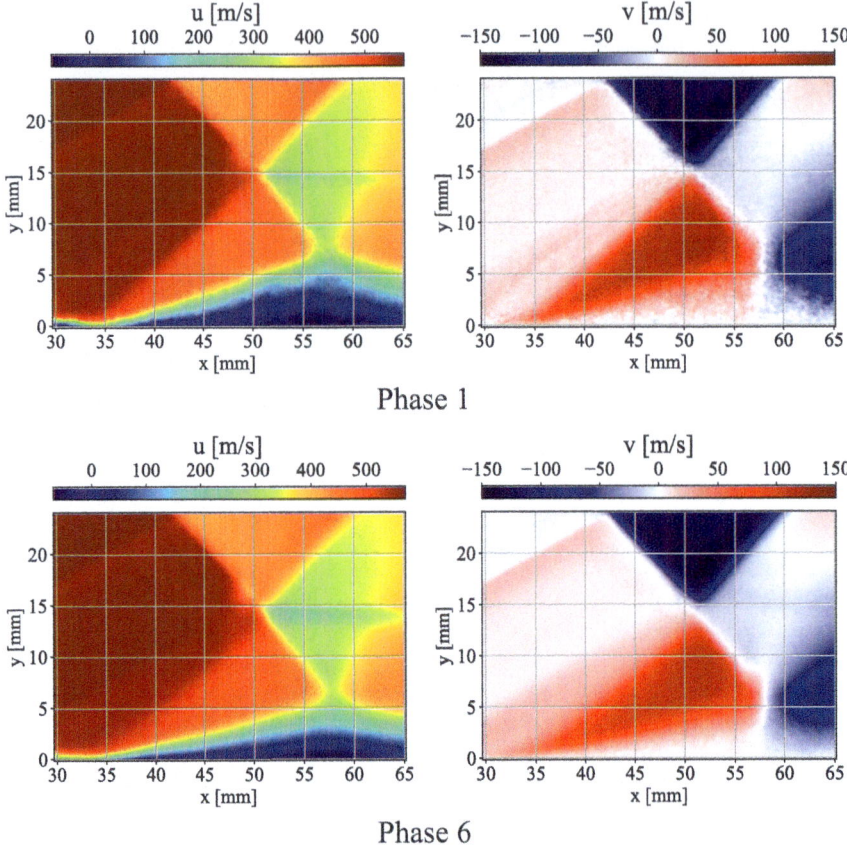

Fig. 15 Phase averages: u (left) and v (right) components [17]

findings from other studies on compressible shear layers [5, 21]. Phase 2 demonstrates a lower initial shear layer growth rate than the other phases, but reaches a similar vorticity thickness after the impinging shock wave. In contrast, Phase 3 shows a higher growth rate upstream of the shock, but a relatively low vorticity thickness downstream. These results indicate clear differences in entrainment strength between phases and suggest that instabilities of varying size are transported downstream depending on the phase.

We further corroborated these findings with an analysis of root mean square (RMS) velocity fluctuations, which indicated that the nondimensionalized u and v RMS velocity components did not reach values of 0.2 and 0.15, respectively. Hence, a fully developed turbulent separated shear layer, as described in [2], was not achieved in this interaction. Values of u'_{rms}/u_∞ were relatively high (0.2–0.3) in the shear layer before the main shock impingement and started to diminish to 0.15 after this

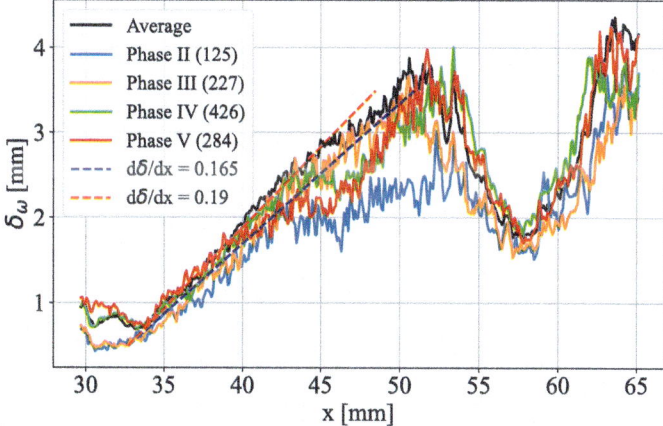

Fig. 16 Vorticity thickness of main (middle) phases. Number of snapshots indicated in brackets [17]

point. Conversely, the values for the transverse component v'_{rms}/u_∞ were relatively low upstream of the main impingement (0.07–0.08), with vertical mixing abruptly increasing to 0.10 afterwards.

3.4 Unsteady Analysis

In this subsection, we analyze the unsteady behavior captured by the high-speed Schlieren imaging to understand the dynamic mechanisms of the flow field. To evaluate the upstream propagation velocity of the laminar separation bubble, the position of the separation shock over time is plotted in Fig. 17. The position of the laminar separation shock is tracked using the point of maximum gradient in the high-speed Schlieren images with a vertical knife edge, measured at the point of emanation on the flat plate. The dashed blue lines indicate a propagation velocity of 66 m/s. The average travel distance of the laminar separation shock is approximately 11.5 mm, and represents a characteristic length scale for the shock oscillation mechanism, since this is the distance over which the laminar separation shock travels before the laminar section of the separation bubble collapses, causing a shock movement.

The separation shock position over time qualitatively matches previous findings from isentropic Mach number distributions (Fig. 9). Both show the periodic upstream movement of the laminar separation shock, with a dominant forward motion before retreating back to the point of turbulent separation. As the laminar separation shock shock vanishes, a new laminar separation shock forms and begins to grow upstream. The pronounced upstream movement signature is due to the steeper detachment during the upstream growth phase of the separation bubble, whereas the collapse exhibits

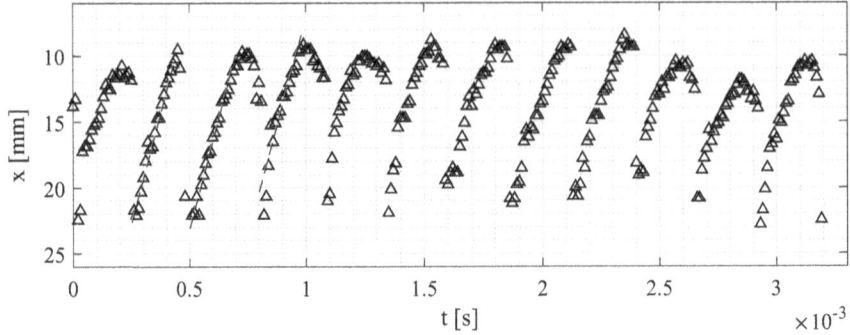

Fig. 17 Separation shock position extracted from high-speed Schlieren. Dashed blue lines indicate a propagation velocity of 66 m/s [16]

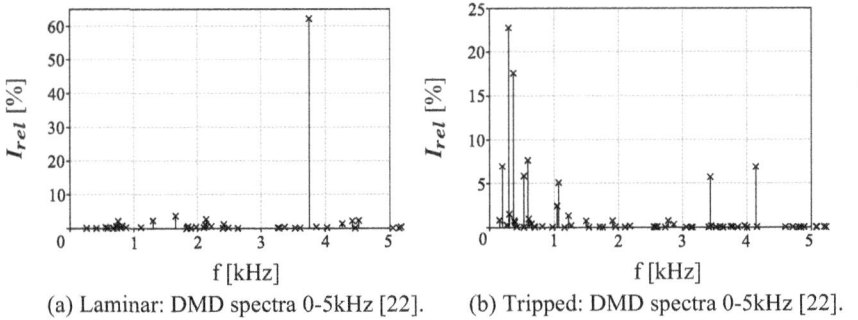

Fig. 18 a Laminar: DMD spectra (0–5 kHz) [16]. **b** Tripped: DMD spectra (0–5 kHz) [16]

a concave (tangential) behaviour. This behavior has been observed in previous studies [15] and can also be seen in the LES simulations of the canonical case [16], such as in Fig. 24.

The spatially and temporally coherent modes of the shock unsteadiness were evaluated using sparsity-promoting optimized dynamic mode decomposition (DMD) [22, 23]. This variant of the standard DMD [20] employs backpropagation and stochastic gradient descent techniques from machine learning to optimize the eigenvalues and eigenvectors of the DMD operator. This method offers improved robustness against noise while promoting sparsity, which aids in identifying the most significant dynamics and filtering out less relevant or spurious modes. For each case, the optimal rank truncation was determined using the singular value hard thresholding algorithm [8]. The relative importance (I_{rel}) represents the contribution of each DMD mode to the overall dynamics.

The DMD spectra of dominant modes for the laminar case is presented in Fig. 18a. For the laminar case, a single dominant mode at 3748 Hz was identified, accounting for 62% of the dynamic behavior of the Schlieren sequence. The associated mode shape is shown in Fig. 19a, where the Schlieren intensity captured by the mode is

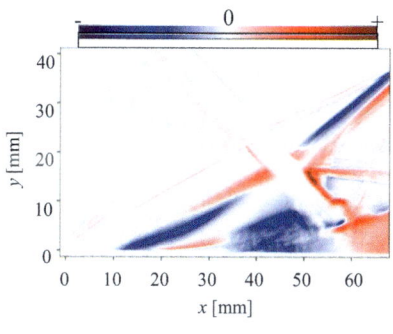
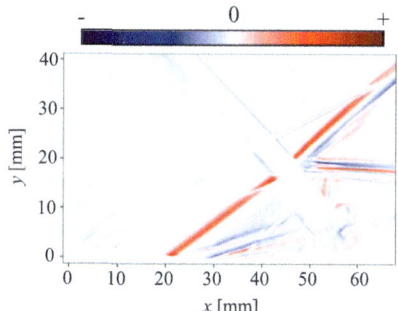

(a) Laminar case: Dominant DMD mode shape (I_{rel} = 62%) at 3748 Hz [22].

(b) Tripped case: Dominant DMD mode shape cyclic temporal evolution (I_{rel} = 23%) at 293 Hz [22].

Fig. 19 a Laminar case: Dominant DMD mode shape (I_{rel} = 62%) at 3748 Hz [16]. **b** Tripped case: Dominant DMD mode shape (I_{rel} = 23%) at 293 Hz [16]

scaled by the relative intensity I_{rel} of the mode. This mode is linked to the periodic appearance of the laminar separation shock, explained by the upstream growth of the laminar portion of the separation bubble. As the separation bubble expands upstream, the laminar part carries instabilities along the separated shear layer. Since the laminar part is extremely thin, when it reaches a certain length (the characteristic length scale of the oscillation frequency), Kelvin-Helmholtz instabilities on the laminar shear layer suppress the upstream laminar section of the separation bubble through entrainment, effectively cutting itself off and causing its collapse. The advection of these instabilities (especially turbulence from the collapse of the upstream laminar section) over the bulk separation bubble also partly suppresses the entire separation, resulting in significant movement of the reflected shock wave.

The mode shape reveals a strong correlation between the laminar separation shock, the separation bubble, and the reflected shock. Additionally, the mode shape suggests that the unsteadiness from the incident shock wave does not influence this mode of shock oscillation, as the incident shock is not visible in the dominant mode shape.

For the tripped case, the DMD spectra is shown in Fig. 18b. Unlike the laminar case, the tripped case exhibits less well-defined oscillatory behavior, with multiple significant modes identified. The dominant mode in the tripped case has a frequency of 293 Hz, an order of magnitude lower than in the laminar case. The associated mode shape (Fig. 19b) also shows movement of the reflected shock and turbulent separation shock, though with a relatively small shock travel distance, as was also observed earlier in the standard deviation comparison (Fig. 13).

There is a notable behavioral difference between the dominant modes of the laminar and tripped cases. In the laminar case, the cyclic colormap of the mode shape repeats from the upstream foot of the separation shock to the reflected shock, indicating a delay (or phase difference) in movement between the x- and y-directions [15].

In contrast, the tripped case exhibits shock movement in the same phase from the point of emanation to the end of the separation shock (with red spanning across the reflected shock in the direction of the reflected shock, as seen in Fig. 19b), suggesting that the separation shock and reflected shock are more rigid compared to the laminar case.

This behavioral difference can be explained by the fact that, in the laminar case, the reflected shock in this mode has a strong relationship with the separation bubble. The separation bubble affects the flow direction with a delay, due to downstream-advecting entrainment eddies originating from periodic upstream instabilities, which oscillate in their x-position relative to the separation bubble and thus vary in entrainment strength. This is supported by the observation that, in the tripped case, the dominant mode shows little connection to the separation bubble, and thus the mechanism necessary for delayed or rolling behavior in the reflected shock wave movement is absent.

Higher frequency spectra revealed a high-frequency mode at approximately 160 kHz in the laminar case. Although this mode was generally associated with turbulent structures, the shedding structures in the separation bubble were larger in scale and exhibited a more coherent nature in the laminar interaction compared to the turbulent case.

3.5 Upstream Effects

The origin of the shock oscillation mechanism can be traced to the upstream laminar section of the separation bubble. This thin and elongated section of the separation bubble grows upstream, carrying instabilities along the shear layer. Once it reaches a critical length, turbulence from these instabilities cut off this upstream elongated section of the separation bubble and cause its collapse. The eddies generated from this upstream collapse then advect downstream, leading to the partial suppression of the entire separation bubble [15, 16]. Consequently, we also focus on a detailed upstream analysis.

The shifting transition location in the laminar case was previously observed in LES calculations [15]. To capture the transition location and the various phases of the oscillation mechanism in greater detail, the flow field was captured instantaneously using 20 ns light bursts.

In Fig. 20a, we zoom in on the upstream section of the separation bubble, where both the separation shock from the laminar interaction and the stronger compression wave region, emanating from the sudden turbulent thickening of the separation bubble, can be seen. This region is referred to as the turbulent separation shock in Fig. 20a. Depending on the phase of the oscillation, this region alternates between displaying densely spaced compression waves and a single shock wave, as will be further illustrated in Fig. 20a and b and spatial modal analysis of PIV snapshots

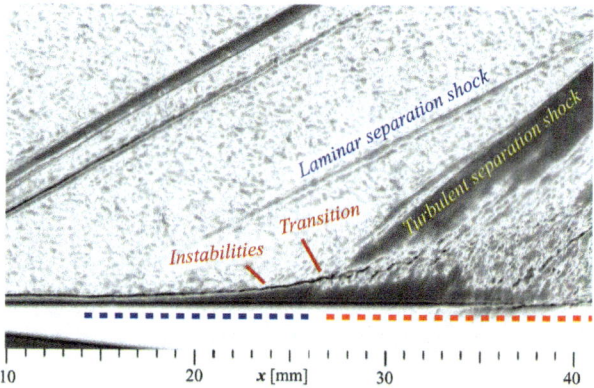

(a) Growth phase of laminar part (blue). Laminar separation shock at x = 14mm, transition at x = 27mm.

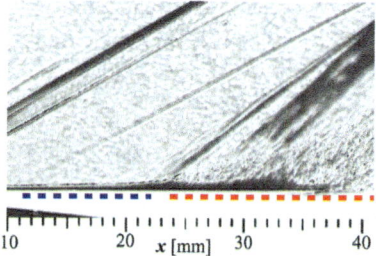

(b) Collapse of laminar part (blue). Laminar separation shock at x = 11mm vanishing. Turbulence (red) at x = 24mm.

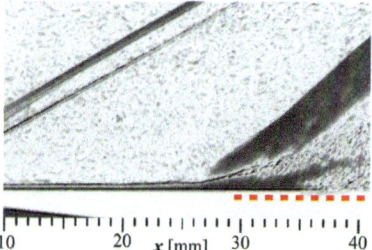

(c) Collapsed laminar part of separation bubble. No visible laminar separation shock. Turbulence (red) at x = 30mm.

Fig. 20 Highly separated transitional SBLI at different stages: **a** Growth phase, **b** Initial collapse, **c** Complete collapse [16]

(Fig. 21). The transition location on the shear layer is also clearly visible, as the relatively featureless laminar shear layer breaks down into turbulent structures, followed immediately by the aforementioned shock wave.

Figure 20b captures an instant when the laminar section (shown in blue) of the separation bubble has begun to collapse. It appears flatter than in Fig. 20a, as was also noted during the discussion of LES results in Sects. 2.3 and 3.4. This can also be seen in volumetric numerical shadowgraph images of the canonical case (Fig. 24). At this stage, the separation shock from the laminar region begins to vanish due to the turbulence generated on the shear layer, which suppresses the separation bubble and cuts off its upstream part. For a brief moment afterward, the interaction behaves as a turbulent one until the upstream turbulent structures have advected downstream of the original laminar region of the separation bubble.

Figure 20c depicts the phase when the originally laminar portion of the separation bubble has vanished. The turbulent separation shock emanating from the turbulent of

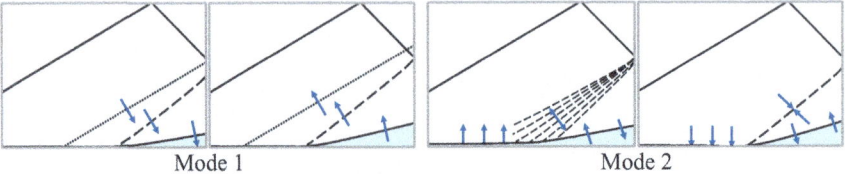

Fig. 21 Upstream analysis: effect of 1st and 2nd modes on flow features

the separation, shown in red, dominates. Notably, between these two shadowgraphs, the transition location shifts from 24 to 30 mm—a finding confirming the shifting transition location expected from LES simulations. In Fig. 20b, during the phase where the upstream section of the separation bubble begins to vanish, the flatter nature of the separation bubble's upstream portion causes the turbulent separation shock region to appear as multiple shock waves or a compression wave region.

Using the decomposition method outlined by Berkooz [1] on PIV results, we further characterize these upstream effects by examining the coherent spatial modes that dominate the unsteady flow field. We focus on the first two modes, as they are the most significant.

The growth of the separation bubble results in an outward movement of the compression waves caused by turbulent thickening, as well as the separation shock from the upstream (laminar) edge of the separation bubble. Figure 21 illustrates this effect. In this upstream analysis, the distinction between the first and second POD modes on the upstream compression waves is more pronounced than in the downstream (separation bubble) spatial modal analysis of Sect. 3.7. The first mode shape of the upstream region is of similar type to the dominant mode seen in the separation bubble region Fig. 26).

The second POD mode reveals shear layer undulation along with boundary layer thickening. This mode is associated with a horizontal spreading of compression waves, which occurs due to the upstream transition on the shear layer in the upstream growth phase of the separation bubble. This spreading of the compression waves is more pronounced in the v component mode shape.

A distinct difference is observed between the upstream effects of the second POD mode and the first. In the first mode, the most upstream separation compression waves originate from the laminar part of the separation bubble. In contrast, the upstream edge of the compression wave region in the second POD mode originates from instabilities that cause thickening of the shear layer. This difference is illustrated schematically by comparing the impact on flow features of the first mode with that of the second mode in (Fig. 21).

The described phenomena are also visible in instantaneous PIV snapshots (Fig. 22). The upstream phase of the oscillation, where a laminar separation shock exists and a spread-out compression wave region is visible, is shown in the left of Fig. 22. In the right of Fig. 22, the laminar section is temporarily suppressed, and the separation shock exhibits the typical form of a turbulent separation shock.

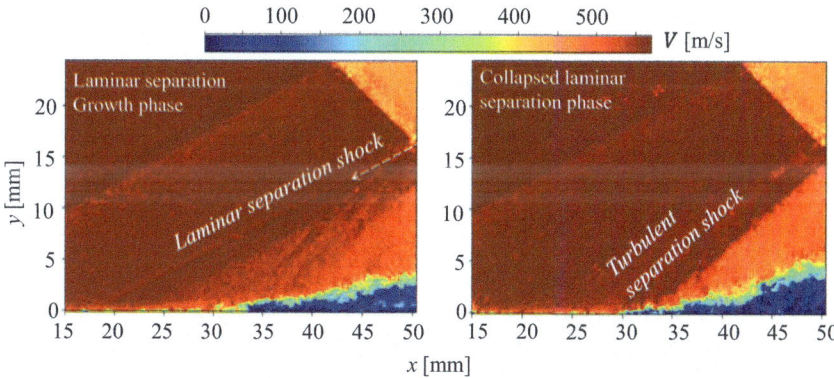

Fig. 22 Example PIV snapshots showing states of oscillation. Laminar separation shock (arrow) can be seen for upstream states [17]

Spark light shadowgraphs [16] provided insights into the instabilities on the shear layer or the transition location, complementing the PIV snapshots at the different oscillation phases. Figure 20a (a) showed the phase where the laminar part of the separation bubble is growing upstream, and the instabilities are visible on the separated shear layer. Figure 20a (b) showed the phase where the laminar section has begun to collapse, while Fig. 20a (c) showed the collapsed state. LES simulations [16] in the next section corroborate and confirm these observations.

3.6 Numerical Investigation

An LES simulation based on the final experimental setup was conducted (Fig. 23) using the Rolls-Royce *HYDRA* code, employing second-order schemes and implicit time-stepping. Despite computational limitations, such as mesh-constrained turbulence scales, the simulation includes a sensitivity study on turbulence [16]. Different turbulence levels (TI = 0.07, 0.11, 0.21%) were tested, and TI = 0.21% suppressed shock oscillation, while TI = 0.11% allowed it, with reduced amplitude compared to a clean inflow. Volumetric numerical shadowgraphs were compared to experiments.

Figure 24 shows volumetric numerical shadowgraphs of the upstream region from an LES simulation, details of which are provided in Sect. 3.6. In Fig. 24: Snapshot (1) shows the suppressed state, which begins to recover by snapshot (2). Between snapshots (2) and (4), the laminar separation shock propagates upstream, originating from the laminar section of the separation bubble. In snapshot (5), the laminar part of the separation bubble collapses. The downward movement of instabilities, which cut off the separation bubble region through entrainment, is clearly visible. Snapshots (5) and (6) show the coalescence of the compression wave region

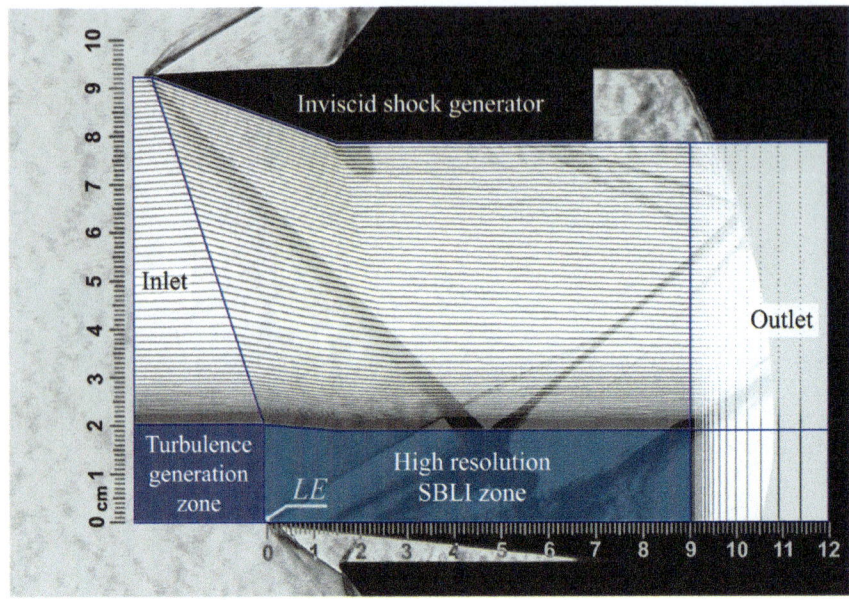

Fig. 23 Final simulation domain for experimental comparison [16]

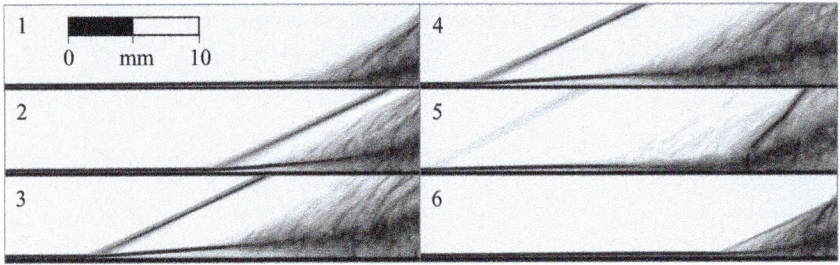

Fig. 24 Upstream zoomed view capturing growing and shrinking upstream part of separation bubble ($\Delta t = 10.35\text{e-}05$ s). Numerical volumetric shadowgraph of the LES simulation with TI = 0.07%

into a single shock wave as it retracts to the turbulent base of the separation bubble. By snapshot (6), the oscillation cycle restarts with the formation of a new laminar separation shock.

Shock oscillation frequencies extracted from the DMD were sensitive to turbulence levels: 2672 Hz for TI = 0.11% and 2430 Hz for TI = 0.07%, with increased free-stream turbulence promoting the earlier formation of shear layer instabilities. Strouhal numbers (0.075) were matching, with discrepancies in absolute frequency likely stemming from 3D effects of the experiment (e.g., end wall separation).

Given that the growth phase of the laminar section of the separation bubble dominates the shock oscillation, with the growth velocity consistent between the experi-

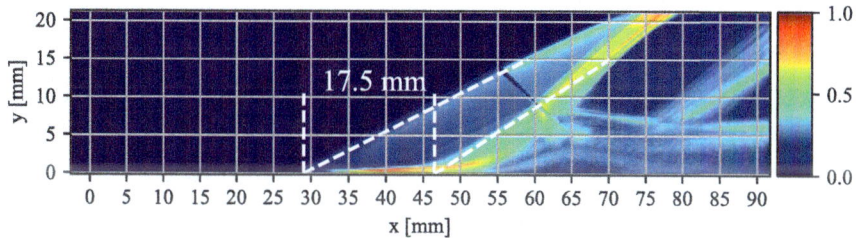

Fig. 25 Standard deviation of numerical volumetric shadowgraph of the LES simulation with TI = 0.07%

Table 2 Strouhal frequencies for experiment and numerical simulations

	Experiment	Numerical TI = 0.07%	Numerical TI = 0.11%
L [mm]	11.5	17.5	15
f [Hz]	3728	2430	2672
St [–]	0.078	0.077	0.073

ment and the numerical simulations, as well as across different cases exhibiting this oscillation mechanism [15], the characteristic length scale for the shock oscillation mechanism should be based on the laminar separation shock travel distance as in Fig. 25. To consistently identify this characteristic length for the Strouhal frequency, the standard deviation of Spark light shadowgraphs from the experiment is used for comparison with that of the numerical shadowgraphs. A comparison of the resulting Strouhal numbers are shown in Table 2.

3.7 Separation Bubble Spatial Behaviour

Next, we analyze the spatial modal characteristics of the separation bubble using PIV results at the mid-plane section to better understand the variation in separation bubble size, which could not be quantified through the DMD analysis of the high-speed Schlieren imaging. As before, the focus is on the first two modes. This first mode shows the main separation bubble size variation, the u and v components of the mode shapes are analysed in detail by Nel et al. [17]. The modal analysis identified that the disappearance of the Mach stem is linked with a larger separation bubble. This was also clear in the phase averages (Sect. 3.3), where for the first phase, displaying the largest separation bubble, the slip line was barely visible in comparison to the 6th phase, which displayed the smallest separation bubble. This was confirmed by evaluating the effect of the first mode using the time coefficients. The effect of the first mode is schematically shown in Fig. 26. The negative u velocity component and positive v component in the separation shock and reflected shock region indicate an

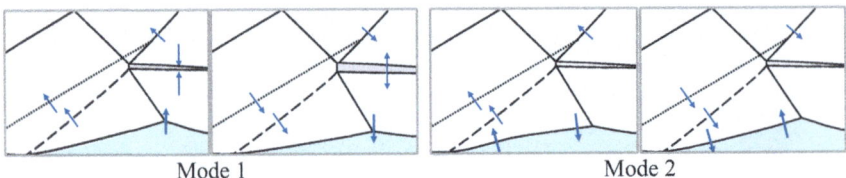

Fig. 26 Downstream analysis: effect of first and second modes on flow features

upstream movement of these shock waves when the separation bubble has grown. The first mode confirms that the Mach stem generally exists for a shallow separation bubble.

The second most dominant POD mode, is the first mode shape which is associated with an undulation of the separated shear layer. Its effect on the flow structure is illustrated in Fig. 26. Although the v component shape is nearly identical to that of the first mode, the impact on the most upstream compression waves is more pronounced in the second mode. This is consistent with the observation that the negative u component in the second mode extends further upstream than in the first mode. Unlike the first mode, the second mode is not associated with variations in the size of the slip line or Mach stem. Both modes influence the laminar separation shock, compression waves from turbulent thickening, and the reflected shock.

3.8 Full Span Fan Demonstration Case

In this section, we discuss preliminary results that demonstrate the relevance of the canonical case to the application of a highly loaded 3D transonic fan, and discuss the influence of the 3D fan geometry on the oscillation mechanism. This study, based on a full-span 3D fan, uses the NASA Rotor 67 geometry. The Reynolds number was reduced to 350,000 to match the low Reynolds cascade case, and the loading was significantly increased to simulate the highly loaded conditions experienced by the real fan under altitude excitation (Fig. 2). It is important to note that without increasing the loading to create a large separation, no significant shock oscillation occurs for Rotor 67, as determined during preliminary tests. For this preliminary study, we focus on the blade, and the endwall flow is not accurately resolved. The mesh over the blade has Δx^+ and Δz^+ values of 50, with a Δy^+ of 1.

A laminar oncoming boundary layer, as well as a case with a step are considered. The step is incorporated close to the leading edge on the suction side between 45 and 95% of the span. It is observed that the pre-shock Mach number, separation, and shock structure are similar to those of the highly loaded fan application at the condition where altitude excitations are present (Fig. 2). This also holds when compared to the quasi-2D cascade simulations [13]. From the standard deviation of the relative Mach number (Fig. 27), it is seen that the laminar case exhibits a significant movement of

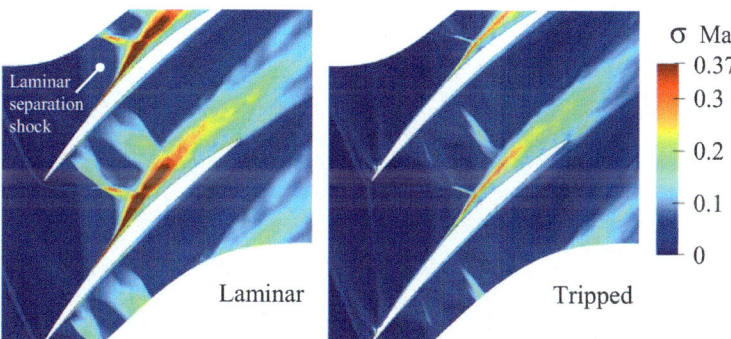

Fig. 27 Full span case: standard deviation of relative Mach number at 55% of the span for the laminar case (left) and tripped case (right)

the separation bubble shear layer and passage shock wave movement, in comparison to the tripped case. Furthermore, a large upstream extent of the laminar separation shock is observed for the laminar boundary layer case, similar to quasi-2D cascade simulations.

Figure 28 shows the flow structure by density gradient at 55% span, along with the blade suction side pressure gradient contours to illustrate the upstream position of the laminar separation shock in the laminar case (left). The tripped case is shown on the right. In this case, the source of the shock oscillation occurs between 45% and 80% span, as turbulence from the upstream tip flow suppresses the growth of a laminar separation shock near the top of the blade, while a weaker interaction is present below 45% span. This is evident when examining the flow field, where the laminar section of the separation bubble begins to grow upstream (accompanied by a clearly visible laminar separation shock) before being suppressed by its own shear layer instabilities. A laminar separation shock can be seen emanating from the blue dotted line at the section cut of the density gradient in Fig. 28. Red dashed lines are provided for positional reference between the snapshots.

The moving transition location on the blade in the laminar case is clearly visible. Additionally, it is worth noting that the growth phase of the laminar section of the separation bubble in the 3D fan case exhibits a 3D nature: the growth occurs in both upstream and downward directions, with the collapse phase also happening in a downward (toward the hub) and downstream direction. This refers not to the direction of advecting turbulence, but to the fact that the mechanism is significantly out of phase when considering different sections along the span. This can be explained by the varying blade loading along the span.

From the standard deviation of the static pressure distribution on the suction side, it was observed that boundary layer tripping reduces pressure deviations by more than 50%. Furthermore, from the pressure distribution along the span it was noted that although the source of the shock oscillation is limited to a region closer at mid span, the effect of introducing a step is present from the hub to the tip, with the oscillation affecting the entire shock structure.

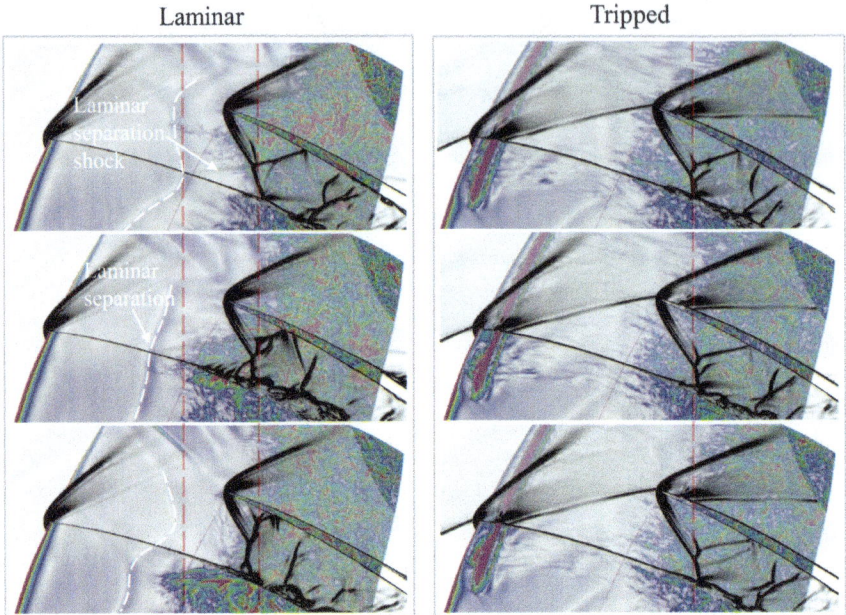

Fig. 28 Full span case: snapshots showing the growth of laminar separation shock and movement of passage shock structure of laminar case (left). Step (tripped) case is shown in the right. Section taken at 55% of the span

4 Conclusions and Outlook

Numerical studies on various cases, including compressor cascades, have demonstrated that the shock oscillation mechanism of interest is primarily driven by the conditions at the shock-wave/boundary-layer interaction, rather than being reliant on compressor-like geometry. The issue of shock oscillation at low Reynolds numbers, where a laminar oncoming boundary layer is present, stems from a strong transitional SBLI on a transonic fan at altitude. This can be simplified to a highly separated transitional oblique shock-wave/boundary-layer interaction, with transition occurring on the shear layer of the upstream laminar section of the separation bubble. Furthermore, it became clear that the existing experimental cascades were too noisy to effectively study the oscillation mechanism observed at altitude or in the idealized, quiet quasi-2D LES simulations designed to replicate altitude conditions. As a result, a canonical research configuration was defined and implemented in a physical experiment, specifically designed to demonstrate and validate this shock oscillation mechanism, which is eliminated when the boundary layer is turbulent. The study primarily focused on the canonical research configuration, analyzing the details of the shock oscillation mechanism with regard to dynamics and spatial behavior. Large Eddy Simulations (LES) were used to complement the experimental findings and validate the shock oscillation mechanism observed in the simulations.

4.1 Conclusions: Shock Oscillation Mechanism

The investigation into the dynamics of highly separated transitional shock-wave/boundary-layer interactions (SBLIs) contributes to efforts aimed at mitigating shock-induced unsteadiness in transonic fans operating at altitude, where such interactions can lead to structural excitation and potential fan blade fatigue.

The goal was to examine the shock oscillation mechanisms present for a laminar oncoming boundary layer and demonstrate that this mechanism is eliminated in the turbulent (tripped boundary layer) case, while also validating the oscillation mechanism observed in Large Eddy Simulations (LES). The study utilized high-speed Schlieren imaging, spark light shadowgraphy, and surface oil flow visualization to capture the dynamic behavior, instantaneous effects, and flow topology of the SBLI. Experiments were conducted using a 19° shock generator in a supersonic wind tunnel at Mach 2.3 with a total pressure of 2.8 bar. High-speed Schlieren imaging at 100 kHz revealed the unsteady separation bubble, reflected shock movement, and the periodic appearance of an upstream laminar separation shock as the laminar part of the separation bubble periodically vanishes. Spark light shadowgraphy, with 20 ns light bursts, provided insights into the location of transition and flow instabilities, while surface oil flow visualization offered three dimensional insights of the interaction region separation bubble. These experimental methods were complemented by LES simulations, which captured the oscillation mechanisms in greater detail. In the laminar boundary layer case, the oscillation mechanism is characterized by a periodically collapsing laminar section of the separation bubble, driven by instabilities in the upstream separated shear layer. The thin and elongated laminar section moves upstream, carrying Kelvin-Helmholtz instabilities along the shear layer. When a critical length is of this section of the separation bubble is reached, turbulence generated by the instabilities suppresses the upstream laminar section of the separation bubble, causing its collapse. Additionally, the increased turbulence during collapse of the upstream section temporarily entrains into the bulk separation bubble, causing a separation suppression and resulting in an oscillation of the reflected shock. A key finding is the identification of a characteristic length scale for the oscillation mechanism, based on the distance between the maximum upstream position of the laminar separation shock and the turbulent separation shock. The calculated Strouhal number of approximately 0.075 aligns well with LES results, validating the numerical approach for resolving the oscillation mechanism. In contrast, the turbulent boundary layer exhibits a different oscillation mechanism, characterized by dominant frequencies significantly lower than in the laminar case. The reflected shock movement in the turbulent SBLI is more rigid and less correlated with the separation bubble movement. The use of a transition strip to trip the boundary layer stabilizes the shock, as indicated by the significantly reduced travel distance of the reflected shock in the turbulent case compared to the laminar case. LES simulations provided detailed insights into these oscillation mechanisms, validating the experimental findings concerning the shock oscillation mechanism. The results also indicated that the shock oscillation mechanism is highly sensitive to free stream turbulence levels. Higher turbulence levels lead to an earlier destabilization of the laminar shear layer, resulting in a higher oscillation frequency as the travel distance of the laminar separation shock decreases. This demonstrates a clear dependency on upstream flow conditions.

The oscillation mechanism source is an interplay and balance of three facts of nature, namely:

- Laminar SBLI's have a larger upstream influence than turbulent SBLI's (separation growth mechanism).
- Kelvin-Helmholtz instabilities develop on laminar shear layers, leading to transition.
- Turbulence has the capability to suppress separation.

4.2 Conclusion: Spatial Modal Analysis

We conducted a statistical analysis of PIV data on a highly separated oblique shock-wave/boundary-layer interaction at Mach 2.3 with a 19° flow deflection angle in order to study provide quantitative insights into the shock oscillation mechanism, with a focus on the size variation of the separation bubble and upstream effects. This was especially important to capture and confirm the flow structure at the mid-plane section and quantify the size variation of the separation bubble, which could not be achieved with Schieren imaging.

The separation bubble shear layer does not reach a fully developed turbulent state, as indicated by root mean square (RMS) velocity fluctuations. Instantaneous snapshots revealed states where a laminar separation shock is visible, as well as states where only a single turbulent separation shock exists. This helped understand Schlieren data which showed the signature of a varying slip line, confirming the presence of an alternating Mach stem feature at the mid-plane section. For a vertically grown separation bubble, there is a reduced slip line region thickness (shorter Mach stem), a more distinct upstream reach of the separation shock, and an upstream positioning of the reflected shock, as demonstrated by phase-averaged results and the analysis of the first POD mode of the separation bubble (downstream view). Differences in entrainment strength between the phases suggest that instabilities of varying sizes are transported downstream depending on the phase. This has been confirmed through the analysis of the shear layer vorticity thickness, which aligns with the observation that the shear layer is not fully developed into a turbulent state. Proper Orthogonal Decomposition (POD) was used to extract coherent structures of the PIV data, and their effects on the flow structure were analyzed. The first mode of the separation bubble region is related to the separation bubble size and also shows an effect on the slip line region size (Mach stem height increases with shallow separation). The second mode is associated with shear layer undulations. Both the first and second modes in the separation bubble region influence the shock wave positioning. The upstream modal analysis revealed a first mode analogous to that of the separation bubble region, while the second mode emphasized the spreading of compression waves due to upstream boundary layer transition.

4.3 Conclusions: Numerical Settings

Large Eddy Simulations are able to resolve the oscillation mechanism. We have shown that there is little difference between the oscillation dynamics when comparing a DNS resolution solution ($\Delta x^+ = \Delta z^+ = 7$, $\Delta y^+ = 1$ and a high resolution LES mesh. Furthermore, the transonic channel investigation, as well as numerous preliminary tests, had shown that the shock oscillation mechanism is not present when the mesh is too coarse. We summarise the grid requirements for resolving the oscillation mechanism in Table 3. The WALE and SIGMA sub-grid scale models both performed well. The Smagorinsky model performed poorly, vastly over-predicting the upstream influence and causing excessive smearing across the SBLI.

Furthermore, it was shown that Reynolds averaged methods cannot be used to resolve the oscillation mechanism of interest, with artificial shock oscillations occurring due to a combination of the transition model and multi grid cycles.

4.3.1 Conclusions: Shock Oscillation Mechanism on 3D Fan

The full span fan simulation demonstrated the relevance of the canonical case to the highly loaded fan application. The primary source of shock oscillation was located between 45 and 75% blade span, with tip clearance disturbances suppressing the shock oscillation near the tip. Below 45% blade span, the interaction is too weak, preventing the shock oscillation mechanism from occurring. The boundary layer tripping significantly reduced shock unsteadiness, with pressure deviations on the blade reduced by more than 50%. The shifting transition location was clearly shown for the laminar case, with the tripped case exhibiting a stable transition location. The behavior of the laminar separation section was three-dimensional, often exhibiting an undulating structure with the oscillation mechanism occurring out of phase across different sections of the blade.

Table 3 Grid requirements

Requirement	Δx^+	Wall Δy^+	Δz^+	Far Δy^+
Mechanism present	<60	1	<60	<60
Mechanism resolved	14	1	14	34

4.4 Outlook

Future studies should include LES simulations of a 3D fan case at the original Reynolds number. Swept geometries can be simulated alongside a baseline case to investigate the effects of cross-flow instabilities on the shock oscillation mechanism. With regard to canonical investigations, future work should consider applying high-speed PIV to the laminar baseline case for comparison with various tripped configurations, offering additional spatio-temporal insights.

Acknowledgements The authors express gratitude to Ferry Schrijer and Bas van Oudheusden for their supervision and assistance during the experimental campaign at the High Speed Aerodynamics Laboratory of the Technical University of Delft, which was an invaluable part of this work. The author participated in the TEAMAero project, which has received funding from the European Union's Horizon 2020 research and innovation programme under grant agreement No 860909.

References

1. Berkooz, G., Holmes, P.J., Lumley, J.: The proper orthogonal decomposition in the analysis of turbulent flows. Ann. Rev. Fluid Mech. **25**, 539–575 (2003). https://doi.org/10.1146/annurev.fl.25.010193.002543
2. Bigillon, F., Niño, Y., García, M.: Measurements of turbulence characteristics in an open-channel flow over a transitionally-rough bed using particle image velocimetry. Exp. Fluids **41**, 857–867 (2006). https://doi.org/10.1007/s00348-006-0201-2
3. Denton, J.D., Xu, L.: The effects of lean and sweep on transonic fan performance. In: Turbo Expo 2002: Power for Land, Sea, and Air, vol. 5, Parts A and B. ASME, pp. 23–32 (2002). https://doi.org/10.1115/GT2002-30327
4. Doerffer, P., Flaszynski, P., Dussauge, J.P., Babinsky, H., Grothe, P., Petersen, A., Billard, F.: Transition location effect on shock wave boundary layer interaction: experimental and numerical findings from the TFAST project. In: Notes on Numerical Fluid Mechanics and Multidisciplinary Design, vol. 144, pp. 229–296. Springer, Cham (2021). https://doi.org/10.1007/978-3-030-47461-4_6
5. Schrijer, F., Sciacchitano, A., Scarano, F.: Spatio-temporal and modal analysis of unsteady fluctuations in a high-subsonic base flow. Phys. Fluids **26**, 086101 (2014). https://doi.org/10.1063/1.4891257
6. Gomar, A., Gourdain, N., Dufour, G.: High-fidelity simulation of the turbulent flow in: a transonic axial compressor. In: 9th European Conference on Turbomachinery: Fluid Dynamics and Thermodynamics, ETC 2011—Conference Proceedings, vol. 1 (2011)
7. Grothe, P., Flaszynski, P., Szwaba, R., Piotrowicz, M., Kaczynski, P., Tartinville, B., Hirsch, C., Hergt, A.: Internal flows-compressors. In: Doerffer, P., et al. (eds.) Transition Location Effect on Shock Wave Boundary Layer Interaction: Experimental and Numerical Findings from the TFAST Project, pp. 229–296. Springer, Cham (2021). https://doi.org/10.1007/978-3-030-47461-4_6
8. Gavish, M., Donoho, D.L.: The optimal hard threshold for singular values is $4/\sqrt{3}$. IEEE Trans. Inf. Theory **60**(8), 5040–5053 (2014). https://doi.org/10.1109/TIT.2014.2323359
9. Hah, C., Bergner, J., Schiffer, H.-P.: Tip clearance vortex oscillation. vortex shedding and rotating instabilities in an axial transonic compressor rotor (2008). https://doi.org/10.1115/GT2008-50105

10. Hah, C.: Large eddy simulation of transonic flow field in NASA Rotor 37. In: 47th AIAA Aerospace Sciences Meeting including the New Horizons Forum and Aerospace Exposition (2009). https://doi.org/10.2514/6.2009-1061
11. Hergt, A., Klinner, J., Wellner, J., Willert, C., Grund, S., Steinert, W., Beversdorff, M.: The present challenge of transonic compressor blade design. J. Turbomach. **141**(9), 091004 (2019). https://doi.org/10.1115/1.4043329
12. Joo, J., Medic, G., Philips, D.A., Bose, S.T.: Large-eddy simulation of a compressor rotor (2014)
13. Nel, P.L., Janke, C., Vasilopoulos, I., Swoboda, M., Schreyer, A.M., Hady, A., Flaszyński, P.: Effect of transition on self-sustained shock oscillations in highly loaded transonic rotors. AIAA J. **62**(6), 2063–2075 (2024). https://doi.org/10.2514/1.J063378
14. Nel, P.L., Grothe, P., Swoboda, M., Pirozzoli, S., Weiss, J.: Towards understanding and resolving natural shock oscillation in transonic compressors. In: Tuovinen, T., Periaux, J., Knoerzer, D., Bugeda, G., Pons-Prats, J. (eds.) Advanced Computational Methods and Design for Greener Aviation, pp. 75–93. Springer International Publishing, Cham (2024). https://doi.org/10.1007/978-3-031-61109-4_6
15. Nel, P.L., Schreyer, A.M., Schrijer, F., van Oudheusden, B., Swoboda, M.: Research configuration to study shock oscillation mechanism in highly loaded transonic fans (Manuscript submitted)
16. Nel, P.L., Schreyer, A.M., Schrijer, F., van Oudheusden, B., Janke, C., Vasilopoulos, I., Swoboda, M.: Shock oscillation mechanism of highly separated transitional shock-wave/boundary-layer interactions. AIAA J. **63**(5), 1703–1715 (2025)
17. Nel, P.L., Schreyer, A.-M., Schrijer, F., van Oudheusden, B., Swoboda, M. Highly separated transitional shock-wave/boundary-layer interactions: a spatial modal study. Phys. Fluids **36**(11), 116114 (2024)
18. Papadogiannis, D., Garnaud, X.: Unstructured large eddy simulations of the transonic compressor Rotor 37 (2017). https://doi.org/10.2514/6.2017-3612
19. Priebe, S., Wilkin, D., II, Breeze-Stringfellow, A., Mousavi, A., Bhaskaran, R., d'Aquila, L.: Large eddy simulations of a transonic airfoil cascade. In: Proceedings of the ASME Turbo Expo 2022: Turbomachinery Technical Conference and Exposition. Volume 10A: Turbomachinery—Axial Flow Fan and Compressor Aerodynamics, Rotterdam, Netherlands, June 13–17, 2022 (2022). V10AT29A010. ASME. https://doi.org/10.1115/GT2022-80683
20. Schmid, P., Sesterhenn, J.: Dynamic mode decomposition of numerical and experimental data. J. Fluid Mech. **656** (2008). https://doi.org/10.1017/S0022112010001217
21. Smits, A.J., Dussauge, J.P.: Turbulent Shear Layers in Supersonic Flow. Springer (2006)
22. Weiner, A., Semaan, R.: A robust dynamic mode decomposition methodology for an airfoil undergoing transonic shock buffet. J. Fluid Mech. (2022). https://doi.org/10.2514/1.J062546
23. Weiner, A., Semaan, R.: Backpropagation and gradient descent for an optimized dynamic mode decomposition (2023). arXiv preprint, physics.flu-dyn, arXiv:2312.12928
24. Li, Z., Yaping, J., Zhang, C.: Quasi-wall-resolved large eddy simulation of transitional flow in a transonic compressor rotor. Aerosp. Sci. Technol. **126**, 107620 (2022). ISSN 1270-9638, https://doi.org/10.1016/j.ast.2022.107620

Open Access This chapter is licensed under the terms of the Creative Commons Attribution 4.0 International License (http://creativecommons.org/licenses/by/4.0/), which permits use, sharing, adaptation, distribution and reproduction in any medium or format, as long as you give appropriate credit to the original author(s) and the source, provide a link to the Creative Commons license and indicate if changes were made.

The images or other third party material in this chapter are included in the chapter's Creative Commons license, unless indicated otherwise in a credit line to the material. If material is not included in the chapter's Creative Commons license and your intended use is not permitted by statutory regulation or exceeds the permitted use, you will need to obtain permission directly from the copyright holder.

The manufacturer's authorised representative in the EU is Springer Nature Customer Service Centre GmbH, Europaplatz 3, 69115 Heidelberg, Germany. If you have any concerns regarding our products, please contact ProductSafety@springernature.com

Printed and bound by CPI Group (UK) Ltd, Croydon, CR0 4YY

26/03/2026

02078989-0002